Advances in

CELL CULTURE

VOLUME 4

CONTRIBUTORS TO THIS VOLUME

Salvatore DiMauro

Karl Heinz Glätzer

C. A. Heckman

Norbert Kociok

Albert Leibovitz

Winfrid Liebrich

Armand F. Miranda

Tiziana Mongini

Paul J. Price

J.-M. Quiot

Shaul Reuveny

Lynne P. Rutzky

C. Vago

A. Vey

Advances in CELL CULTURE

Edited by

KARL MARAMOROSCH

Robert L. Starkey Professor of Microbiology
Waksman Institute of Microbiology
Rutgers University
New Brunswick, New Jersey

VOLUME 4

1985

ACADEMIC PRESS, INC.

(Harcourt Brace Jovanovich, Publishers)

Orlando San Diego New York London
Toronto Montreal Sydney Tokyo

ACADEMIC PRESS, INC.
Orlando, Florida 32887

United Kingdom Edition published by
ACADEMIC PRESS INC. (LONDON) LTD.
24–28 Oval Road, London NW1 7DX

ISSN 0275-6358
ISBN 0-12-007904-6

PRINTED IN THE UNITED STATES OF AMERICA

85 86 87 88 9 8 7 6 5 4 3 2 1

CONTENTS

Hybridoma Technology

PAUL J. PRICE

Morphogenesis of Mitochondria during Spermiogenesis in *Drosophila* Organ Culture

WINFRID LIEBRICH, KARL HEINZ GLÄTZER, AND NORBERT KOCIOK

Effects of Mycotoxins on Invertebrate Cells *in Vitro*

J.-M. QUIOT, A. VEY, AND C. VAGO

Microcarriers in Cell Culture: Structure and Applications

SHAUL REUVENY

The Establishment of Cell Lines from Human Solid Tumors

ALBERT LEIBOVITZ

CONTRIBUTORS TO VOLUME 4

Numbers in parentheses indicate the pages on which the authors' contributions begin.

SALVATORE DIMAURO (1), *Department of Neurology, Columbia University, College of Physicians and Surgeons, and the H. Houston Merritt Clinical Research Center for Muscular Dystrophy and Related Diseases, New York, New York 10032*

KARL HEINZ GLÄTZER (179), *Institut für Genetik, Universität Düsseldorf, D-4000 Düsseldorf, Federal Republic of Germany*

C. A. HECKMAN (85), *Department of Biological Sciences, Bowling Green State University, Bowling Green, Ohio 43403*

NORBERT KOCIOK (179), *Institut für Genetik, Universität Düsseldorf, D-4000 Düsseldorf, Federal Republic of Germany*

ALBERT LEIBOVITZ (249), *Arizona Cancer Center, University of Arizona College of Medicine, Tucson, Arizona 85724*

WINFRID LIEBRICH (179), *Institut für Genetik, Universität Düsseldorf, D-4000 Düsseldorf, Federal Republic of Germany*

ARMAND F. MIRANDA (1), *Department of Pathology, Columbia University, College of Physicians and Surgeons, and the H. Houston Merritt Clinical Research Center for Muscular Dystrophy and Related Diseases, New York, New York 10032*

TIZIANA MONGINI (1), *Department of Neurology, Columbia University, College of Physicians and Surgeons, and the H. Houston Merritt Clinical Research Center for Muscular Dystrophy and Related Diseases, New York, New York 10032*

PAUL J. PRICE (157), *Hybridoma Sciences, Inc., Atlanta, Georgia 30084*

J.-M. QUIOT (199), *Centre de Recherches de Pathologie Comparée, INRA, CNRS, USTL, 30380 Saint-Christol, France*

SHAUL REUVENY[1] (213), *Department of Biotechnology, Israel Institute for Biological Research, Ness-Ziona 70450, Israel*

LYNNE P. RUTZKY (47), *Department of Surgery, The University of Texas Medical School, Houston, Texas 77030*

C. VAGO (199), *Centre de Recherches de Pathologie Comparée, INRA, CNRS, USTL, 30380 Saint-Christol, France*

A. VEY (199), *Centre de Recherches de Pathologie Comparée, INRA, CNRS, USTL, 30380 Saint-Christol, France*

[1]Present address: New Brunswick Scientific Co., Inc., 44 Talmadge Rd., Edison, New Jersey 08818.

PREFACE

Volume 4 of *Advances in Cell Culture* continues the wide coverage of this serial publication. Chapters are devoted to basic aspects such as cell shape and growth control *in vitro,* morphological and cytochemical changes occurring during muscle culture of human myopathies, and the biology of cultured human colon tumor cells. Among the important practical topics are the structure and application of microcarriers and hybridoma technology. One chapter is devoted to the establishment of cell lines from human solid tumors. The use of invertebrate organ culture for the study of spermiogenesis and the study of mycotoxin action on invertebrate cells are presented in two chapters. Readers will find stimulating ideas, new concepts, and fresh approaches to problems in the current contributions. Facets of cell culture that may have immediate, as well as long-range economic potential, are presented. The contributions reflect the thinking and accomplishments of those who are in the forefront of the broad field of cell culture today. The depth and sophistication of the articles indicate current strength in the diverse areas of *in vitro* research.

In this volume, a biographical sketch has been devoted to George Gey, who had a profound impact on cell culture and on numerous scientists who became inspired by him.

KARL MARAMOROSCH

George Gey

DREAMS OF CELLULAR PHYSIOLOGY IN TISSUE CULTURE: GEORGE GEY'S BEQUEST

In 1923, Dr. John L. Yates, a graduate of The Johns Hopkins University School of Medicine and Chief of Surgery at Columbia Hospital in Milwaukee, Wisconsin, and whose interest was in cancer of the breast, wrote to Dr. Warren Lewis of the Carnegie Institution at The Johns Hopkins Medical School asking for scientists trained in tissue culture to come to establish a tissue culture facility in Milwaukee. At this time, George Gey, who had worked part time in this field, was a first-year medical student at Johns Hopkins. Chance would have it that George's funds for medical school had been depleted before finishing his second year, and Dr. Warren Lewis suggested that he take the position at Columbia Hospital. In September 1923, George transferred to Milwaukee. He asked Margaret Koudelka, at that time the operating room supervisor, to assist him in sterilization in his new laboratory, which consisted of two small rooms, one a sterile room and the other an office.

George obtained the services of two Milwaukeeans, whom he trained in tissue culture and in handling laboratory animals and laboratory technical design. From 1923 to 1926, George Gey built a laboratory and a more than strong friendship with Margaret Koudelka. In 1926, George married Margaret, and the two initiated a scientific career which took them all over the country and blessed them with two wonderful children, George, Jr., and Frances.

It was clear that George was going to return to Johns Hopkins to complete his medical school training, and after six years in Milwaukee, he did indeed return to Baltimore with his new wife where he was to return to his medical studies and simultaneously develop a tissue culture laboratory in the Carnegie Institution. Dr. Joseph Bloodgood and Dr. Warren Lewis facilitated this joint endeavor. From the ground level George and Margaret physically built their laboratory on the first floor of the Carnegie Institution building from an old janitor's quarters of three rooms and a bathroom. A cubicle and an incubator were moved into the space in which the janitor's bathroom had been torn out. It was cleaned and painted, and a drying oven, an autoclave, a small sink, and a laboratory work table were personally installed by Dr. George Gey. Dr. Dean Lewis, the Professor of Surgery, Dr. Woodrich Williams, the Professor of Obstetrics and Gynecology, and Dr. Bloodgood joined in providing the biological resources necessary to obtain tissue culture specimens for this work. In 1931, Bill

Fitzwilliams and Tom Stark, two of George Gey's original technicians in Milwaukee, were hit by the economic depression, and quickly moved to Baltimore to continue to assist in the development of this pioneering tissue culture endeavor. At the same time, Dr. and Mrs. Dean Lewis invited Dr. Gey and Mrs. Gey to conduct summer experiments in Salisbury Cove, Maine, which would require moving the tissue culture equipment to conduct this work. The laboratory was in an old farmhouse. Margaret and George lived in a tent on the grounds throughout the summer. In 1932, Carl Koudelka, Margaret's brother from Milwaukee joined them. Later, Tom Stark was to join Dr. Wilton Earl at the National Institutes of Health, but Bill Fitzwilliams stayed on with the Gey's for forty-three years.

George received his medical degree from The Johns Hopkins Medical School in 1933. It was in this year that he, in the laboratory, increased the growth of cells by converting from the old Maximow Hanging Drop methods to the now famous "roller tube" technique of tissue culture that he pioneered. This was the method that was used later to establish the HeLa cell.

In 1939, Dr. George Korner became director of the Carnegie Laboratories, and Dr. Gey moved his laboratories to the Department of Surgical Pathology with Dr. Warfield Firor and Dr. Bloodgood. This new move into the Department of Surgical Pathology meant new construction. George Gey did the construction himself. He built three cubicles of brick, cement, glass, and stainless steel with incubators, animal rooms, cleanup and washing rooms to mold a tissue culture system that attracted fellows and scientists from all over the world to learn the Gey tissue culture techniques. The laboratory remained in the Surgical Pathology section until the new Director of the Department of Surgery, Dr. Alfred Blaloch, was appointed. Then the laboratory moved into the Blaloch Building. The Gey laboratory was given the thirteenth floor, and work began again to build new and expanded laboratories. These laboratories were occupied until George Gey's death on November 8, 1970. It was in this laboratory that the HeLa cell was established. Numerous other contributions to medical science were made from these portals, and many scientists were trained for work in their own institutions world wide. It was in these laboratories that the BeWo Trophoblast cell line was established by Dr. Gey's last fellow, another Milwaukeean, Roland A. Pattillo, who had come to Hopkins in 1965.

On February 1, 1951, a thirty-one-year-old black woman who complained of intermenstrual spotting appeared in the Gynecologic Outpatient Department of The Johns Hopkins Hospital. Eight days later, the resident gynecologist who saw her obtained a cervical biopsy on the patient, Henrietta Laacks, made it available to Dr. George Gey, who established the immortality of a cancer victim dead since 1951 in the tissue culture laboratory. This was the first human cell line to be established, and was widely distributed throughout the world for scientific research. It formed the groundwork on which poliovirus could be grown and led to the ultimate solution of the poliomyelitis crippling disease entity. It formed the basis for a wider and wider cell biological research system which spread all over the world. It provided the basic scientific investigative tool that brought together many nations in the joint effort to conquer human disease through *in vitro* techniques.

The HeLa cells rapid and sometimes uncontrollable growth characteristics made it a ready source of human cells for investigation of almost every facet of human biology. By the same token, however, this rapid and almost uncontrollable growth generated enormous problems when contamination of many other cell lines around the world led to the discoveries of Gartler and Nelson-Reese that HeLa cells contaminated many cell lines thought to be other entities. In addition, the HeLa cell line was carried for many years as an epidermoid carcinoma of the cervix only to have Dr. Howard Jones, a gynecologist who was the first to see Henrietta Laacks, along with the Hopkins pathologist, Dr. Donald Woodruff, review the original tumor from which the HeLa cell was derived and found it to be predominantly an adenocarcinoma of the cervix.

Initially, in the 1930s and 1940s, Dr. Gey's laboratory, with the assistance of fellows that included Dr. Georgeanna Seeger Jones, embarked on tissue culture efforts to establish hydatidiform moles, choriocarcinomas, and normal trophoblastic tissue in continuous culture. This cell type which secretes hCG would have hormone marker which could be detected in the culture media if a contamination with another cell type inadvertently occurred. This anticipated the problem of contamination that occurred in the HeLa cell systems. It was not until twenty-five years later that Dr. Roland Pattillo, in Dr. Gey's laboratory, established the BeWo cell line from a trophoblastic tumor which had been transplanted to the hamster cheek pouch by Dr. Roy Hertz at the National Institutes of Health.

This cell line, a human chorionic gonadotropin producer, continues to produce hCG in continuous culture eighteen years after its establishment. The "pregnancy hormone" secreted by these cells is detected by radioimmunoassay—rabbit, rat, mouse, frog, and other assays. Like HeLa, these cells have been distributed all over the world, but unlike HeLa, the marker for hCG is continuously produced and serves to identify this cell line. Since the establishment of the BeWo line, and because the BeWo line had been maintained in hamster cheek pouches for a number of years, it was deemed necessary to establish cell lines directly from patients so that no animal transplantation intermediary would be present. To this end, the JaR cell line was established when Pattillo and his laboratory moved from Johns Hopkins to The Medical College of Wisconsin in Milwaukee. Multiple trophoblastic cell lines, as well as other cell lines of the cervix, breast, and ovary have been applied in medical research throughout the world. The First International Trophoblast Congress was held in Nairobi, Kenya, East Africa in 1982, which resulted in a book, "Human Trophoblast Neoplasms" published by Plenum Press in 1984.

The pursuit of dreams of cellular physiology and tissue culture systems answers the quests of Dr. George Gey over and over again in laboratories all over the world, truly George Gey's bequest to human biology.

ROLAND A. PATTILLO
The Medical College of Wisconsin
Milwaukee, Wisconsin 53226
with the assistance of
Margaret Gey

CONTENTS OF PREVIOUS VOLUMES

ADVANCES IN CELL CULTURE, VOL. 4

HUMAN MYOPATHIES IN MUSCLE CULTURE: MORPHOLOGICAL, CYTOCHEMICAL, AND BIOCHEMICAL STUDIES

Armand F. Miranda,* Tiziana Mongini,† and Salvatore DiMauro†

Departments of *Pathology and †Neurology
Columbia University, College of Physicians and Surgeons
and the H. Houston Merritt Clinical Research Center
for Muscular Dystrophy and Related Diseases
New York, New York

I. Introduction

A. *Myogenesis*

During ontogeny, myogenic stem cells derived from mesenchyme proliferate to produce myoblasts that line up and fuse spontaneously to form multinucleated syncytia (myotubes) (see Fischman, 1972; Bischoff, 1978; Hauschka *et al.*, 1982). These postmitotic muscle straps differentiate further into cross-striated myofibers (for review see Fischman, 1982) but attain full maturity only after they are innervated by motor neurons. This entire process can be reproduced in culture, making it an ideal system for studying the cellular and molecular processes of myogenesis. The induction of genes coding for myofibrillar proteins and the correlation between cell fusion and initiation

ISBN 0-12-007904-6

of synthesis of muscle-specific gene products are now being studied extensively (see Hastings and Emerson, 1982; Nguyen *et al.*, 1983). The most commonly used muscle cell strains are derived from avian or rodent embryos or neonates (Konigsberg, 1963; Bischoff and Holtzer, 1969; Miranda and Godman, 1973), although several permanent rodent cell lines are also commonly employed (Richler and Yaffe, 1970; Yaffe, 1973; Hauschka *et al.*, 1979). For our studies, we use cultures derived from human muscle biopsies, which grow and differentiate similarly to cultures derived from embryonic sources.

B. *Muscle Satellite Cells and Regeneration*

Regeneration is the process by which tissue, injured mechanically or by disease, is replaced by newly formed tissue with similar structural and functional characteristics. Regeneration of muscle is quite efficient, unless damage is very extensive in which case "repair" may occur due to ingress of connective tissue cells and formation of scar tissue (see Godman, 1957; Murray, 1965; Carlson *et al.*, 1979). Regeneration can also be reproduced *in vitro,* and this forms the basis for obtaining differentiating muscle cultures derived from mature muscle (Fig. 1). In this system, dormant stem cells, named "muscle satellite cells" by Mauro (1961) who first identified them ultrastructurally in frog skeletal muscle, were shown to be a source of myoblasts which fuse to form new muscle fibers during muscle regeneration (Bischoff, 1974, 1975; Snow, 1977, 1978, 1979). Regeneration of muscle following injury is a recapitulation of myogenesis occurring during ontogeny (Carlson *et al.*, 1979; Mong *et al.*, 1982; Fig. 1).

Although human muscle cultures obtained from muscle biopsies of adults grow and differentiate remarkably well (Pogogeff and Murray, 1946; Fig. 1), relatively little use was made of this *in vitro* system to study muscle diseases until the 1970s (Askanas and Engel, 1975; Witkowski, 1977; Miranda *et al.*, 1979a; Yasin *et al.*, 1981). However, in recent years numerous *in vitro* studies of genetic and acquired human myopathies have yielded important information that would have been difficult to obtain from *in vivo* investigations. With the use of tissue cultures, many questions concerning the pathobiology of neuromuscular diseases can be posed.

1. In genetic muscle diseases of unknown etiology, the myopathologic changes may be due to abnormality of the muscle fibers or to "sick motor neurons." Can these diseases be classified more accurately using muscle culture?

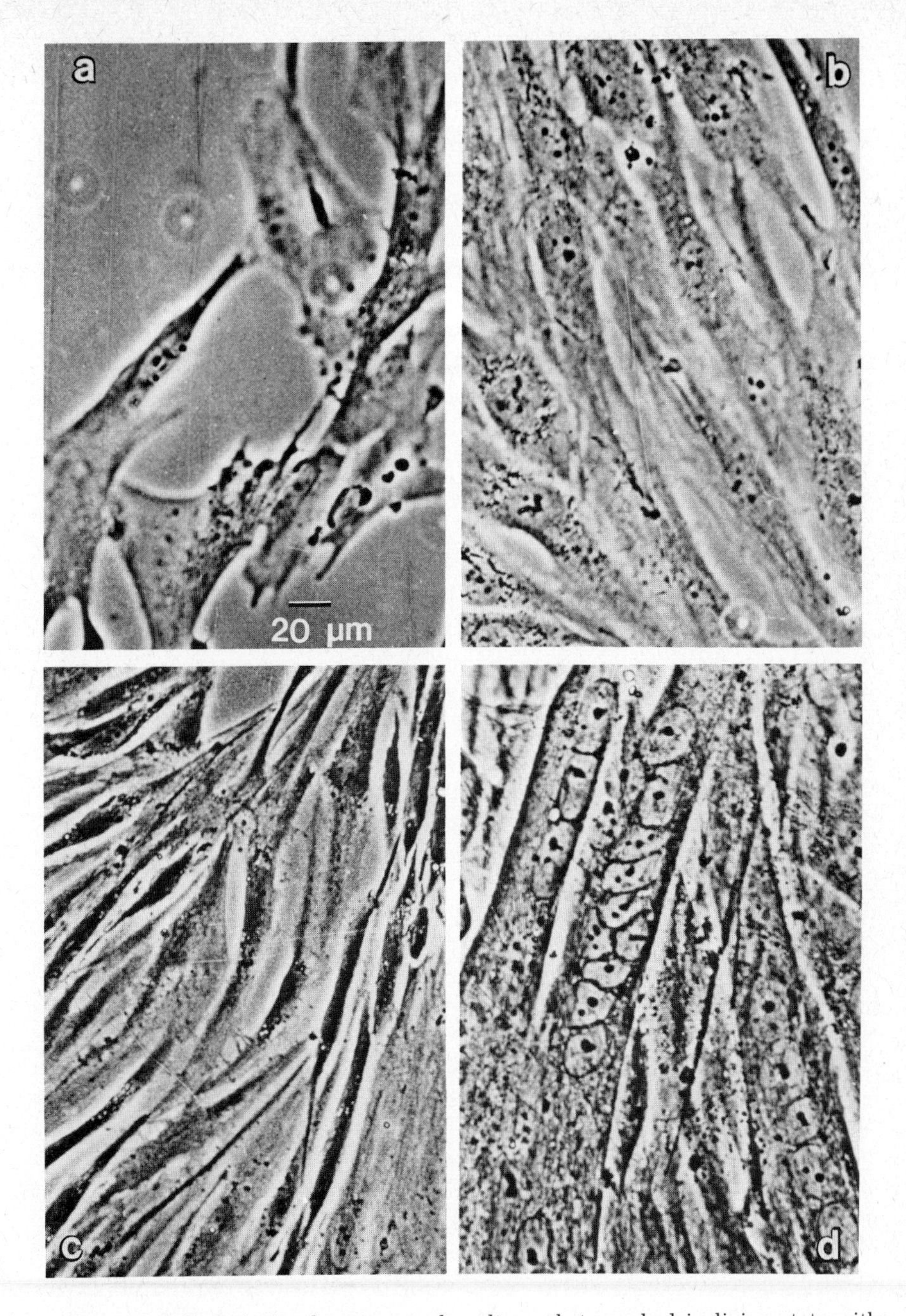

FIG. 1. Differentiating human muscle culture photographed in living state with phase optics. (a) After 48 hours muscle satellite cells began to proliferate. (b) At 120 hours the culture was almost confluent. (c) At 144 hours the cells lined up into rows and began to fuse. (d) At 192 hours prominent multinucleated myotubes were seen.

2. Some genetic myopathies are not clinically expressed until later in postnatal life. Can muscle cultures be used to pinpoint the stage of muscle differentiation at which these diseases become fully expressed?

3. Since nerve factors and innervation play an important role in muscle maturation, can nerve–muscle cocultures help elucidate the role of nerve–muscle interactions in the pathogenesis of genetic myopathies?

4. Nonhereditary muscle diseases may be due to infectious agents (such as viruses), myotoxic chemicals, or dysfunction of tissues other than muscle (e.g., abnormal immunologic response). Can muscle cultures be used to identify infectious agents in such "acquired myopathies" or to study the effects of potentially myotoxic agents, myotropic viruses, and autoantibodies?

5. Since differential diagnosis is sometimes difficult in myopathies of unknown etiology, can muscle cultures be used as a diagnostic tool?

In this article, we will discuss muscle culture systems that are presently used to study human disease. We will analyze in more detail genetic metabolic myopathies due to specific enzyme defects because, besides clarifying aspects of pathogenetic mechanisms, these investigations can also help elucidate the functional significance of specific enzymes in normal muscle metabolism. Finally, we will consider the value of human muscle culture as an experimental tool to learn something more about normal muscle regeneration and the pathogenesis of nonhereditary (acquired) myopathies.

C. The Muscle Culture System

Even though human muscle cultures are readily grown from fetal muscle, a more important source of muscle to study genetic or acquired myopathies *in vitro* is represented by biopsies obtained in the course of diagnostic evaluation. The methods for growing human muscle are similar to the standard techniques originally developed for avian and rodent embryonic muscle, although several studies have indicated that for human muscle culture the need for a collagenous substrate and for embryo extract in the medium appears less stringent. Witkowski *et al.*, (1976) outlined the various techniques and nutrient media employed to grow human muscle *in vitro*. More recently, several investigators have concentrated on cultures derived from muscle clones, which would eliminate contaminating fibroblasts and facilitate comparative studies between normal and genetically deficient muscle (Blau and Webster,

1981; Yasin *et al.*, 1981; Ionasescu and Ionasescu, 1982). The media and substrate more commonly used at the present time are outlined in Table I. Most of these procedures were adapted for growing mature human muscle. Hauschka (1974) analyzed the requirement for *in vitro* cultivation of *fetal* human muscle in detail.

To evaluate critically the data obtained on neuromuscular diseases in muscle culture, it is important to review briefly the various methods of culture.

1. Explants

Most early studies were carried out on explant cultures in which small pieces of muscle freed from fat and connective tissue are placed in culture dishes in appropriate media. After 4–6 days, myogenic stem cells begin to migrate out of the explant. These cells proliferate and ultimately fuse into multinucleate myotubes. After removal of the initial explants, the cultures are harvested by brief enzymatic digestion or by scraping with a rubber policeman. The explants can be regrown for a second time with similar results.

This method yields good differentiating myotubes, but the degree of differentiation may vary considerably between different cultures, even when they are derived from the same individual (Herrmann *et al.*, 1960; Miranda and Mongini, 1983, for review). Moreover, particularly in short-term cultures some of the explanted muscle may remain intact and such remnants may contaminate the culture samples, thus leading to erroneous biochemical data. A reexplantation method to reduce contamination with fibroblasts has been reported by Askanas and Engel (1975).

Muscle explants can also be cultured in the presence of fetal spinal cord of rodents ("organotypic") in which the human muscle becomes innervated by motorneurons. In contrast to noninnervated explant cultures which tend to deteriorate after 4 or 5 weeks, nerve–muscle cultures can be maintained for several months. Myotube formation in such cocultures often occurs within intact sarcolemmas of the explanted fibers, and this results in highly organized regenerating muscle bundles which can be analyzed morphologically, histochemically, electrophysiologically, and even biochemically (Crain *et al.*, 1970; Peterson and Crain, 1979; Miranda *et al.*, 1983a). Innervated cultures are more mature than uninnervated ones and are therefore a more suitable system for studying neuromuscular diseases that are not fully expressed in less mature, aneurally grown muscle (Miranda *et al.*, 1982, 1983a).

TABLE I

SOME MEDIA AND SUBSTRATES USED FOR HUMAN MUSCLE CULTURE

Type of culture	Dissociation and other procedures	Medium and adjuvants[a]		Substrate	Reference
Clonal	Collagenase	Ham F-10	79	Gelatin or plastic	Hauschka (1974)
		HS	15		
		EE	5		
Explant	Reexplantation	(a) Dulbecco	56	Human plasma/gelatin	Askanas and Engel (1975)
		M199	19		
		HPS	15		
		(b) Dulbecco	63		
		M199	21		
		FBS	10		
		EE	5		
		(c) Dulbecco[b]	45		
		Eagle	45		
		HS	10		
		Glucose (a–c)	0.6% (w/v)		
Monolayer	From explants: cell outgrowth detached with trypsin	(a) M199	87.5	Gelatin	Witkowski *et al.* (1976)
		HS	10		
		EE	2.5		
		(b) M199[b]	88.75		
		HS	10		
		EE	1.25		
		Glucose (10 m*M* Hepes buffered; a, b)	0.6%(a,b)		

Nerve–muscle	13–14 day fetal mouse spinal cord and human muscle fibers	Eagle	50	Reconstituted rat tail collagen	Peterson (1978)
		EE	10		
		HPS	30		
		Hanks' BBS	7		
		Glucose	0.6%		
Clonal, monolayer	From explants:cell outgrowth detached with trypsin/EDTA preplating	(a) Eagle	82	Plastic or gelatin-coated	Miranda (1974); Miranda *et al.* (1979a)
		FBS	15		
		Nonessential amino acids	1		
		Vitamins	1		
		Na-pyruvate (100 ×; Gibco)	1		
		(b) Eagle[b]	90		
		FBS	10		
Clonal	Trypsin/collagenase	(a) Dulbecco	100	Gelatin	Yasin *et al.* (1981)
		FBS	20		
		EE	2		
		(b) Dulbecco[b]	100		
		FBS	10		
		EE	2		
Clonal	Trypsin/EDTA pre-plating, prior to cloning	(a) Ham F12	75	Denatured calfskin collagen	Blau and Webster (1981)
		FBC	20		
		EE	5		
		(b) Dulbecco[b]	98		
		HS (conditioned)	2		

[a] Additional formulations are discussed by Hauschka *et al.* and Yasin *et al.* All media (except Peterson's) also contained antibiotics. Numbers refer to relative volumes, sometimes approximated (see references for details). Abbreviations: HS, horse serum; FBS, fetal bovine serum; HPS, human placental serum; M199, medium 199; BSS, balanced salt solution; EE, chick embryo extract.

[b] These media are used for maintenance or to promote differentiation.

2. *Mass Monolayers*

The most commonly employed human muscle culture system involves the growth of trypsinized cells in monolayers and the removal of contaminating nonmuscle cells (e.g., fibroblasts) by various methods. One advantage of this system over explant cultures is that cells can be planted at preselected densities and different culture dishes can be seeded with the same number of cells, thus allowing experimentation in duplicate or even multiple identical cultures.

Removal of contaminating nonmuscle cells can be carried out before myoblast fusion by "preplating," a technique that takes advantage of the fact that dissociated fibroblasts adhere more rapidly to the growth substrate than myoblasts. After incubation of the trypsinized cells for 15 minutes or longer, the myoblast-enriched supernate can be transferred to another culture vessel for further growth and differentiation (Yaffe, 1973; Miranda *et al.*, 1979a; Fig. 1). Another method to select for myoblasts is treatment with fungal metabolites, cytochalasins B or D, which cause selective detachment of myoblasts in cultures derived from embryonic avian muscle (Sanger, 1974). Since the effects of these drugs are fully reversible upon removal, this method is both efficient and reliable. However, cytochalasin selection has not yet been tested adequately in regenerating human muscle cultures.

Pure myotube cultures can be obtained by application of "DNA poisons" such as cytosine arabinoside (ara C) which become incorporated into replicating cellular DNA. Since conventional DNA synthesis no longer occurs in myotubes, replicating mononuclear cells can be selectively removed with ara C (Fishbach and Cohen, 1973). Again, the use of such agents has not been adequately tested in human muscle cultures and, moreover, it has not been completely excluded that such poisons are not harmful to postsynthetic cells, such as myotubes. Fambrough *et al.* (1982) obtained monoclonal antibodies that selectively kill mononuclear cells in myotube cultures in the presence of complement. This novel approach might become a method of choice for eliminating "contaminating fibroblasts" in differentiating muscle cultures.

3. *Clones*

Perhaps the most rational approach to obtain "pure" muscle cultures is to grow the cells at clonal density and isolate myogenic clones for further propagation. Several investigators have already applied this method successfully to obtain human muscle cells for biochemical analyses (Blau and Webster, 1981; Yasin *et al.*, 1981; Ionasescu and Ionasescu, 1982). The use of muscle clones however, still requires that suitable markers of differentiation, such as creatine kinase isozyme transition and myoblast fusion index, be available to determine the degree of differentiation (in cultures from controls and patients). This is because actively proliferating myogenic cells lose muscle cell characteristics and acquire fibroblastic features. This has been demonstrated in clonal cultures from embryonic chick (Abbott *et al.*, 1974). Moreover, individual cell clones may exhibit differences in degree of myoblast fusion and differentiation while still retaining muscle cell characteristics, as reported for clonal cultures of human origin (Yasin *et al.*, 1982).

4. *Muscle Grafts*

Implantation of muscle in a suitable animal host can be considered an *in vivo* method of culture and has already been applied successfully in human muscle studies (O'Steen, 1963; Yarom *et al.*, 1981). To avoid tissue rejection, athymic nude mice were used as hosts to study regeneration of biopsied human muscle (Wakayama *et al.*, 1980; Wakayama and Ohbu, 1982). The *in vivo* culture model might turn out to be a suitable alternative to nerve–muscle cocultures since cross-species innervation can occur in heterologous muscle grafts.

5. *Permanent Cell Lines*

Because muscle cells derived from primary tissue do not have an unlimited life span, permanent cell lines established from normal and diseased human muscle would represent an ideal source of unlimited amounts of cultured muscle cells for biochemical and genetic analyses. Apparently immortal muscle cell lines from mouse and rat are already widely used for experimental studies (Section I,A). Human muscle cells transformed with wild-type simian virus 40 (SV40) have also been obtained, but these cultures are only of limited value for studying myopathies because these transformants lose the ability to differentiate after several transfers (Miranda *et al.*, 1983b). Using temperature-sensitive SV40 mutants (TsA-SV40), it might be feasible to obtain transformed human muscle cell lines that retain the capacity to differentiate. The transformed cultures would be grown initially at 33°C, permissive to TsA-SV40, and the transformed cultures would later be switched to about 40°C, nonpermissive to the transforming virus. TsA-SV40 mutants have already been used successfully to obtain differentiating human cultures from human placenta (Chou, 1978). Transfection of human epidermal cells by Ts-SV40 DNA has been applied successfully as well: These cells were shown to differentiate normally at elevated temperature (Banks-Schlegel and Howley, 1983).

Another method for studying genetic myopathies *in vitro* is the use of somatic muscle cell hybrids obtained by fusions between cells of a permanent rodent muscle cell line and human myoblasts. Since rodent–human hybrids tend to lose human chromosomes upon continuous cultivation, cell colonies can be selected that synthesize human proteins coded on chromosomes or fractions of chromosomes that have been retained. Fusions between L6 rat myoblasts and human myoblasts have already yielded hybrids that of the human genome only retain a functional X chromosome, thus facilitating the gene mapping of normal cells and cells with X-linked genetic deficiencies (e.g., Duchenne muscular dystrophy) (Quinn *et al.*, 1981; Walsh *et al.*, 1982).

II. Hereditary Diseases of Muscle

Sequential studies of human muscle are limited by the difficulty to perform repeated muscle biopsies. As an alternative, animal models have been used because several neuromuscular diseases in animals resemble human disorders and may contribute to a better understanding of the pathogenesis of hereditary human myopathies. However, animal and human disorders may have different etiologies, and to understand the causes of human myopathies, it remains essential to analyze human tissues. Since *in vivo* experimentation is obviously limited or impossible in humans, the muscle culture system is the only other approach that makes such studies feasible. In this section, we will review the studies carried out in human muscle cultures and the results obtained to date. Some of these findings are still controversial and others have not been confirmed by repeated analysis in other laboratories. The use of uninnervated and innervated human muscle cultures could contribute significantly toward understanding the developmental pathobiology of lesions caused by genetic mutations.

A. *Metabolic Myopathies*

The metabolic myopathies are hereditary diseases in which gene defects cause specific muscle enzyme deficiencies. Some of these disorders are due to defects of muscle-specific enzyme subunits which may be present exclusively in striated muscle or may be found also in other tissues, but in smaller amounts. Most metabolic myopathies identified to date are due to deficiencies of enzymes of glycogen metabolism (Fig. 2). The few biochemically identified disorders of lipid metabolism will be discussed separately. In the one disorder of nucleotide metabolism,

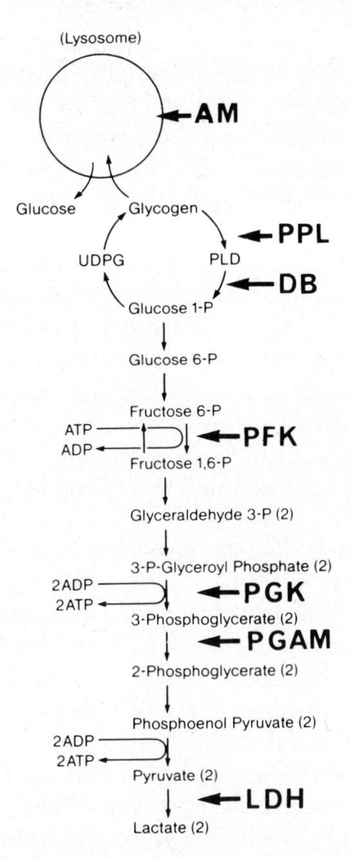

FIG. 2. Pathways of glycogen metabolism and glycolysis. Arrows indicate the sites of the metabolic blocks due to genetic deficiencies of AM (acid maltase), PPL (phosphorylase), DB (debrancher enzyme), PFK (phosphofructokinase), PGK (phosphoglycerate kinase), PGAM (phosphoglycerate mutase), and LDH (lactate dehydrogenase).

adenylate deaminase (AMPD) deficiency, it is still uncertain whether the clinical symptoms are directly related to the enzyme defect (Engel *et al.*, 1964; Fishbein *et al.*, 1978; Shumate *et al.*, 1979).

1. Disorders of Glycogen Metabolism

a. Phosphorylase (PPL) Deficiency. PPL catalyzes the first reaction in the degradation of glycogen through the stepwise removal of 1,4-glucosyl residues from the peripheral branches of glycogen with liberation of glucose 1-phosphate. The partially degraded glycogen that results from the action of phosphorylase (phosphorylase-limit-dextrin), is further broken down hydrolytically by the debrancher enzyme (Fig. 2).

In humans and other vertebrates, PPL occurs in multiple tissue-specific forms which are under separate genetic control (Yunis *et al.*, 1962; Davis *et al.*, 1967; Will *et al.*, 1974). During myogenesis and muscle regeneration, PPL isozymes undergo developmentally controlled transitions and in mature muscle a single molecular form, known as muscle phosphorylase (PPL-M; PPL-III), is found almost exclusively (DiMauro *et al.*, 1978).

An autosomal recessive genetic deficit of PPL-M (McArdle disease) causes exercise intolerance, cramps, appearance of myoglobin in urine (myoglobinuria), and, in about one-third of the patients, fixed weakness (DiMauro and Bresolin, 1984). The muscle biopsy typically shows subsarcolemmal glycogen accumulation and lack of histochemically or biochemically detectible PPL activity. The disease is usually diagnosed in young adults and the course is benign. More recently, however, a lethal form of PPL-M deficiency has been reported (DiMauro and Hartlage, 1978).

PPL-M deficiency was the first hereditary metabolic myopathy to be studied in muscle culture (Roelofs *et al.*, 1972). The results were surprising because histochemically detectable phosphorylase activity was found in myotube cultures from patients. Moreover, regenerating areas of muscle taken by repeat biopsy of the same site also had phosphorylase activity, showing that this phenomenon was not due to an artifact of the *in vitro* cultivation. This puzzling observation was confirmed in subsequent more detailed studies, and it was also shown that total PPL activity was normal and that there was no glycogen accumulation in cultured myotubes from patients (Meienhofer *et al.*, 1977; Sato *et al.*, 1977; DiMauro *et al.*, 1978).

Several hypotheses have been proposed to explain the "mysterious" reappearance of PPL in muscle cultures from patients with genetic lack of PPL-M activity, including the reexpression or reactivation of the PPL protein. However, a more likely explanation is that the PPL activity in muscle cultures is due to a different PPL isozyme. Immunochemical and electrophoretic studies confirmed the presence of non-muscle ("fetal") isozymes in myotube cultures from patients, which differed from mature muscle PPL: one of the two or three bands of PPL activity present in fetal muscle comigrated with the single band found in electropherograms of cultures from normal individuals and from patients with PPL-M deficiency. Also, PPL activity of muscle cultures could not be precipitated by anti-PPL-M antibodies (Sato *et al.*, 1977; DiMauro *et al.*, 1978). However, more recent studies with specific anti-PPL-M antibodies showed that PPL-M protein is already present in myotube cultures from normal individuals (Davidson *et al.*, 1983b). It is not yet known at which stage of myogenesis PPL-M deficiency is expressed. Perhaps more mature cultured muscle straps obtained by *in vitro* innervation of myotubes from patients will confirm the isozyme hypothesis, reveal the disappearance of PPL activity, and show characteristic myopathologic changes (e.g., glycogen accumulation).

b. Debrancher Enzyme (DB) Deficiency. DB catalyzes the second step in glycogen degradation by acting on partially degraded glycogen (phosphorylase-limit-dextrin, PLD). DB has two catalytic functions, apparently shared by a single protein. In the first step maltotriosyl units of PLD are transferred from donor to acceptor chains by a transferase. In the second step, a glucosidase catalyzes the removal of α-1-6 glucosyl units (see Brown, 1984; Fig. 2).

DB deficiency causes marked accumulation of PLD in muscle. The disease is transmitted by autosomal recessive inheritance, and most patients lack both transferase and hydrolase activities. However, in a few patients only one but not the other catalytic function was defective (Brown, 1984).

Muscle cultures from two patients with DB deficiency involving both transferase and hydrolase functions also showed complete lack of DB activity. The myoblasts proliferated and fused normally, but electron microscopic studies showed abnormally large glycogen deposits both in mononuclear cells and myotubes (Fig. 3). Most of the glycogen was free in the cytoplasm, although some membrane-bound lysosome-like structures filled with glycogen were also present (Miranda *et al.*, 1981a; Miranda and Mongini, 1984). Membrane-bound glycogen is also found in muscle biopsies from patients and may represent an attempt by the cell to degrade the excessive glycogen through a lysosomal pathway (DiMauro *et al.*, 1979; Miranda *et al.*, 1981a).

Unlike PPL-M deficiency, DB deficiency is expressed in all tissues (Illingsworth, 1961; Justice *et al.*, 1970; DiMauro *et al.*, 1979; Miranda *et al.*, 1981a). The full phenotypic expression of DB deficiency in tissues other than muscle and in muscle cultures at all stages of differentiation indicates that the enzyme does not occur in multimolecular forms and is not developmentally controlled.

c. Phosphofructokinase (PFK) Deficiency. PFK catalyzes the conversion of fructose 6-phosphate to fructose 1,6-bisphosphate with utilization of one molecule of ATP, a major rate-limiting step in the glycolytic pathway (Fig. 2). The enzyme is a tetramer, and in humans there are three tissue-specific subunits under separate genetic control: muscle (M), liver (L), and platelet (P) (also called F for fibroblast). These subunits can hybridize randomly to yield as many as 15 homo-and heterotetramers (Vora, 1982, for review). In humans, the genetic loci coding

MF
Gly
05 µm
a

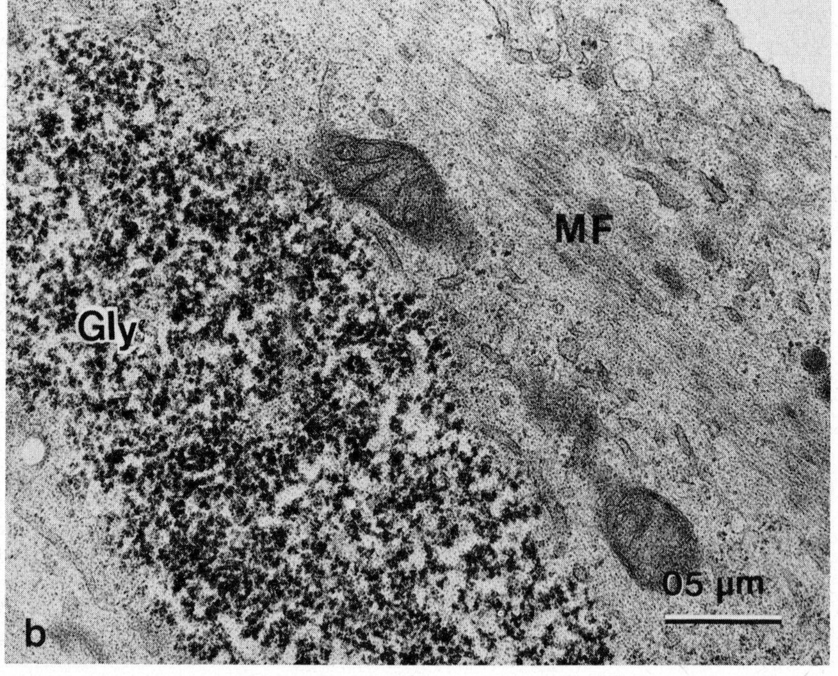
MF
Gly
05 µm
b

for the M, L, and P (or F) subunits of PFK have been assigned to chromosomes 1, 21, and 10, respectively (Weil *et al.*, 1980; Vora and Francke, 1981; Vora *et al.*, 1982, 1983).

Not only are PFK isozyme patterns different in different tissues, but during myogenesis or muscle regeneration there are isozyme transitions from a pattern composed mostly of nonmuscle PFK-L and -P to one composed exclusively of the muscle-specific PFK-M. Thus, in normal mature muscle PFK-M_4 is the only isozyme present. PFK isozymes can be studied electrophoretically or immunochemically using antibodies against the different subunits. However, electrophoretic separation does not clearly distinguish all the isozymes, and a cleaner separation of PFK-M_4, -L_4, -P_4, and their heterotetramers can be achieved by affinity chromatography (Vora *et al.*, 1980).

Glycogenosis type VII or Tarui disease (Tarui *et al.*, 1965; Layzer *et al.*, 1967) is due to a genetic deficiency of the PFK-M subunits. PFK-M deficiency is usually associated with exercise intolerance and hemolysis, although a few patients showed either predominantly myopathic or predominantly hemolytic features (see Vora, 1982).

PFK-M deficiency has been studied in differentiating muscle cultures. As in the case of PPL-M deficiency, the cultures showed histochemically detectable PFK activity and normal or near-normal total PFK activity, and there was no evidence of abnormal glycogen accumulation (Miranda *et al.*, 1982; Davidson *et al.*, 1983a; Fig. 4). However, using ion-exchange column chromatography and selective immunologic precipitation of the enzymatically active fractions with monoclonal antibodies, it was shown that muscle cultures from patients lack the PFK-M_4 peak, and the "normal" activity was due to the presence of isozymes containing PFK-P and -L subunits. In normal control muscle cultures, PFK-M containing isozymes represented only about 15% of the total activity, thus explaining the "reappearance" of almost normal enzyme activity in myotube cultures from patients. These studies also show that the presence of nonmuscle PFK isozymes in immature muscle can compensate for the lack of PFK-M because there was no abnormal accumulation of glycogen in cultures (Davidson *et al.*, 1983a). The presence of PFK-L-containing isozymes was further documented by indirect immunocytochemistry using monospecific antibody

FIG. 3. (a) Electron micrograph of a muscle section from a patient with debrancher enzyme deficiency showing massive intermyofibrillar glycogen (Gly) deposits. (b) Glycogen accumulation is also seen in the patient's muscle in culture. MF, myofibrils. Reproduced with publisher's permission (a) from DiMauro *et al.* (1979) (*Ann. Neurol.*).

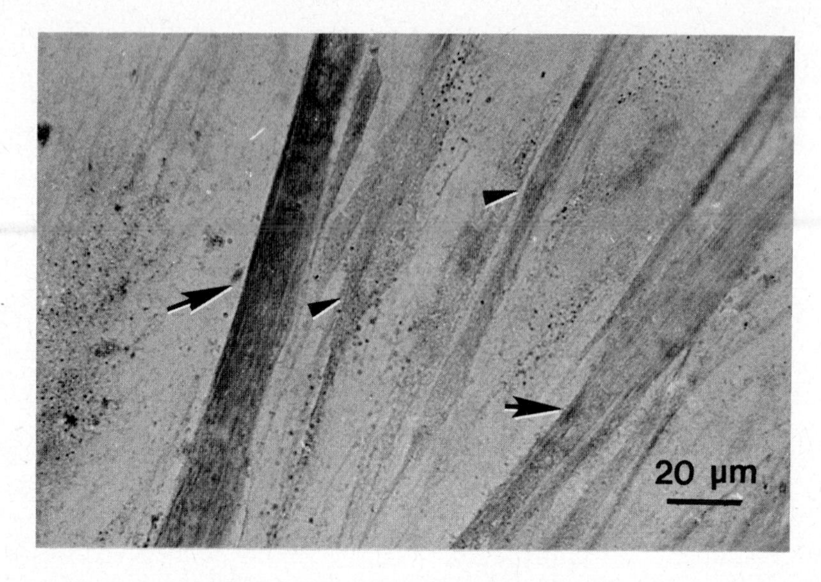

FIG. 4. Muscle culture from a patient with PFK-M deficiency (Tarui disease) showing prominent histochemical staining of PFK activity in multinucleated myotubes (arrows) and in some mononuclear cells (arrowheads). Normal control cultures (not shown) had similar staining patterns.

against subunit L and a second fluorescein-tagged antimouse IgG. In the 3 patients studied, PFK-M deficiency was due to synthesis of an inactive PFK-M subunit rather than to lack of enzyme synthesis because immunologically PFK-M-cross-reacting protein was present in myotubes (Davidson *et al.*, 1983a). Two other studies showed partial or complete lack of PFK activity in muscle cultures from patients, but these studies did not include detailed isoenzyme analysis (Askanas and Engel, 1979; Zanella *et al.*, 1982). Since more advanced muscle maturation can be achieved in cultures (e.g., by innervation; Miranda *et al.*, 1983a), full phenotypic expression of PFK-M deficiency may be reproduced *in vitro*.

d. Phosphoglycerate Kinase (PGK) Deficiency. PGK catalyzes the conversion of 3-phosphoglyceroyl phosphate to 3-phosphoglycerate with generation of one molecule of ATP (Fig. 2). In humans, only two molecular forms of PGK have been described: PGK 1 is found in most tissues, including muscle at all stages of differentiation, and the gene locus is on the long arm of the X chromosome (Chen *et al.*, 1971). On the other hand, PGK 2 appears to be restricted to spermatogenic cells, and the gene locus coding for this enzyme has not yet been assigned. Several PGK mutations have been described in humans. Some of these cause hemolytic anemia and brain dysfunction while others are clinically silent (see DiMauro *et al.*, 1983). In two PGK variants, single amino acid substitutions have been demonstrated (Fujii *et al.*, 1980, 1981). A newly identified clinical variant was characterized by recurrent myoglobinuria after strenuous physical exercise without hemolytic anemia (DiMauro *et al.*, 1982b; Bresolin *et al.*, 1983). Muscle cultures and skin fibroblast cultures from this patient, a 15-year-old boy, showed decreased PGK activity, although the cultured cells appeared morphologically normal.

The expression of PGK deficiency in muscle cultures is not surprising, since PGK appears to exist as a single molecular form at all stages of myogenesis and is also found in skin fibroblasts. This was also demonstrated *in vivo* during myogenesis of normal intact fetal muscle (DiMauro *et al.*, 1982b; Miranda *et al.*, 1982). Although the enzyme defect was reproduced in muscle cultures, there was no abnormal accumulation of glycogen. Glycogen concentration was also normal in the muscle biopsy of the patient. This was probably due to the presence in both intact and cultured muscle of residual enzyme activity, which may have allowed near normal glycogen degradation (DiMauro *et al.*, 1982b). Even though only two PGK isozymes have been documented to date, there is evidence that other PGK enzyme forms may exist in mammalian tissues (Hiremath and Rothstein, 1982). In addition, skin fibroblasts from the patient with myoglobinuria, transformed by simian virus 40 (SV40), showed an increase of PGK activity approaching normal values. Electrophoresis of these transformed cells revealed an additional band of PGK activity that was not visible in electropherograms of untransformed fibroblast from patient or controls. However, it is not yet certain whether this represents a "true isozyme" or whether the difference in electrophoretic mobility is due to posttranslational modification of the PGK molecule (N. Bresolin, A. F. Miranda, and S. DiMauro, unpublished).

e. Phosphoglyceromutase (PGAM) Deficiency. The conversion of 3-phosphoglycerate to 2-phosphoglycerate is catalyzed by PGAM (Fig. 2), a dimeric enzyme composed of a "muscle" (M) subunit and a "brain" (B) subunit. The M and B subunits, which are under separate genetic control, form the homodimeric isozymes MM and BB, but can also hybridize under physiologic conditions yielding the heterodimer isozyme MB that migrates electrophoretically between the slow PGAM-MM and the faster BB isozyme (Omenn and Cheung, 1974; Omenn and Hermodson, 1975). Within the past 2 years, 3 patients have been identified with exercise-induced recurrent myoglobinuria (DiMauro *et al.*, 1981, 1982a; Bresolin *et al.*, 1983; J. A. Mendell and S. DiMauro, unpublished) due to genetic lack of PGAM-MM activity. Muscle cultures were obtained from 2 patients. Differentiating myotubes were morphologically normal and showed no evidence of glycogen accumulation. Total PGAM activity was also normal, suggesting the presence of the nonmuscle PGAM isozyme. This was confirmed by electrophoretic studies on cellulose acetate showing only a single band corresponding to PGAM-BB (DiMauro *et al.*, 1981, 1982a; Bresolin *et al.*, 1983; Miranda and Mongini, 1984; Fig. 5). Long-term uninnervated cultures from normal muscle (Fig. 5 showed both PGAM-BB (70%) and PGAM-MM (30%), suggesting that PGAM-M deficiency may be at least partially expressed *in vitro* after prolonged cultivation (5–6 weeks) (Davidson *et al.*, 1983b). Full phenotypic expression of PGAM-M deficiency may be demonstrated in cocultures of human muscle innervated by fetal mouse motor neurons because in normal muscle cultures under these conditions conversion from PGAM-BB to PGAM-MM is virtually complete (Miranda *et al.*, 1983a).

f. Lactate Dehydrogenase (LDH) Deficiency. The final step of anaerobic glycolysis (i.e., reduction of pyruvate to lactate) is catalyzed by

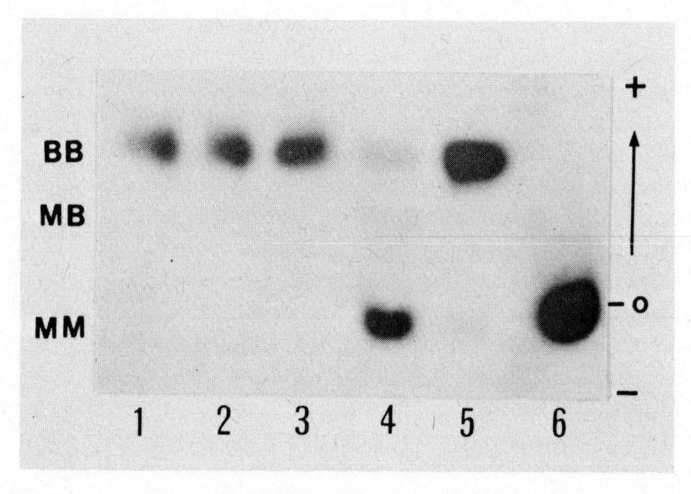

FIG. 5. Cellulose acetate electrophoresis stained for PGAM activity. Lane 1: Muscle culture from a patient with PGAM deficiency showing only the BB isozyme, similar to normal control muscle cultures (lanes 2,3). Lane 4: Fetal muscle (22-weeks gestation) already shows a more advanced stage of PGAM isozyme differentiation; 3 bands are present, PGAM-MM predominates. Lane 5: Muscle from a patient with PGAM deficiency shows only the BB isozyme; PGAM-MM is present as a very faint band only in undiluted 10% muscle homogenates. Lane 6: Normal control muscle homogenate, diluted 1:1000 shows prominent PGAM-MM activity. Reproduced with publisher's permission from DiMauro *et al.* (1981) (*Science*).

LDH, a tetrameric enzyme (Fig. 2). Five isozymes of LDH are present in vertebrate tissues, resulting from random tetramerization of 2 subunits, LDH-M and LDH-H, which are encoded on chromosomes 11 and 12 (Boone *et al.,* 1972; Mayeda *et al.,* 1974; Miranda *et al.,* 1979a for review). The 5 isozymes, 2 homotetramers, and 3 heterotetramers (M_4, M_3H_1, M_2H_2, M_1H_3, H_4) can be separated electrophoretically. During myogenesis there is a gradual transition from a pattern with a prevalence of the anodic LDH-H isozymes to a pattern where the cathodic LDH-M isozymes predominate. This conversion also occurs in muscle culture, even in uninnervated myotubes (Miranda *et al.,* 1979a).

Deficiency of the LDH-H subunit was found serendipitously in one Japanese family and did not appear to cause any symptoms (Kitamura *et al.,* 1971). Deficiency of LDH-M was identified in another Japanese kindred and was associated with exercise-induced myoglobinuria (Kanno *et al.,* 1980). LDH deficiencies have not yet been studied in muscle cultures, but since the LDH isozyme pattern of cultured myotubes is almost indistinguishable from mature skeletal muscle, full phenotypic expression of LDH-M deficiency might be demonstrable in muscle culture. Hybridization studies of cultured cells from patients might reveal whether these genetic LDH deficiencies are due to defect of structural or regulatory gene loci.

2. *Disorders of Lipid Metabolism*

Carnitine palmitoyltransferase (CPT) deficiency is the only hereditary disorder of lipid metabolism in which the deficient enzyme has been identified (DiMauro and Melis-DiMauro, 1973; DiMauro and Papadimitriou, 1984). Another disorder of lipid metabolism, a triglyceride storage myopathy (TSM) of unknown etiology, has been described in 3 patients with excessive neutral lipid accumulation in all tissues, including muscle (Chanarin *et al.,* 1975; Miranda *et al.,* 1979b; Angelini *et al.,* 1980). Although these cases were sporadic, the phenotypic expression of TSM in fibroblast and muscle cultures suggests that this is a genetic disease, but persistent infection by a "hidden" virus should also be considered. No enzyme defect has yet been identified in TSM, but since the lipid droplets in the sarcoplasm are not membrane-bound, the deficient enzyme is likely to be extralysosomal. Carnitine deficiency, both systemic and myopathic, is another disorder of lipid metabolism affecting muscle in which the genetic lesion(s) remains to be identified (Karpati *et al.,* 1975; Engel and Rebouche, 1982). All three disorders have been studied in skin fibroblasts and muscle cultures and the findings are presented below.

a. Carnitine Palmitoyltransferase (CPT) Deficiency. CPT (CPT I), loosely bound to the outer face of the inner mitochondrial membrane, catalyzes the synthesis of palmitoylcarnitine from carnitine and palmitoyl-CoA, an essential metabolic step to facilitate transport of long-chain fatty acid residues across the inner mitochondrial membrane (CPT I). Carnitine is subsequently removed by a second CPT (CPT II) which is tightly bound to the inner aspect of the inner mitochondrial membrane. Although CPT I and CPT II perform separate catalytic functions, it is not yet known whether they are encoded on different gene loci. Alternatively, CPT may be a single protein in which different catalytic functions are dictated by differences in localization or are due to posttranslational modifications of the enzyme molecule (DiMauro and Papadimitriou, 1984). Layzer *et al.*, (1980) found that both CPT I and II had half normal activity in subfractions of mitochondria isolated from cultured fibroblasts of one patient. However, the enzyme defect was not consistently demonstrated in other studies of fibroblast cultures using whole-cell homogenates (DiDonato *et al.*, 1978, 1981; Pula *et al.*, 1981). Different enzyme assays were used in different studies and the discrepancies were probably due to technical reasons. When CPT was measured in the same fibroblast preparation by the isotope exchange assay of Norum (1964) and by the "backward" reaction (i.e., measuring release of radiolabeled carnitine from added palmitoyl[^{14}C]carnitine; Solberg, 1974) the decrease of activity was found only by the backward reaction (see DiMauro and Papadimitriou, 1984). In contrast to fibroblast cultures, CPT activity was undetectable by the isotope exchange method in myotube cultures from one patient. Neither cultured fibroblasts nor myoblasts or myotubes from CPT-deficient patients showed abnormal lipid accumulation (Miranda *et al.*, 1982). This is not surprising, since lipid storage is also absent in most muscle biopsies (DiMauro and Papadimitriou, 1984).

b. Triglyceride Storage Myopathy (TSM). A distinct disorder of lipid metabolism, characterized by chronic myopathy, scaly skin rash (ichthyosis), and fatty stools (steatorrhea), was identified in 3 unrelated patients. There was moderate to large increase of lipid droplets in muscle biopsies and in several other tissues such as liver, leukocytes, and cells of the gastrointestinal tract (Chanarin *et al.*, 1975; Miranda *et al.*, 1979a; Angelini *et al.*, 1980; Fig. 6). In muscle and fibroblast cultures, histochemistry showed abnormal deposits of neutral lipid, and thin-layer chromatography indicated that the excess neutral lipids were almost exclusively triglycerides (Chanarin *et al.*, 1975; Miranda *et al.*, 1979b; Fig. 6). Quantitation of the triglycerides in fibroblasts showed a 20- to 40-fold increase over control cultures (Miranda *et al.*,

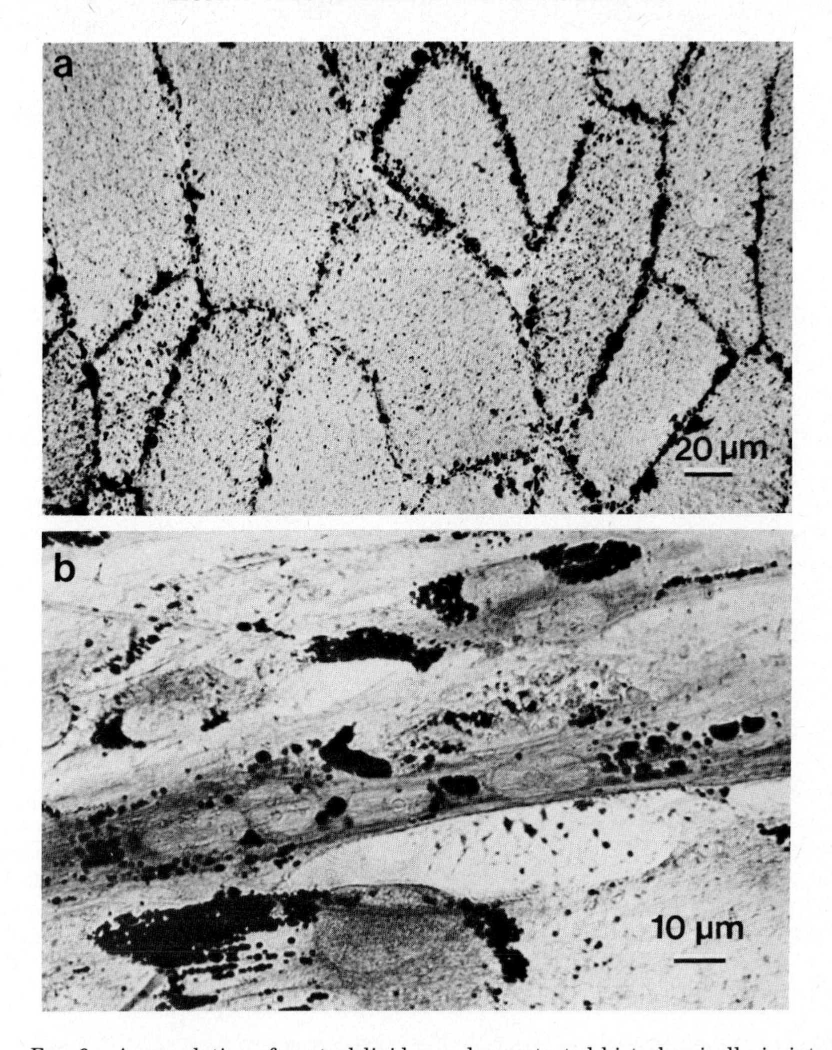

FIG. 6. Accumulation of neutral lipid was demonstrated histochemically in intact muscle (a) and muscle culture (b) from a patient with triglyceride storage myopathy. Reproduced with publisher's permission from Miranda *et al.* (1979b) (*Muscle and Nerve*).

1979b; Fig. 7). The lipid storage in the myotube cultures worsened with increasing culture age and could become severe enough to obscure the nuclei. Electron microscopic analysis of muscle cultures showed numerous lipid globules, not limited by membranes, suggesting that they were extralysosomal. Wolman disease is a genetic lysosomal disorder of lipid metabolism due to acid lipase deficiency (Kyriakides *et al.*, 1972). The clinical picture of Wolman disease does not include

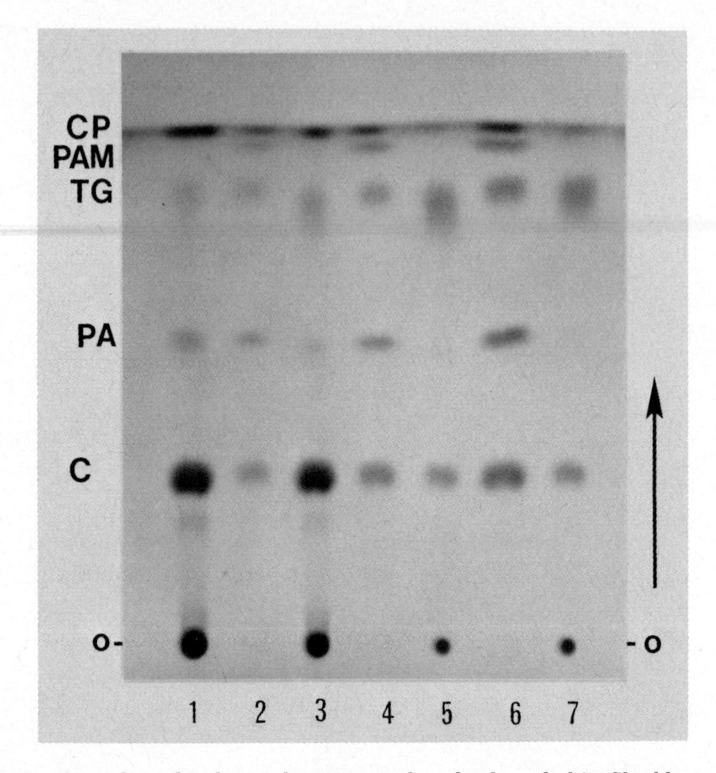

FIG. 7. Ascending thin-layer chromatography of cultured skin fibroblast extracts from normal individuals and a patient with triglyceride storage myopathy. Lanes 1 and 3 are lipid extracts from normal controls (concentrated 10-fold). Lanes 5 and 7 are lipid extracts from the patient (unconcentrated). Lanes 2, 4, and 6 are standards of neutral lipid and fatty acid mixture. The increase in neutral lipid in the patient's cells is due to triglyceride. C, Cholesterol; PA, palmitic acid; TG, triglyceride; tripalmitin; PAM, palmitic acid methyl ester; CP, cholesterol palmitin. Reproduced with publisher's permission from Miranda *et al.* (1979b) (*Muscle and Nerve*).

myopathy, and electron microscopic studies show neutral lipid-laden membrane-bound structures corresponding to lysosomes. In contrast to Wolman disease, tissues from patients with TSM showed normal acid lipase activity. Additional evidence that TSM is not a variant of Wolman disease came from cocultivation studies of TSM and Wolman fibroblasts showing disappearance of the stored lipid in Wolman fibroblasts but no change in the lipid content in TSM cells. Similarly, cocultivation of control fibroblasts with TSM or Wolman cells "normalized" the Wolman cells but had no effect on the TSM fibroblasts (Miranda *et al.,* 1979b). Similar results were obtained by exposing Wolman or TSM fibroblast to culture medium previously applied ("conditioned") to normal or TSM cells. These results are explained by the fact that lysosomal enzymes are excreted into the medium and can be transferred from normal to deficient cells, thus correcting the metabolic defect *in vitro*. The persistence of lipid storage in TSM fibroblasts in these conditions further indicates that the biochemical defect does not involve a lysosomal enzyme. Since utilization of [^{14}C]palmitate by fibroblasts was normal in 2 patients, but the radiolabel was retained longer than normal, a defect of cytoplasmic triglyceride *degradation* was proposed (Chanarin *et al.*, 1975; Miranda *et al.*, 1979b). In the third patient, however, impaired *utilization* of [^{14}C]palmitate was reported and a defect of β-oxidation was suggested (Angelini *et al.*, 1980). The phenotypic expression of TSM in all tissues and in muscle cultures at all stages of differentiation suggests that the missing enzyme exists as a single molecular form.

c. Carnitine Deficiency (CD). In systemic carnitine deficiency (SCD) the concentration of the amino acid carnitine is decreased in muscle, liver, plasma, and sometimes also in heart and kidney (Karpati *et al.*, 1975; Engel and Rebouche, 1982). The primary biochemical lesion has not yet been identified. Studies of intact tissues indicated that neither synthesis nor degradation of carnitine are impaired.

Tissue culture studies investigated whether CD might be due to defective cellular transport of carnitine. However, it was found that skin fibroblasts and differentiating muscle cultures from patients grew normally and showed no significant difference in the uptake of radiolabeled L-carnitine added to the medium when compared with control cultures or cultures from patients with a variety of neuromuscular diseases (Rebouche and Engel, 1982). It is conceivable that more mature muscle cultures might reveal an impairment of carnitine transport *in vitro*, but these studies have not yet been done.

Tissue culture studies of hereditary metabolic myopathies illustrate several points.

1. Hereditary metabolic defects may not be expressed in uninnervated cultured muscle. In some disorders (e.g., debrancher enzyme, phosphoglycerate kinase, and acid maltase deficiency) enzyme activities in myotube cultures are absent or decreased, but in others (e.g., phosphorylase, phosphofructokinase, or phosphoglyceromutase deficiencies) enzyme activities are normal, and there is no detectable increase of glycogen.

2. A general pattern has emerged from these studies. Genetic defects of enzymes that exist in a single molecular form in most tissues are also expressed in "immature" muscle cultures because the same

enzyme is present in muscle at all stages of differentiation. On the other hand, genetic defects of muscle-type isoenzyme subunits that are developmentally controlled are not manifest in aneural cultures from patients because these cultures still express isoenzyme patterns characteristic of fetal muscle. These immature isozymes are under separate genetic control from the deficient enzymes which are characteristic of mature muscle.

3. All metabolic myopathies in which the deficient enzymes have been identified are due to defects of the *mature* muscle-type isozymes. Defects of isoenzymes present in immature muscle and absent in mature muscle have not yet been described. However, I cell disease, a genetic disorder of lysosomal processing, is expressed in immature but not in mature muscle and might be one example of "fetal muscle isozyme myopathy" (Section II, A, 3, *b*).

4. Hereditary metabolic myopathies in which the biochemical defect is not known and which are not apparently expressed in muscle cultures may well be due to genetic defects of mature muscle isozymes.

3. *Lysosomal Disorders Affecting Muscle*

a. Acid Maltase (AM:α-Glucosidase) Deficiency. AM is a lysosomal hydrolase that catalyzes the degradation of glycogen, acting on α-1-4 and on α-1-6 glycosyl linkages (Engel, 1984a; Fig. 2). The enzyme is present in all tissues in a single molecular form which is coded on chromosome 17 (D'Ancona *et al.,* 1979). Normal muscle does not contain many lysosomes, but lysosomal glycogen degradation is nevertheless crucial for normal muscle function, since deficiency of AM (AMD) in its most severe form (Pompe disease) is generally lethal within the first year of life. AM deficiency of later onset is a more benign but crippling disease (see Engel, 1984a).

AMD has been studied in tissue cultures derived from patients with both infantile and later-onset forms. The disease can be identified prenatally in cultured amniotic cells and is also expressed in muscle cultures at all stages of differentation (Askanas *et al.,* 1976; Askanas and Engel, 1979; Miranda *et al.,* 1982). This is not surprising because AM exists in a single molecular form, and there are no developmentally regulated isoenzyme transitions during myogenesis. AM activity is completely lacking in cultures from patients with infantile AM, but some residual activity is present in muscle cultures from individuals with late-onset AMD. Residual activity is also present in muscle biopsies and cultured skin fibroblast from patients with late-onset AMD (Askanas *et al.,* 1976; Mehler and DiMauro, 1977). Cultured cells typically contain many glycogen-filled lysosomes together with pools of extralysosomal glycogen (Askanas *et al.*, 1976; Askanas and Engel, 1979; Miranda *et al.*, 1982). It is not yet known whether AMD is due to lack of AM synthesis or to synthesis of inactive enzyme protein. Immunochemical studies using anti-AM antibodies failed to show cross-reacting material (CRM) in cultured muscle from two infantile AMD patients (Miranda *et al.*, 1984; Beratis *et al.*, 1983, for review). This may be due to synthetic failure or to the fact that the enzyme is altered to such an extent that immunologic properties are lost (see Beratis *et al.*, 1983; Miranda *et al.*, 1984). In agreement with the finding that cells from late-onset AMD patients show residual enzyme activity, some immunocytochemical staining of lysosomes with AM antibody could also be demonstrated in muscle cultures from 2 late-onset AMD patients, but not in cultures of Pompe patients (Miranda *et al.*, 1984). It is not clear why the clinical expression of AMD is so heterogeneous, since both infantile and late-onset forms appear to be due to defects of the same gene locus as suggested by studies of somatic hybridization of skin fibroblasts. Shanske *et al.* (1974), Reuser *et al.* (1978), and Hays *et al.* (1979) found that somatic cell hybridization with fibroblasts from infantile and late-onset AMD patients did not correct the enzyme defect. More work is needed to elucidate the mechanisms by which AMD affects muscle function. Since the disease is expressed in tissue culture, including long-lived rapidly growing SV40 transformed cell lines (Miranda *et al.*, 1981b), such cultured cells are an excellent source of tissue for more detailed molecular analyses.

b. Mucolipidosis Type II (Inclusion Body or I Cell Disease). I cell disease is a genetic metabolic disorder in which the proper processing of several acid hydrolases is defective (Hickman and Neufeld, 1974). Deficiency of an acetylglucosamine transferase was recently identified (Haselik *et al.*, 1981; Reitman *et al.*, 1981). Patients with I cell disease usually die within the first decade of life and the cells from affected tissues are filled with lysosomal inclusion bodies, hence the name of the disease. Many, but not all, lysosomal enzyme activities are decreased within the cells and increased in body fluids (Leroy and Demars, 1967; Leroy *et al.*, 1972; Hickman and Neufeld, 1974; Shanske *et al.*, 1981).

Cultured fibroblasts also show characteristic lysosome-laden cells with decreased activities of several acid hydrolases such as β-galactosidase, β-hexosaminidase, α-mannosidase: these same enzymes are increased in the medium of cultures (Leroy and Demars, 1967; Leroy *et al.*, 1972). I cell disease does not cause a myopathy, and muscle biopsies do not show increased lysosomes or decreased activity of those hydrolases that are deficient in other tissues (Shanske *et al.*, 1981). However, when muscle from one patient with I cell disease was grown in culture, typical inclusion bodies were seen in unfused cloned myoblasts and persisted in fused myotubes (Shanske *et al.,* 1981). Fusion of myoblasts was not as efficient as in control muscle grown simultaneously. This was not due to increased activities of lysosomal enzymes in the medium because exposure of normal cultures to I cell muscle culture medium had no detrimental effect on myoblast fusion. Conversely, repeated changes of medium in I cell cultures failed to improve fusion efficiency, even when conditioned medium from normal controls was used. The I cell myotubes were also affected at early stages of differentiation and showed numerous inclusions. However, accumulation of lysosomes was no longer visible in more mature, cross-striated myotubes. This suggested that lysosomal processing may be different in immature and mature muscle. An alternative hypothesis is that the defective enzyme occurs in multimolecular isozyme forms that are developmentally regulated during myogenesis. While all known metabolic myopathies are due to deficiencies of mature muscle isozymes, genetic defects of isozymes characteristic of immature muscle must also exist. These disorders may go undetected either because they are lethal in utero or because they are "self-curing" as the muscle attains full maturity. I cell disease may be such a disorder but this still remains to be proved. The expression of I cell disease in cultured myotubes is not an artifact of the *in vitro* system because electron microscopic studies of a muscle biopsy from a patient showed normal ultrastructure of the muscle fibers, whereas the satellite cells, considered to be dormant muscle stem cells were filled with lysosomal inclusions (Shafiq and Kula, unpublished; Shanske *et al.,* 1981; Miranda *et al.,* 1982).

4. A Disorder of Nucleotide Metabolism

*Adenylate Deaminase (AMPD).*AMPD catalyzes the formation of IMP through deamination of AMP, the first step in the adenine nucleotide cycle (Engel *et al.,* 1964; Fishbein *et al.,* 1978; Shumate *et al.,* 1979). The enzyme exists in multiple molecular forms in vertebrate tissues. During muscle development and regeneration, there are isozyme transitions from nonmuscle forms to the muscle-type AMPD, which predominates in normal mature muscle (Sammons and Chilson, 1978; DiMauro *et al.,* 1980; Miranda *et al.,* 1982).

An apparently frequent mutation affects the muscle-type isozyme causing muscle AMPD deficiency (Fishbein *et al.,* 1978). However, studies of muscle culture from one patient showed presence of normal AMPD activity in myotubes, which appeared to be due to reinitiation of nonmuscle "fetal" AMPD isozyme synthesis in immature cultured myotobes. In rat muscle cultures, muscle-type AMPD isozyme has been demonstrated immunocytochemically and biochemically in close association with the myosin region of the sarcomere (Ashby and Frieden, 1978; Ashby *et al.,* 1979). In normal, uninnervated human muscle cultures muscle-type AMPD has not yet been demonstrated but if present, it should represent only a small fraction of the total enzyme activity because normal and patient's myotubes had similar AMPD activities (DiMauro *et al.,* 1980).

B. Myopathies of Unknown Etiology

Although several genetic myopathies have been attributed to specific enzyme defects for most hereditary neuromuscular disorders, the genetic error remains unknown. Among the genetic myopathies of unknown etiology are the muscular dystrophies, which include the rapidly progressive X-linked form (Duchenne dystrophy) and myotonic dystrophy, an autosomal dominant multisystem disorder (Rowland and Layzer, 1979; Engel, 1984b; Harper, 1984; Munsat, 1984, for review).

The "mitochondrial myopathies" are also a mixed bag of disorders in which the mode of inheritance is poorly understood, perhaps because some of these disorders may be due to mutations of the mitochondrial genome. Muscle biopsies usually show large accumulations of mitochondria, some of which may contain highly organized paracrystalline inclusions (Morgan-Hughes, 1984, for review). In a few patients with these morphological abnormalities, defects of respiratory chain components (e.g., cytochrome *b* and cytochrome oxidase) have been demonstrated, (DiMauro *et al.,* 1983; Morgan-Hughes, 1984).

1. Muscular Dystrophies

The muscular dystrophies comprise a heterogeneous group of hereditary myopathies of unknown etiology, characterized by muscle degeneration followed by replacement of muscle with connective tissue and fat. The most severe and most common form of dystrophy is the Duchenne muscular dystrophy (DMD), an X-linked disorder in which progressive proximal weakness starts in childhood and usually leads to death in the second or third decade of life. In a more slowly progressive variant (Becker dystrophy), also transmitted as an X-linked recessive trait, weakness starts later, the course is slower, and death occurs in the fourth or fifth decade. Other forms of dystrophy, affecting certain muscle groups more than others, have autosomal recessive inheritance, such as limb-girdle dystrophy, or autosomal dominant inheritance, such as facioscupulohumeral (FSH) dystrophy. Even though DMD is X-linked, a few affected girls showed X-chromosome translocations, with a breakpoint at or near the Xp 21 region (Canki *et al.*, 1979; Zatz *et al.*, 1981). This suggested that the DMD locus might be in that area of the X chromosome. Recombinant DNA studies have now confirmed that the gene coding for DMD is indeed near this region (Davies *et al.*, 1983).

Although DMD has been extensively studied in muscle culture, the findings have been conflicting (see Miranda and Mongini, 1983). In earlier studies using the explanation culture technique, it was reported that DMD cultures proliferated more rapidly at early stages *in vitro* and that the smaller than normal myotubes failed to develop normal cross-striations (Geiger and Garvin, 1957; Kakulas *et al.*, 1968; Bateson *et al.*, 1972). However, these findings were not confirmed by others, who reported normal myogenesis *in vitro* (see Miranda and Mongini, 1983). The abnormalities described in DMD cultures could be attributed to two factors: (1) severe connective tissue infiltration in the biopsies from which the cultures were derived, and (2) heterogeneity of the myotube cultures, with different degrees of differentation from explant to explant, even when cultures were derived from the same individual (Witkowski, 1977; Miranda and Mongini, 1984, for review). To eliminate these problems, more recent studies used monolayer muscle cultures obtained from protease-dissociated mononuclear cells from biopsies and grew identical replicate cultures (Witkowski *et al.*, 1976; Yasin *et al.*, 1977; Miranda *et al.*, 1979a). However, the results were variable even with these methods. Thompson *et al.* (1977) found a peculiar clustering and piling up of cells in DMD muscle cultures, but similar cell clumps were also seen occasionally in cultures from controls or patients with other neuromuscular disorders, suggesting that this growth pattern, even if related to the primary DMD lesion, is not a sensitive morphologic maker of the disease (Ecob-Johnston and Brown, 1981; Thompson *et al.*, 1981).

Electron microscopic studies also failed to show specific alterations in cultured myotubes from DMD patients, and freeze-fracture replica of myotube membranes did not show the differences in membrane particle density and distribution that had been described in DMD muscle biopsies (Osame *et al.*, 1982). Moreover, binding of the plant lectin concanavalin A (Con A) to sarcolemma of cultured myotubes was normal, in contrast to the abnormal Con A binding pattern observed in biopsied muscle (Heiman-Patterson *et al.*, 1982).

More recent clonal analysis by Blau *et al.* (1983) showed that DMD myoblasts had "impaired proliferative capacity," concluding that there might be a basic defect of DMD satellite cells. Hauschka *et al.* (1979) noted a similar limited proliferative capacity of presumptive myoblasts from dystrophic mice (strain 129 Re/J), but felt that this phenomenon might reflect terminal proliferative senescence of dystrophic myoblasts, since satellite cells of dystrophic degenerating muscle are continuously stimulated to proliferate. It would be important to determine the mechanisms of premature senescence of dystrophic myoblasts to assess whether this phenomenon represents a basic defect of the muscle stem cells or simply a secondary manifestation resulting from continuous regeneration of the dystrophic muscle. Clonal analysis of muscle from DMD carriers and muscle from patients with other disorders that show extensive regeneration (e.g., polymyositis and dermatomyositis) might help resolve this question.

Several biochemical abnormalities have been reported in DMD cultures, such as decreased intracellular creatine kinase (CK) levels and altered CK isozyme pattern with a relative increase of CK-BB isozyme, suggesting impaired muscle maturation in DMD. However, when cultures were derived from pure myoblast populations obtained from myogenic clones, both CK levels and isozyme profiles were normal and there was no detectable leakage of CK in the culture medium (Blau *et al.*, 1982). Adenylate cyclase (AC), a membrane marker enzyme, was decreased in intact DMD muscle and could not be stimulated normally by addition of catecholamines. In monolayer muscle cultures from DMD patients, basal AC activity was higher than in controls and stimulation by catecholamines was also impaired (Mawatari *et al.*, 1976; Schonberg *et al.*, 1977). However, these findings should be confirmed in clonal muscle cultures.

The lack of specific morphologic abnormalities in conventional DMD cultures may indicate that uninnervated cultures are not mature enough to express the DMD phenotype. A less likely explanation may be that expression of the disorder requires interaction of muscle with other affected tissues, such as "sick motor neurons." Preliminary studies of DMD cultures innervated with normal fetal mouse neurons and grown for several weeks have failed to reveal any morphologic abnormalities (Witkowski and Dubowitz, 1975). Longer-term innervated DMD cultures (6 months or older) did show some degenerative changes, such as premature loss of cross-striations and filamentous "cytoplasmic bodies," but these preliminary studies must be confirmed in cultures from more patients (Peterson and Crain, 1979; Peterson *et al.*, 1979).

The primary genetic lesion in DMD will probably be identified at the primary structure of DNA, using hybridization techniques. However, muscle cultures remain the method of choice to study the factors determining the pathologic expression of this devastating myopathy.

2. Myotonic Dystrophy

Myotonic dystrophy (myotonic atrophy) is an autosomal dominant disease of unknown etiology. In muscle from patients characteristic abnormalities of membrane function include decreased resting membrane potentials (RMPs) and trains of repetitive action potentials of varying amplitude and frequency ("myotonic discharges") upon mechanical or electrical stimulation (see Harper, 1979; Harper, 1984, for review).

It is controversial whether myotonic dystrophy can be reproduced in uninnervated muscle cultures from patients. One group of investigators found decreased RMPs, decreased action potential amplitudes, and repetitive discharges in some myotubes after anodal break excitation (Merickel *et al.*, 1982). However, other investigators observed that in each culture, whether derived from patients or from controls, there was a difference in the degree of differentiation between broad, flattened myotubes that showed RMPs of −30 mV, and cylindrical-appearing myotubes showing RMPs of −50 mV (Tahmoush *et al.*, 1983). Moreover, only the cylindrical phase-bright myotubes generated action potentials when hyperpolarized at −80 mV. When only the excitable myotubes were studied, there was no difference in RMPs or action potential amplitudes between cultures from patients and controls. Repetitive discharges following anodal break excitation could be obtained in both types of cultures. It was suggested that the genetic lesion might not be expressed in poorly differentiated cultured myotubes and that the findings of Merickel *et al.* (1982) may have been due to improperly matched cultured myotubes from controls and patients (Tahmoush *et al.*, 1983). If this is the case, it may be feasible to reproduce myotonic dystrophy *in vitro* using more mature myotube cultures obtained by cocultivation with fetal mouse motor neurons (Peterson and Crain, 1979; Miranda *et al.*, 1983a).

3. Mitochondrial Myopathies

This is a heterogeneous group of muscle disorders associated with abnormal mitochondria, sometimes containing paracrystalline inclusions. The etiology is still unclear. In a few patients, disorders of the respiratory chain involving deficiencies of cytochrome oxidase or cytochrome *b* were documented in intact muscle, but these specific disorders have not yet been studied in muscle culture. Carnitine palmitoyltransferase (CPT) deficiency, the first metabolic disorder of mitochondria, was discovered in 1973 (DiMauro and Melis-DeMauro, 1973). CPT deficiency has been described above (see Section II, A, 2, *a*).

Askanas and Engel (1977, 1979) and Askanas *et al.* (1978) studied muscle cultures from several patients with abnormal mitochondria in their muscle biopsies. They found that the cultures grew and differentiated normally, but that many myotubes showed bizarre or enlarged mitochondria which contained abnormal electron-dense granular material. Treatment of normal human muscle culture with an uncoupler of oxidative phosphorylation (2,4-dinitrophenol) yielded similar ultrastructural changes. However, paracrystalline inclusions, which are a common feature in biopsies from patients with mitochondrial myopathies, have not yet been reproduced in cultured myotubes. The modes of inheritance in mitochondrial disorders is sometimes difficult to assess, since at least some of these myopathies may be due to abnormal gene products coded for in mitochondrial rather than nuclear DNA (DiMauro *et al.*, 1983; Morgan-Hughes, 1984). Systematic studies of these syndromes may, therefore, yield important information regarding the molecular regulation of mitochondrial metabolism.

4. *Other Rare Myopathies of Unknown Etiology*

A rare, X-linked recessive neuromuscular disease affecting several boys in three generations of a Dutch family has been described (Askanas and Engel, 1979). All the affected infants had smaller than normal muscle fibers with central nuclei and increased sarcoplasmic glycogen deposits. Muscle from two affected children were studied in culture and showed unusually rapid cell proliferation even upon repeated transfers of the cultures. The nuclei of the cultured myotubes were unusually large, with prominent nucleoli and increased number of sarcoplasmic ribosomes. The cultured myoblasts fused normally, but failed to develop cross-striations.

Vacuolar "cabbage body" myopathy (VCM) is a slowly progressive vacuolar myopathy described in a 32-year-old man. A muscle biopsy from this patient showed numerous lysosome-like inclusions which were also seen in muscle culture. Several lysosomal enzymes were increased, including acid maltase, but no specific enzyme defect was identifed. Because the morphological abnormalities were reproduced in muscle culture, a hereditary cause seemed likely, although an infectious etiology (e.g., viral) cannot be excluded (Barth *et al.*, 1975; Askanas and Engel, 1979).

III. Acquired Diseases of Muscle

A. Myasthenia Gravis

Myasthenia gravis (MG) is an acquired autoimmune disease characterized by muscle weakness due to a defect in cholinergic neuromuscular transmission. The motor end plates of patients with MG have decreased number of functional acetylcholine receptors (AChRs), resulting in decreased miniature end plate potentials (MEPs). Binding of α-bungarotoxin (α-BT), a specific marker of AChRs, is also reduced. MG patients produce autoantibodies against AChRs: binding of antibodies to the receptors appears to accelerate the loss of surface AChRs (Engel, 1984c, for review). Muscle cultures have been useful in understanding the pathobiology of MG. Bevan *et al.* (1977) showed that application of serum from MG patients to fetal human myotube cultures decreased the sensitivity to iontophoretically applied ACh. This effect appeared to be due to humoral antibodies because it was reduced considerably by removal of IgG. It was suggested that decreased ACh sensitivity was due to increased degradation of AChRs rather than to direct inactivation by anti-AChR antibodies. Similar findings were obtained in rat muscle cultures. Anwyl *et al.* (1977) found gradual disappearance of ACh sensitivity in rat myotubes exposed to serum from MG patients. In another study, Appel *et al.* (1977) found that binding of radiolabeled α-BT to cultured rat myotubes was reduced after exposure to MG sera. However, when ACh rather than α-BT was used to block the receptors, binding of ACh was similar to controls when measured electrophysiologically. Since α-BT is a large molecule (MW 8000) when compared with ACh (MW 150), the partial blockade by α-BT could be due to steric hindrance (Appel *et al.*, 1977). When MG serum-treated cultures were exposed to low temperature or metabolic poisons, known to affect AChR turnover, the accelerated disappearance of ^{125}I-labeled α-BT was slowed down, suggesting that the decreased sensitivity to ACh of myotubes exposed to MG sera was due to increased internalization and/or degradation of the cell surface AChRs (Appel *et al.*, 1977). Marked increase in AChR degradation was found by Kao and Drachman (1977) who exposed fetal rat myotube cultures to immunoglobulin fractions of MG sera: there was acceleration of ^{125}I-labeled α-BT release in the culture medium when compared with control cultures treated with immunoglobulin fractions from normal individuals. In another study, the same investigators found that the increase of AChR degradation is preceded by formation of AChR–antibody complexes, since divalent myasthenic (MG) $F(ab)'_2$ fragments increased AChR degradation whereas monovalent Fab fragmenns did not. Although there is good evidence that in MG impairment of AChR function is due to circulating anti-AChR antibodies, there is no direct correlation between clinical severity and AChR antibody titers. In tissue culture, however, there is good correlation between increased AChR degradation and blockade and clinical status of patients from whom sera were obtained. This conclusion was based on studies in which rat muscle cultures were exposed to immunoglobulin fractions from 59 patients and 15 controls (Drachman *et al.*, 1978). "AChR degradation" was measured by labeling the AChRs with ^{125}I-labeled α-BT and studying the effect of serum immunoglobin fractions from patients and control on the release of ^{125}I derived from degraded AChR–antibody complexes. AChR blockade was studied by applying medium containing immunoglobulin fractions from patients and controls to cultures at 4°C (to eliminate internalization and degradation of the AChRs) and labeling the remaining unblocked AChRs with ^{125}I-labeled α-BT. The values obtained with cultures exposed to each patient's immunoglobulin were subtracted from the values obtained in similar cultures treated with control immunoglobulin fractions. The investigators found that 91% of the patients' sera caused increased AChR degradation and that the relative rates corresponded well with the clinical status of the patients: 88% of the patients' sera blocked the AChRs. Both criteria together increased the correlation with severity of the symptoms to 98%. These studies support the hypothesis that both AChR degradation and blockade by autoimmune antibodies contribute to cause the clinical manifestations of MG (Drachman *et al.*, 1978). However, other mechanisms, such as complement-associated immune injury, may also be involved in the pathogenesis of MG (Engel, 1984c).

B. Inflammatory Myopathies

The etiology of other acquired myopathies is poorly understood. Subchronic and chronic polymyositis (PM) and dermatomyositis (DM) may be due to immunologic injury, but the mechanisms that trigger the abnormal autoimmune responses in patients are not yet known. Acute inflammatory myopathy (acute myositis), often with severe muscle pain (myalgia), is sometimes seen in individuals, mostly children, following respiratory viral infections such as influenza B (Gamboa *et al.*, 1979, for review). Biopsies from these patients are not usually available for tissue culture because the disease tends to be self-limiting. Several muscle biopsies from patients with PM and DM showed virus-like particles in muscle fibers and other cells such as endothelial cells of blood capillaries. It was also shown, however, that at least some of these "virus-like inclusions" are artifactual. For instance, glycogen arranged in paracrystalline arrays may resemble picornaviruses (Collins and Gilbert, 1977). Therefore, direct proof for a viral etiology has to rely on the demonstration of viral products and virus isolation. Attempts to isolate viruses from patients with myositis have not been very rewarding. Cocultivation studies of muscle cells from patients with PM and DM with cell lines (e.g., HeLa, Vero) known to support lytic infection of several viruses showed no evidence of virus production (Hashimoto *et al.*, 1971). Attempts to stimulate proliferation of "hidden viruses" by treatment of PM and DM muscle cultures with virus-inducing agents such as dimethyl sulfoxide and iododeoxyuridine have also been unsuccessful (A. F. Miranda, unpublished). Despite these negative findings there is enough indirect evidence to indicate that at least some of the inflammatory myopathies are caused by viral injury. Influenza B virus antigen was documented in the muscle biopsy of a patient with a polymyositis-like syndrome and the virus was isolated. Moreover, electron microscopic studies of this patient's muscle showed elongated structures budding off the muscle cell surface and resembling influenza B (Gamboa *et al.*, 1979). Unfortunately, muscle from this patient was not available for tissue culture.

The presence of inflammatory lesions in muscle from patients with PM and DM suggests that these diseases may be due to infectious agents. However, subacute and chronic muscle necrosis may be due to autoimmune responses rather than to direct toxicity of infectious agents. Experimental studies in which tissue cultures were exposed to lymphonode cells from rats previously sensitized with rabbit or monkey muscle showed myonecrosis and other cytopathic effects on cultured embryonic rat myotubes (Kakulas, 1966; Currie, 1971). However, similar muscle culture studies using peripheral lymphocytes from patients have yielded inconsistent results. Using morphologic criteria together with a radioactive chromium release assay to evaluate sarcolemmal membrane damage, Haas (1980) found no cytotoxicity in fetal rat muscle cultures exposed to blood monocytes from patients with PM or DM. In contrast, Kakulas *et al.* (1971) found evidence of myonecrosis in similar studies using human rather then rat fetal muscle cultures. When peripheral blood lymphocytes from PM and DM patients were exposed to minced human muscle, they produced a "lymphokine lymphotoxin" that caused cytopathic effects in human fetal muscle cultures (Johnson *et al.*, 1972). Iannacone *et al.* (1982) found no evidence of cell-mediated cytotoxicity when blood lymphocyte and monocyte preparations from 9 children with DM were applied directly to fetal human muscle cultures. To explain these negative findings, Iannacone and co-workers suggested that the immunologic target might be endothelium rather than muscle since focal deposits of immunoglobulins were found in subendothelial areas of blood vessels of intact muscle from patients (Whitaker and Engel, 1973). It is also conceivable that the immunologic target is the cell surface of mature muscle, which is absent or poorly developed in immature cultured muscle. Studies of uninnervated and innervated muscle cultures from patients with PM and DM exposed to their own peripheral blood monocyte preparations may yield more conclusive data regarding the etiology of inflammatory myopathies and the possible mechanisms of immunologic injury by autoantibodies.

IV. Experimental Viral Myopathies

In humans, sequential studies of diseased muscle are difficult or impossible to carry out *in vivo*. Because mature muscle can regenerate remarkably well in culture, acquired myopathies can be studied *in vitro*. Moreover, application of potential myotoxic agents to human muscle cultures can provide insight into the pathogenic mechanisms of muscle injury. Muscle cultures may also be useful in the evaluation of experimental therapeutic agents. However, so far most experimental studies in muscle cultures have concentrated on the effects of virus infection. These are discussed below.

a. Influenza. There is good indirect evidence that influenza virus can cause acute myositis, particularly in children, following respiratory infection (Gamboa *et al.,* 1979). It is unclear, however, whether the muscle symptomatology is caused by direct virus infection of the muscle, by virally induced "toxic" effects, or by immunologic injury. It is also not clear why most individuals do not develop postinfluenzal myositis. Influenza B virus (FluB) was demonstrated in a man with a polymyositis-like syndrome by electron microscopic, immunocytochemical, and virus isolation studies. It is not certain, however, whether viral infection was the primary cause of the disease or whether an unrelated acquired muscle disease increased susceptibility to FluB (Gamboa *et al.,* 1979).

Both embryonic chicken muscle and regenerating human muscle were found to be susceptible to influenza A virus (FluA) (O'Neill and Kendall, 1975; Armstrong *et al.,* 1978; Miranda *et al.,* 1978; Fig. 8). However, while the virus caused lysis and cell death, usually within 72

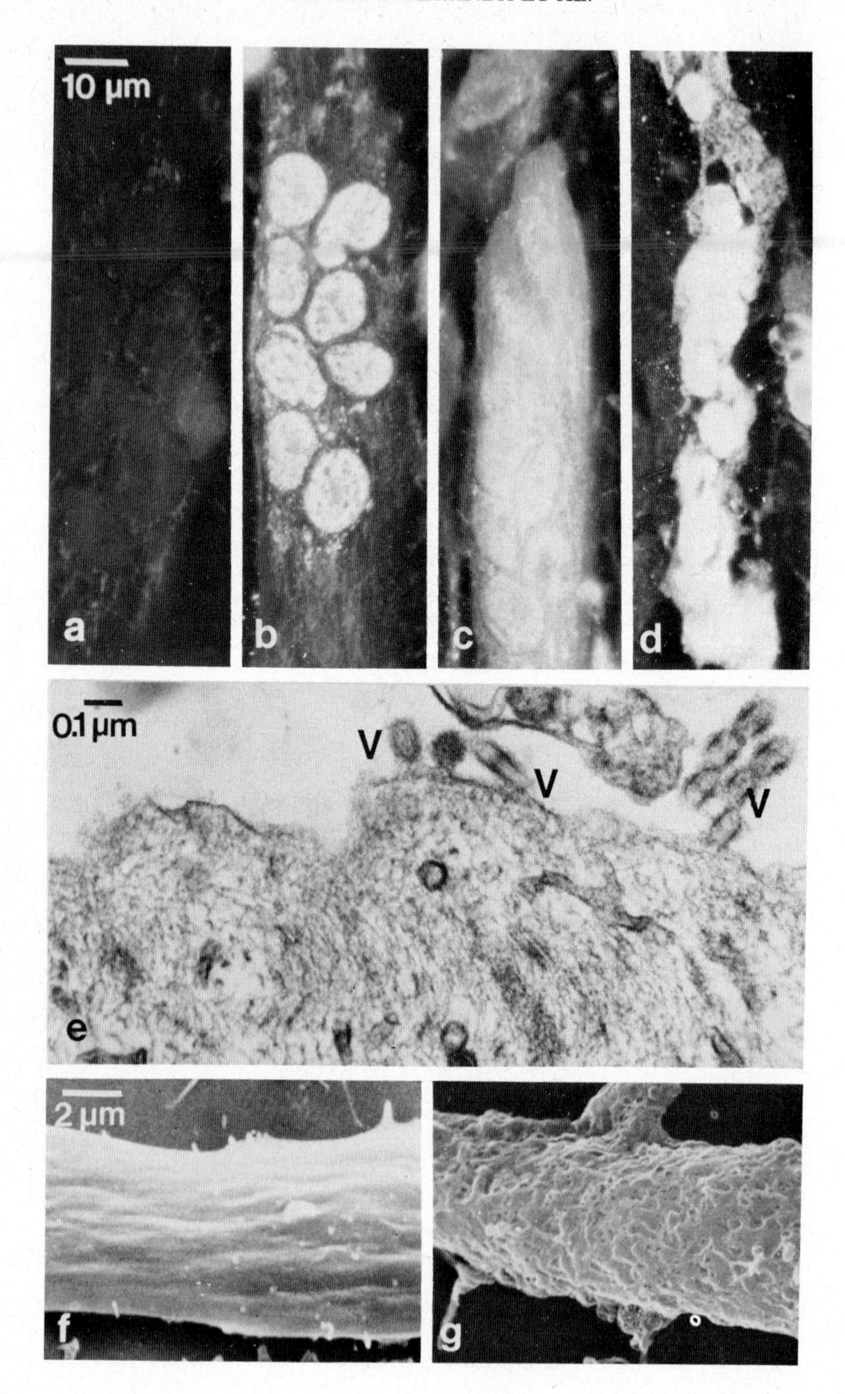
10 μm
a
b
c
d
0.1 μm
V
V
V
e
2 μm
f
g

hours, in multinucleated myotubes, immunocytochemical studies of clonal human muscle cultures showed that unfused myoblasts were considerably less susceptible to the infection and showed little or no cytopathic effects, even after 72 hours or longer (Armstrong *et al.*, 1978; Miranda *et al.*, 1978).

In vivo studies in rodents also showed that skeletal muscle is differentially susceptible to FluA virus at different stages of differentiation. Normal mature muscle appeared to be refractory, while denervated and regenerating muscle were susceptible (see Gamboa and Hays, 1984). It would appear from these studies that whenever muscle regeneration is induced in patients following mechanical injury or by disease, they may be at risk for developing influenzal myositis when exposed to the virus. The mechanisms responsible for the differential viral susceptibility of muscle at various stages of differentiation are not understood, and the nature of the influenza virus receptor in muscle has not yet been studied in detail. Valderrama *et al.* (1981) found that application of α-neurotoxin (α-NT), an AChR-blocking agent, to human muscle cultures diminished the susceptibility of myotubes to influenza A virus and that this was not due to a direct toxic effect of α-NT on the virus. It was suggested that AChRs and FluA receptors may be similar, or that the inhibition of infection by α-NT was due to steric hindrance. The fact that myotubes are more susceptible to FluA than unfused myoblasts supports the former hypothesis because myotubes, but not myoblasts, have abundant extrajunctional ACh surface receptors. It is interesting to note in this respect that rabies virus was shown to bind specifically to cholinergic end plate regions of intact muscle (Lentz *et al.*, 1982).

b. Coxsacki. Both Coxsackie A and B viruses (CA, CB) have been implicated in some cases of acute or chronic inflammatory myopathies (see Gamboa and Hays, 1984). Studies of rat and mouse muscle cultures showed that, like FluA virus, CA readily infected immature myotubes,

---

FIG. 8. Normal human muscle cultures experimentally infected with the WSN strain of influenza A virus (FluA) (a–d). Myotubes stained immunocytochemically with fluorescein-tagged antibodies against FluA: uninfected control does not fluoresce (a); at 18 hours postinfection (pi) viral antigen is present mostly in myotube nuclei (b); at 48 hours pi, both nuclei and cytoplasm are intensely fluorescent (c); at 72 hours infected myotubes are moribund or dead (d). (e) Transmission electron micrograph of an infected myotube 48 hours pi showing budding, spherical, or elongate virions (v) at the surface membrane. (f) Scanning electron micrograph of uninfected control myotube showing a rather smooth surface. (g) Myotube 48 hours pi showing an apparently "leaky" surface membrane with crater-like lesions. Reproduced with publisher's permission from (a–d) Armstrong *et al.* (1978) (*J. Neurol. Sci.*) and (e–g) Miranda *et al.* (1978) (*J. Neurol Sci.*).

while myoblasts were refractory to the infection. As myotubes differentiated further, however, they became refractory again (Goldberg and Crowell, 1971; Landau *et al.*, 1972; Schultz and Crowell, 1980). Studies of intact rodent muscle indicated that mature muscle was resistant to CA infection, but denervated muscle was susceptible (see Gamboa and Hays, 1984). The differential susceptibility in cultures did not appear to be due to differences in virus receptor density at different stages of myogenesis since CA virus attachment was similar on myoblasts and myotubes (Schultz and Crowell, 1980). Furthermore, recent preliminary studies (Andrew *et al.*, 1982) showed that, unlike FluA, there was no detectable relationship between AChRs and CA receptors because application of virus did not diminish the ability of the muscle cells to bind radiolabeled α-bungarotoxin.

c. *Measles.* The Edmundson strain of measles virus also appears to be myotropic, as reported by Askanas *et al.* (1981) who studied the appearance of viral antigen immunochemically and electron microscopically. Cytopathic effects in infected cultures were also reported, but viral isolation was not performed.

d. *Vaccinia and Herpes.* These two DNA viruses may also be associated with myositis, but have not yet been studied in human muscle cultures. Although studies of cultured chick muscle showed that there is no differential susceptibility of muscle cells in culture, both myoblasts and myotubes showed similar cytopathic effects (Cox and Kendall, 1978).

There are only two reports in which viral-like particles were detected in human muscle cultures and in biopsied muscle from which they were derived. Three patients with a nonspecific vacuolar myopathy showed reovirus-like particles in cultured muscle cells examined electron microscopically. In retrospect, similar "viriods" were also found in biopsied muscle from which these cultures were derived. However, virus isolation and experimental virus transmission to normal cultures or to experimental animals have not yet been done (Askanas *et al.*, 1981).

More detailed analyses of human muscle cultures that are experimentally infected with potentially myotropic viruses may provide more insight into the etiology of acquired inflammatory myopathies.

V. Conclusion

In this article we illustrated the use of regenerating human muscle cultures for studying hereditary and acquired diseases of muscle. Even though some of the data are still controversial, they help illustrate the great potential of this *in vitro* system. If permanent human muscle cell lines from normal individuals and patients with genetic myopathies can be established, these may provide an unlimited supply of tissue for molecular studies. For hereditary myopathies due to developmentally regulated proteins that are expressed only at more advanced stages of muscle differentiation, the use of highly differentiated nerve–muscle cocultures might contribute significantly toward a better understanding of their developmental pathogenesis.

ACKNOWLEDGMENTS

The authors thank Ms. Sharon Foster and Ms. Mary Voss for excellent technical assistance and Ms. Mary Tortorelis for typing and editing the manuscript.

This work was supported by grants from the National Institute of Neurological and Communicative Disorders and Stroke (NS-11766-09 and NS 18446-01), the National Institute of Arthritis, Metabolism and Digestive Diseases (AM-25506), and the Muscular Dystrophy Association. Dr. Mongini was supported by a postdoctoral fellowship from Ordine Mauriziano and U.I.L.D.M. section of Turin, Italy.

REFERENCES

Abbott, J., Schiltz, J. Dienstman, S., and Holtzer, H. (1974). *Proc. Natl. Acad. Sci. U.S.A.* **71,** 1506–1511.

Andrew, C. G., Drachman, D. B., and Narayan, D. (1982). *Soc. Neuro Sci. Abs.* **8,** 417.

Angelini, C., Philippart, M., Borrone, C., Bresolin, N., Cantini, M., and Lucke, S. (1980). *Ann. Neurol,* **7,** 5–10.

Anwyl, R., Appel, S. H., and Narahashi, T. (1977). *Nature (London)* **267,** 262–263.

Appel, S. H., Anwyl, R., McAdams, M. W., and Elias, S. (1977). *Proc. Natl. Acad. Sci. U.S.A.* **74,** 2130–2134.

Armstrong, C. L., Miranda, A. F., Hsu, K. C., and Gamboa, E. T. (1978). *J. Neurol. Sci.* **35,** 43–57.

Ashby, B., and Frieden, C. (1978). *J. Biol. Chem.* **253,** 8728–8735.

Ashby, B., Frieden, C., and Bischoff, R. (1979). *J. Cell Biol.* **81,** 361–373.

Askanas, V., and Engel, W. K. (1975). *Neurology* **25,** 58–67.

Askanas, V., and Engel, W. K. (1977). *In* "Pathogenesis of Human Muscular Dystrophies" (L. P. Rowland, ed.), pp. 856–874. Excerpta Medica, Amsterdam.

Askanas, V., and Engel, W. K. (1979). *In* "Handbook of Clinical Neurology" (P. J. Vinken and G. W. Bruyn, eds), Vol. 40, pp. 183–196. North Holland, Amsterdam.

Askanas, V., Engel, W. K., DiMauro, S., Mehler, M., and Brooks, B. R. (1976). *New Engl. J. Med.* **294,** 573–578.

Askanas, V., Engel, W. K., Britton, D. E., Adornato, B. T., and Eiben, R. M. (1978). *Arch. Neurol.* **35,** 801–809.

Askanas, V., McFarland, H., Engel, W. K., and Rodman, R. B. (1981). *Neurology* **31,** 57 (abs.).

Banks-Schlegel, S. P., and Howley, P. M. (1983). *J. Cell Biol.* **96,** 330–337.
Barth, P. G., van Wijngaarden, G. K., and Bethlem, J. (1975). *Neurology* **25,** 531–536.
Bateson, R. G., Hindle, D., and Warren, J. (1972). *J. Neurol. Sci.* **15,** 183–191.
Beratis, N. G., LaBadie, G. U., and Hirschhorn, K. (1983). *Am. J. Hum. Gen.* **35,** 21–33.
Bevan, S., Kullberg, R. W., and Heinemann, S. F. (1977). *Nature (London)* **267,** 263–265.
Bischoff, R. (1974). *Anat. Rec.* **180,** 645–662.
Bischoff, R. (1975). *Anat. Rec.* **182,** 215–236.
Bischoff, R. (1978). *In* "Membrane Fusion" (G. Poste and G. L. Nicolson, eds), pp. 127–179. Elsevier/North Holland, Amsterdam.
Bischoff, R., and Holtzer, H. (1969). *J. Cell Biol.* **41,** 188–200.
Blau, H. M., and Webster, C. (1981). *Proc. Natl. Acad. Sci. U.S.A.* **78,** 5623–5627.
Blau, H. M., Webster, C., Chiu, C-P. Guttman, S., Adornato, B., and Chandler, F. (1982). *In* "Muscle Development: Molecular and Cellular Control". (M. L. Pearson and H. F. Epstein, eds.), pp. 543–556. Cold Spring Harbor Laboratory, Cold Spring Harbor.
Blau, H. M., Webster, C., and Pavlath, G. K. (1983). *Proc. Natl. Acad. Sci. U.S.A.* **80,** 4856–4860.
Boone, C. M., Chen, T. R., and Ruddle, F. H. (1972). *Proc. Natl. Acad. Sci. U.S.A.* **692,** 510–514.
Bresolin, N., Ro W., Reyes, M., Miranda, A. F., and DiMauro, S. (1983). *Neurology* **33,** 1049–1052.
Brown, B. I. (1984). *In* "Myology" (A. G. Engel and B. Q. Banker, eds.). McGraw-Hill, New York (in press).
Canki, N., Dutrillaux, B., and Tivadar, I. (1979). *Ann. Genet.* **22,** 35–39.
Carlson, B. M., Hansen-Smith, F. M., and Davindra, K. M. (1979). *In* "Muscle Regeneration" (A. Mauro, ed.), pp. 501–507. Raven, New York.
Chanarin, I., Patel, A., Slavin, G., Willis, E. J., Andrews, T. M., and Stewart, G. (1975). *Br. Med. J.* **1,** 553–555.
Chen, S. H., Malcolm, L. A., Yoshida, A., and Giblett, E. (1971). *Am. J. Hum. Genet.* **23,** 87–91.
Chou, J. Y. (1978). *Proc. Natl. Acad. Sci. U.S.A.* **75,** 1409–1414.
Collins, D., and Gilbert, E. (1977). *Lab. Invest.* **36,** 91–99.
Cox, N., and Kendall, A. P. (1978). *J. Gen. Virol.* **41,** 635–640.
Crain, S. M., Alfei, L., and Peterson, E. R. (1970). *J. Neurobiol.* **1,** 471–489.
Currie, S. (1971). *J. Path.* **105,** 169–185.
D'Ancona, G. G., Wurm, J., and Croce, C. (1979). *Proc. Natl. Acad. Sci. U.S.A.* **76,** 4526–4529.
Davidson, M., Miranda, A. F., Bender, A., DiMauro, S., and Vora, S. (1983a). *J. Clin. Invest.* **72,** 545–550.
Davidson, M., Mongini, T., Foster, S. A., and DiMauro, S. (1983b). *In Vitro* **19,** 272 (abs.).
Davies, K. E., Pearson, P. L., Harper, P. S., Murray, J. M., O'Brien, T. O., Sarfarazi, M., and Williamson, R. (1983). *Nucleic Acids Res.* **11,** 2303–2312.
Davis, C. H. Schliselfeld, L. H., Wolf, D. P., Leavitt, C. A., and Krebs, E. G. (1967) *J. Biol. Chem.* **242,** 4824–4833.
DiDonato, S., Cornelio, F., Pacini, L., Peluchetti, D., and Rimoldi M. (1978). *Ann. Neurol.* **4,** 465–467.
DiDonato, S., Castiglione, A., and Rimoldi, M. (1981). *J. Neurol. Sci.* **50,** 207–215.
DiMauro, S., and Melis-DiMauro, P. M. (1973). *Science* **182,** 929–931.
DiMauro, S., and Hartlage, P. L. (1978). *Neurology* **28,** 1124–1129.
DiMauro, S., and Bresolin, N. (1984). *In "Myology"* (A. G. Engel and B. Q. Banker, eds.). McGraw-Hill, New York (in press).

DiMauro, S., and Papadimitriou, A. (1984). *In* "Myology" (A. G. Engel and B. Q. Banker, eds.). McGraw-Hill, New York (in press).

DiMauro, S., Arnold, S., Miranda, A. F., and Rowland, L. P. (1978). *Ann. Neurol.* **3,** 60–66.

DiMauro, S., Hartwig, G. B., Hays, A. P., Eastwood, A. B., Franco, R., Olarte, M., Chang, M., Roses, A. D., Fetell, R., Schoenfeldt, R. S., and Stern, L. Z. (1979). *Ann. Neurol.* **5,** 422–436.

DiMauro, S., Miranda, A. F., Hays, A. P., Franck, W. A., Hoffman, G., Schoenfeldt, R. S., and Singh, N. (1980). *J. Neurol. Sci.* **47,** 191–202.

DiMauro, S., Miranda, A. F., Friedman, R., Khan, S., and Gitlin, K. (1981). *Science* **212,** 277–1279.

DiMauro, S., Miranda, A. F., Olarte, M., Friedman, R., and Hays, A. P. (1982a). *Neurology* **32,** 584–591.

DiMauro, S., Dalakas, M., and Miranda, A. F. (1982b). *Ann. Neurol.* **13,** 11–19.

DiMauro, S., Hays, A. P., and Eastwood, A. B. (1983). *In* "Mitochondrial Pathology in Muscle Diseases" (G. Scarlato and C. Cerri, eds.). Piccin Editore, Padova (in press).

Drachman, D. B., Angus, C. W., Adams, R. N., Michelson, J. D., and Hoffman, G. J. (1978). *New Engl. J. Med.* **298,** 1116–1122.

Ecob-Johnston, M. S., and Brown, A. E. (1981). *Exp. Neurol.* **71,** 390–397.

Egger, J., and Wilson, J. (1983). *New Engl. J. Med.* **309,** 142–184.

Engel, A. G. (1984a). *In* "Myology" (A. G. Engel and B. Q. Banker, eds.). McGraw-Hill, New York (in press).

Engel, A. G. (1984b). *In* "Myology" (A. G. Engel and B. Q. Banker, eds.). McGraw-Hill, New York (in press).

Engel, A. G. (1984c). *In* "Myology" (A. G. Engel and B. Q. Banker, eds.). McGraw-Hill, New York (in press).

Engel, A. G., and Rebouche, C. A. (1982), *In* "Disorders of the Motor Unit" (D. J. Schotland, ed.), pp. 643–656. Wiley, New York.

Engel, A. G., Potter, C. S., and Rosevear, J. W. (1964). *Nature (London)* **202,** 670–672.

Fambrough, D. M., Bayne, E. K., Gardner, J. M., Anderson, M. J., Wakshull, E., and Rotundo, R. L. (1982). *In* "Neuroimmunology," pp. 49–89. Plenum, New York.

Fischman, D. A. (1972). *In* "The Structure and Function of Muscle" (G. Bourne, ed.), pp. 75–148. Academic Press, New York.

Fischman, D. A. (1982). *In* "Muscle Development: Molecular and Cellular Control" (M. L. Pearson and H. F. Epstein, eds.), pp. 397–404. Cold Spring Harbor Laboratory, Cold Spring Harbor.

Fishbach, G. D., and Cohen, S. A. (1973). *Develop. Biol.* **31,** 147–162.

Fishbein, W. N., Ambrustmacher, V. W., and Griffin, J. L. (1978). *Science* **200,** 545–548.

Fujii, H., Krietsch, W. K. G., and Yoshida, A. (1980). *J. Biol. Chem.* **255,** 6421–6423.

Fujii, H. Chen, S. H., Akatsuka, J., Miwa, S., and Yoshida, A. (1981). *Proc. Natl. Acad. Sci. U.S.A.* **78,** 2587–2590.

Gamboa, E. G., and Hays, A. P. (1984). *In* "Myology" (A. G. Engel and B. Q. Banker, eds.). McGraw-Hill, New York (in press).

Gamboa, E. T., Eastwood, A. B., Hays, A. P. Maxwell, J., and Penn, A. S. (1979). *Neurology* **29,** 1323–1335.

Geiger, R. S., and Garvin, J. S. (1957). *J. Neuropath. Exp.Neurol.* **16,** 532–543.

Godman, G. C. (1957). *J. Morphol.* **100,** 27–82.

Goldberg, R. J., and Crowell, R. L. (1971). *J. Virol.* **7,** 759–769.

Haas, D. C. (1980). *J. Rheumat.* **7,** 671–676.

Harper, P. S. (1979). "Myotonic Dystrophy," pp. 274–290. Saunders, Philadelphia.

Harper, P. S. (1984). *In* "Myology" (A. G. Engel and B. Q. Banker, eds.). McGraw-Hill, New York (in press).
Haselik, A. A., Waheed, A., and von Figura, K. (1981). *Biochem. Biophys. Res. Commun.* **98,** 761–767.
Hashimoto, K., Robinson, L., Velayos, E., and Niizuma, K. (1971). *Arch. Dermat.* **103,** 120–135.
Hastings, K. E., and Emerson, C. P. (1982). *Proc. Natl. Acad. Sci. U.S.A.* **79,** 1553–1557.
Hauschka, S. D. (1974). *Develop. Biol.* **37,** 329–344.
Hauschka, S. D., Linkhart, T. A., Clegg, C., and Merrill, G. (1979). *In* "Muscle Regeneration" (A. Mauro, ed.), pp. 311–322. Raven, New York.
Hauschka, S. D., Rutz, R., Linkhart, T. A., Clegg, C. H., Merrill, G. F., Haney, C. M., and Lim, R. W. (1982). *In* "Disorders of the Motor Unit" (D. L. Schotland, ed.), pp. 903–923.
Hays, A. P., Miranda, A. F., Mehler, M., and DiMauro, S. (1979). *In Vitro* **15,** 204 (abs.).
Heiman-Patterson, T. D., Bonilla, E., and Schotland, D. L. (1982). *Ann. Neurol.* **12,** 305–307.
Herrmann, H., Konigsberg, U. R., and Robinson, G. (1960). *Proc. Soc. Exp. Biol. Med.* **105,** 217–221.
Hickman, S., and Neufeld, E. F. (1974). *Biochem. Biophys. Res. Commun.* **49,** 992–999.
Hiremath, L. S., and Rothstein, M. (1982). *Biochem. Biophys. Acta* **705** 200–209.
Iannacone, S., Bowen, D. E., and Samaha, F. J. (1982). *Arch. Neurol.* **39,** 400–402.
Illingsworth, B. (1961). *Am. J. Clin. Nutr.* **9,** 683–690.
Ionasescu, V., and Ionasescu, R. (1982). *J. Neurol. Sci.* **54,** 79–87.
Johnson, R. L., Fink, C. W., and Ziff, M. (1972). *J. Clin. Invest.* **51,** 2435–2449.
Justice, P., Ryan, C., Hsia, H. Y. Y., and Krmpotik, E. (1970). *Biochem. Biophys. Res. Commun.* **40,** 1259–1265.
Kakulas, B. A. (1966). *Nature (London)* **210,** 115–118.
Kakulas, B. A., Papadimitriou, J. M., Knight, J. V., and Mastaglia, F. L. (1968). *Proc. Aust. Assoc. Neurol.* **5,** 79–81.
Kakulas, B. A., Shute, G. H., and LeClere A. L. F. (1971). *Proc. Aust. Assoc. Neurol.* **8,** 85–92.
Kanno, T., Sudo, K., Takuchi, I., Kanda, S., Honda, N., Nishimura, Y., and Oyama, K. (1980). *Clin. Chim. Acta* **108,** 267–276.
Kao, I., and Drachman, J. B. (1977). *Science* **196,** 527–529.
Karpati, G., Carpenter, S., Engel, A. G. Watters, J., Allen, J., Rothman, S., Klassen, G., and Mamer, O. A. (1975). *Neurology* **25,** 16–24.
Kitamura, M., Itjima, N., Hashimoto, F., and Hiratsuka, A. (1971). *Clin. Chim. Acta* **34,** 419–423.
Koningsberg, I. R. (1963). *Science* **140,** 1273–1284.
Kyriakides, E. C., Paul, B., and Balint, J. A. (1972). *J. Lab. Clin. Med.* **80,** 810–816.
Landau, B. J., Crowell, R. L., Boclair, C. W., and Zajac, B. A. (1972). *Proc. Soc. Exp. Biol. Med.* **141,** 753–758.
Layzer, R. B., Rowland, L. P. and Ranney, H. M. (1967). *Arch. Neurol.* **17,** 512–523.
Layzer, R. B., Havel, R. J., and McLeroy, M. B. (1980). *Neurology* **30,** 627.
Lentz, T. L., Burrage, T. G., and Smith, A. W. (1982). *Science* **215,** 182–184.
Leroy, J. G., and Demars, R. I. (1967). *Science* **157,** 804.
Leroy, J. G., Ho, M. W., McBrinn, M. C., Zielke, K., Jacob, J., and O'Brien, J. S. (1972). *Pediat. Res.* **6,** 752.
Mauro, A. (1961). *J. Biophys. Biochem. Cytol.* **9,** 493–495.
Mawatari, S., Miranda, A. F., and Rowland, L. P. (1976). *Neurology* **26,** 1021–1026.

Mayeda, K., Weiss, L., Lindahl, R., and Dully, M. (1974). *Am. J. Hum. Genet.* **26,** 59–64.
Mehler, M., and DiMauro, S. (1977). *Neurology* **27,** 178–184.
Meienhofer, M. C., Askanas, V., Proux-Daegelen, D., Dreyfus, J-C., and Engel, W. K. (1977). *Arch. Neurol.* **34,** 799–781.
Merickel, M., Gray, R., Chauvin, P., and Appel, S. (1982). *In* "Oisorders of the Motor Unit" (D. L. Schotland, ed.), pp. 889–898. Wiley, New York.
Miranda, A. F. (1974). *J. Cell. Biol.* **63,** 228 (abs.).
Miranda, A. F., and Godman, G. C. (1973). *Tissue Cell* **5,** 1–22.
Miranda, A. F., and Mongini, T. (1983). *In* "Neuromuscular Diseases" (G. Seratrice *et al.*, eds.), pp. 365–371. Raven, New York.
Miranda, A. F., and Mongini, T. (1984). *In* "Myology" (A. G. Engel and B. Q. Banker, eds.). McGraw-Hill, New York (in press).
Miranda, A. F., Gamboa, E. T., Armstrong, C. L., and Hsu, K. C. (1978). *J. Neurol. Sci.* **36,** 63–81.
Miranda, A. F., Somer, H., and DiMauro, S. (1979a). *In* "Muscle Regeneration" (A. Mauro, ed.), pp. 453–473. Raven, New York.
Miranda, A. F., DiMauro, S., Eastwood, A. B., Hays, A. P., Johnson, W. G., Olarte, M., Whitlock, R., Mayeux, R., and Rowland, L. P. (1979b). *Muscle Nerve* **2,** 1–13.
Miranda, A. F., DiMauro, S., Antler, A., Stern, L. Z., and Rowland, L. P. (1981a). *Ann. Neurol* **9,** 283–288.
Miranda, A. F., Fisher, P. B., Johnson, W. G., Shanske, S., and Khan, S. (1981b). *Neurology* **31,** 58 (abs.).
Miranda, A. F., Shanske, S., and DiMauro, S. (1982). *In* "Muscle Development: Molecular and Cellular Control" (M. L. Pearson and H. F. Epstein, eds.), pp. 515–525. Cold Spring Harbor Laboratory, Cold Spring Harbor.
Miranda, A. F., Peterson, E. R., and Masurovsky, E. B. (1983a). *J. Neuropathol. Exp. Neurol.* **42,** 350 (abs.).
Miranda, A. F., Babiss, L. E., and Fisher, P. B. (1983b). *Proc. Natl. Acad. Sci. U.S.A.* **80,** 6581–6585.
Miranda, A. F., Shanske, S., Hays, A. P., and DiMauro, S. (1984). *Arch. Neurol.* (in press).
Mong, S. F., Hays, A. P., and Miranda, A. F. (1982). *Cell Differentiation* **11,** 141–145.
Morgan-Hughes, J. A. (1984). *In* "Myology" (A. G. Engel and B. Q. Banker, eds.). McGraw-Hill, New York (in press).
Munsat, T. (1984). *In* "Myology" (A. G. Engel and B. Q. Banker, eds.). McGraw-Hill, New York (in press).
Murray, M. R. (1965). *In* "Cells and Tissues in Culture" (E. N. Willner, ed.). Academic Press, New York.
Nguyen, H. T., Medford, R. M., and Nadal-Ginard, B. (1983). *Cell* **34,** 281–293.
Norum, K. (1964). *Biochem. Biophys. Acta* **89,** 95–108.
Omenn, G. S., and Cheung, C. V. (1974). *Am. J. Hum. Genet.* **26,** 393–399.
Omenn, G. S., and Hermodson, M. A. (1975). *In* "Isozymes III: Developmental Biology" (C.M. Markert, ed.), pp. 1505–1018. Academic Press, New York.
O'Neill, M. C., and Kendall, A. P. (1975). *Nature (London)* **253,** 195–198.
Osame, M., Engel, A. G., Rebouche, C. J., and Scott, E. R. (1982). *In* "Disorders of the Motor Unit" (D. L. Schotland, ed), pp. 498–501. Wiley, New York.
O'Steen, W. K. (1963). *Tex. Rep. Biol. Med.* **3,** 369–379.
Peterson, E. R. (1978). *In* "Tissue Culture Association Manual" (V. J. Evans, V. P. Perry, and M. M. Vincent, eds.), Vol. 4, pp. 921–923. Tissue Culture Association, Rockville, Maryland.

Peterson, E. R., and Crain, S. M. (1979). *In* "Muscle Regeneration" (A. Mauro, ed.), pp. 429–441. Raven, New York.
Peterson, E. R., Masurovsky, E. B., Spiro, A., and Crain, S. M. (1979). *Soc. Neurosci. Abs.* **5,** 517.
Pogogeff, I. A., and Murray, M. R. (1946). *Anat. Rec.* **95,** 321–335.
Pula, T. P., Max, S. R., Zielke, H. R. Chacon, M., Baab, P., and Gumbinas M. (1981). *Ann. Neurol.* **10,** 196–198.
Quinn, C. A., Goodfellow, P. N., Povey, S., and Walsh, F. S. (1981). *Proc. Natl. Acad. Sci. U.S.A.* **78,** 5031–5035.
Rebouche, C. J., and Engel, A. G. (1982). *In Vitro* **18,** 495–500.
Reitman, M. L. Varki, A., and Kornfeld, S. (1981). *J. Clin. Invest.* **67,** 1574–1579.
Reuser, A. J. J., Koster, J. F., Hoogeveen, A., and Galjaard H. (1978). *Am. J. Hum. Genet.* **30,** 132–143.
Richler, C., and Yaffe, D. (1970). *Dev. Biol.* **37,** 1–22.
Roelofs, R. I., Engel, W. K., and Chauvin, P. B. (1972). *Science* **177,** 795–797.
Rowland, L. P., and Layzer, R. B. (1979). *In* "Handbook of Clinical Neurology" (P. J. Vinken and G. W. Bruyn, eds.), Vol. 40, pp. 349–414. North Holland, Amsterdam.
Sammons, D. W., and Chilson, O. P. (1978). *Arch. Biochem. Biophys.* **191,** 561–570.
Sanger, J. (1974). *Proc. Natl. Acad. Sci. U.S.A.* **71,** 3621–3625.
Sato, K., Imai, F., Hatayama, I., and Roelofs, R. I. (1977). *Biochem. Biophys. Acta* **78,** 663–668.
Schultz, M., and Crowell, R. L. (1980). *J. Gen. Virol.* **46,** 39–49.
Schonberg, M., Miranda, A. F., Mawatari, S., and Rowland, L. P. (1977). *In* "Pathogenesis of Human Muscular Dystrophies" (L. P. Rowland, ed.), pp. 599–611. Excerpta Medica, Amsterdam.
Shanske, S., Shanske, A., Nitowski, H. M. (1974). *Pediat. Res.* **8,** 395, 1976 (abs.).
Shanske, S., Miranda, A. F., Penn, A. S., and DiMauro, S. (1981). *Pediat. Res.* **15,** 1334–1339.
Shumate, J. B., Katnik, M., Ruiz, D., Kaiser, C., Frieden, M. H., Brooke, M. H., and Caroll, J. E. (1979). *Muscle Nerve.* **2,** 213–216.
Snow, M. H. (1977). *Anat. Rec.* **188,** 201–218.
Snow, M. H. (1978). *Cell Tiss. Res.* **186,** 535–540.
Snow, M. H. (1979). *In* "Muscle Regeneration" (A. Mauro, ed.), pp. 91–99. Raven, New York.
Solberg, H. E. (1974). *Biochim. Biophys. Acta* **360.**
Tahmoush, A. J., Askanas, V., Nelson, P. G., and Engel, W. K. (1983). *Neurology* **33,** 311–316.
Tarui, S., Okuno, G., Ikua, Y., Janak, A. T., Suda, M., and Nishikawa, M. (1965). *Biochem. Biophys. Res. Commun.* **19,** 517–523.
Thompson, E. J., Yasin, R., van Beers, G., Nurse, K. C. E., and Al-Ani, S. (1977). *Nature (London)* **268,** 241–243.
Thompson, E. J., Cavanagh, N. P. C., and Yasin, R. (1981). *Exp. Neurol.* **74,** 940–942.
Valderrama, R., Gamboa, E. T., and Miranda, A. F. (1981). *Trans. Am. Neurol. Assoc.* **106,** 291–293.
Vora, S. (1982). *In* "Current Topics in Biological and Medical Research: Isozymes" (M. C. Ratazzi, J. G. Scandalios, and G. S. Whitt, eds.), Vol. 6, pp. 119–167. A. R. Liss, New York.
Vora, S., and Francke, U. (1981). *Proc. Natl. Acad. Sci. U.S.A.* **78,** 3738–3742.
Vora, S., Seaman, C., Durham, S., and Piomelli, S. (1980). *Proc. Natl. Acad. Sci. U.S.A.* **77,** 62–66.

Vora, S., Durham, S., de Martinville, B., George, U., and Francke, U. (1982). *Somat. Cell Genet.* **8,** 95–104.

Vora, S., Miranda, A. F., Hernandez, E., and Francke, U. (1983). *Hum. Genet.* **63,** 374–379.

Wakayama, Y., and Ohbu, S. (1982). *J. Neurol. Sci.* **55,** 59–77.

Wakayama, Y., Schotland, D. L., and Bonilla, E. (1980). *Neurology* **30,** 740–748.

Walsh, F. S., Quinn, C. A., Yasin, R., Thompson, E. T., and Hurko, O. (1982). *In* "Muscle Development: Molecular and Cellular Control" (M. L. Pearson and H. F. Epstein, eds.), pp. 535–542. Cold Spring Harbor Laboratory, Cold Spring Harbor.

Weil, D., Cottreau, D., van Cong, N., Rebourcet, R., Foubert, C., Gross, M-S., Dreyfus J-C., and Kahn, A. (1980). *Ann. Hum. Genet.* **44,** 11–16.

Whitaker, J. N., and Engel, W. K. (1973). *N. Engl. J. Med.* **289,** 107–108.

Will, H., Krause, E. G., Guski, H., and Wollenberger, A. (1974). *Acta Biol. Med. Germ.* **33,** 149–160.

Witkowski, J. A. (1977). *Biol. Rev.* **52,** 431–476.

Witkowski, J. A., and Dubowitz, V. (1975). *J. Neurol. Sci.* **26,** 203–320.

Witkowski, J., Durbridge, M., and Dubowitz, V. (1976). *In Vitro* **12,** 98–106.

Yaffe, D. (1973). *In* "Tissue Culture: Methods and Applications" (P. F. Kruse and M. K. Patterson, eds.), pp. 106–114. Academic Press, New York.

Yarom, R., Carmy, O., Gordon, G. C., and More, R. (1981). *Hum. Path.* **12,** 623–631.

Yasin, R., van Beers, G., Nurse, K. C. E., Al-Ani, S., Landon, D. N. and Thompson, E. J. (1977). *J. Neurol. Sci.* **32,** 347–360.

Yasin, R., Kundu, D., and Thompson, E. J. (1981). *Cell Differentiation* **10,** 131–137.

Yasin, R., Kundu, D., and Thompson, E. J. (1982). *Exp. Cell Res.* **138,** 419–422.

Yunis, A. A., Fischer, E. H., and Krebs, E. G. (1962). *J. Biol. Chem.* **237,** 2809–2815.

Zanella, A., Mariani, M., Meola, G., Fagnani, G., and Sirchia, G. (1982). *Am. J. Hemat.* **12,** 215–225.

Zatz, M., Vianna-Morgante, A. M. Campos, P., and Diament, A. J. (1981). *J. Med. Genet.* **18,** 442–447.

THE BIOLOGY OF HUMAN COLON TUMOR CELLS IN CULTURE

Lynne P. Rutzky

Department of Surgery
The University of Texas Medical School
Houston, Texas

I. Introduction

Cell culture model systems have been helpful in furthering our understanding of the biology of human cancer. Colon cancer, one of the most prevalent forms of human neoplastic disease, is highly refractory to current treatment regimens. Inadequate early detection techniques, difficulties in evaluating tumor remaining after surgical resection, and ineffective chemo- and radiotherapy regimens all contribute to the high mortality of this disease.

Knowledge of the biology of malignant colonic cells is required be-

ISBN 0-12-007904-6

fore more effective treatment modes can be developed. It is only within the past 10 years that pertinent information has been available on the biochemistry, metabolism, immunogenicity, and physiology of colon cancer cells. This article describes the use of well-characterized human colonic adenocarcinoma cell lines as a model system to study cell biology and to identify neoantigens associated with the disease.

The focus of work in this laboratory has been directed toward the development of a set of human colon cancer cell lines, LS174T and LS180, that are representative models of the disease and appropriate sources of tumor neoantigens. These *in vitro* models provide a large uniform population of neoplastic cells free from detectable microbial contamination, passenger host lymphoid and stromal cells, and necrotic debris. Cells recently released from tumor tissue offer an alternate source of materials. However, this approach is complicated by the presence of heterogeneous cell populations, possible low tumor cell viability, and necrotic cells and debris. Cells harvested in sufficient quantity from one tumor can be stored in liquid nitrogen for later retrieval. While this procedure may be useful for some studies, it probably causes additional damage to the recently dissociated cells, adding more selective pressures to the population. In view of these considerations, human tumor cell lines expressing several properties associated with colon cancer have been developed and characterized in this laboratory. These lines are being used to study the biology of human tumor cells grown either *in vitro* or in immunodeficient animals and to evaluate colon–tumor neoantigens. All of the studies have used serially cultured cells that are below 100 population doublings.

II. Biological Properties of Human Tumor Cell Lines

A. *Background*

Since the 1950s (Gey *et al.*, 1952; Eagle, 1955; Berman and Stulberg, 1956) cell lines have been established from various human tumors. Some of the early tumor cell lines (Fjelde, 1955; Moore *et al.*, 1955; Goldenberg *et al.*, 1966; Pattillo and Gey, 1968) were established in immunosuppressed animals (Toolan, 1953, 1954, 1957). More recently, some cell lines have been established from athymic nude mice that were transplanted with cells isolated from human tumors (Carrel *et al.*, 1976; Koura and Isaka, 1980; Dexter *et al.*, 1981). Feeder layers have been used to initiate cell lines from several human tumors (Epstein and Kaplan), 1979; Smith *et al.*, 1981).

Several early studies often used rapidly growing, poorly differentiated HeLa (Gey *et al.*, 1952) or HeLa-like cells (Nelson-Rees and Flandermeyer, 1976) such as HEp-1, HEp-2 (Fjelde, 1955), Chang Liver (Chang, 1954), and some of the Detroit epithelial cell lines (Berman and Stulberg, 1956). Recently, several cell lines established from human tumors have been found to have HeLa-like properties (Lavappa *et al.*, 1976; Nelson-Rees *et al.*, 1980; Harris *et al.*, 1981) or to possess characteristics suggesting cross-contamination with other cell lines (Nelson-Rees *et al.*, 1980; Leibovitz *et al.*, 1979). These studies indicate the need for the complete characterization and continuous monitoring of cell line integrity in order to evaluate properly the data derived from *in vitro* systems. Data derived from *in vitro* models are only as good as the system being studied.

The number of human colorectal tumor cell lines that display various neoplastic properties has risen dramatically in the past 10 years. To date, 63 cell lines have been described in the literature (Table I). In 1973, Giard *et et al.* (1973) reported that a relatively low percentage of cell lines was established from 200 tumor specimens. Recently, McBain (1983) reported the establishment at a high efficiency of cell lines from human colorectal tumors. Leibovitz (1963) described a glucose-free culture medium, L-15, that was used to successfully establish monolayer cell lines from several types of human tumor tissues including 16 SW-colorectal tumors (Leibovitz *et al.*, 1976). Several other human colorectal tumor cell lines established in monolayer cultures are HuCCL-14 (Yaniv *et al.*, 1978), WiDr (Noguchi *et al.*, 1979), and HCT-15 (Dexter *et al.*, 1979). Novel methods in the development of human colon tumor cell lines include (1) passage of tumor cells in cheek pouches of either immunosuppressed or untreated hamsters (Goldenberg *et al.*, 1976; Goldenberg, 1972), (2) cell fusions induced by virus (Watkins and Sanger, 1977), (3) nude mice heterotransplants (Carrel *et al.*, 1976; Orfeo *et al.*, 1980; van der Bosch *et al.*, 1981), (4) fibroblast feeder layers (Brattain *et al.*, 1981, 1983), and (5) in organ culture systems (Breborowicz *et al.*, 1975; Wolff and Wolff, 1975). One potential problem that may occur when xenogeneic animals or their cells are used to initiate human cell lines is contamination with murine viruses or fusion with mouse cells that may produce cell lines appearing to be of human origin but which contain foreign cell elements (Goldenberg and Pavia, 1981; Bowen *et al.*, 1980). Information derived from studies using these cell lines may be difficult to interpret because of the possible presence of mouse genetic or virus-associated properties.

A survey of the literature indicates that 53 human colorectal tumor

TABLE I

PROPERTIES OF HUMAN COLORECTAL TUMOR CELL LINES[a]

Donor tumor tissue					Cell line					
Cell line	Tissue	Tissue origin	Age	Sex	Dukes' class	PDT (hr)	Chromosome mode	Tumor marker CEA	Genetic signature (43)	Reference
1. RPMI-4788	Colon ACA	M	73	F	ND	12	ND	ND	ND	137
2. GW-39	Colon ACA	P	58	F	ND	ND	ND	CEA positive	ND	50
3. HCT-8	Colon ACA	P	67	M	ND	20	48	290 ng/ml	ND	148
4. HRT-18	Rectum ACA	P	65	M	ND	20	48	32 ng/ml	ND	148
5. HT-29	Colon ACA	P	44	F	ND	12	Triploid	3.70 ng CEA/10^6 cells[b]	0.004	40
6. SK-CO-1	Colon ACA	M	65	M	ND	ND	ND	ND	ND	40
7. SW-48	Colon	P	83	F	C_2	70	47	8 ng CEA/10^6 cells[b]	0.015	84
8. SW-707	Rectum	ND	81	M	ND	47	47	31 ng CEA/10^6 cells[b]	ND	84
9. SW-802	Colon	P	64	F	B_1	58	47	214 ng CEA/10^6 cells[b]	0.037	84
10. SW-480	Colon	P	50	M	ND	50	55	21 ng CEA/10^6 cells[b]	ND	84
11. SW-620	Lymph node	M	51	M	ND	30	54	11 ng CEA/10^6 cells[b]	ND	84
12. SW-403	Colon	P	51	M	C_1	79	66	7500 ng CEA/10^6 cells[b]	0.016	84
13. SW-742	Colon	P	75	M	B_2	273	54	2500 ng CEA/10^6 cells[b]	0.042	84
14. SW-837	Rectum	P	53	M	C_2	169	42;80	1200 ng CEA/10^6 cells[b]	0.007	84
15. SW-948	Colon	P	81	F	C_1	120	76	2000 ng CEA/10^6 cells[b]	0.007	84
16. SW-1083	Perineum rectum	ND	83	M	ND	ND	42;80	ND	ND	84
17. SW-1116	Colon	P	73	M	B_2	163	63	7000 ng CEA/10^6 cells[b]	0.017	84
18. SW-1222	Colon	P	44	M	C_2	ND	44	ND	0.008	[c]
19. SW-1345	Rectum	P	53	F	C_2	51	51	ND	0.008	[c]
20. SW-1398	Colon	P	77	F	B_2	ND	ND	ND	0.10×10^{-4}	[c]
21. SW-1417	Colon	P	53	F	C_2	ND	ND	ND	0.006	[c]
22. SW-1463	Rectum	P	66	F	C_2	ND	ND	ND	0.8×10^{-4}	[c]

23. LoVo	Colon ACA	M	56	M	ND	37	49	54 ng CEA/10^6 cells	ND	27
24. LS174T	Colon ACA	P	58	F	B	22	45	401 ng CEA/10^6 cells[b]	2.4×10^{-4}	145
25. LS180	Colon ACA	P	58	F	B	72	45	775 ng CEA/10^6 cells[b]	2.4×10^{-4}	145
26. Co-115	Colon ACA	P	77	F	ND	36	49	30 ng CEA/ml	ND	13
27. HT-55	Rectum ACA	P	54	F	ND	75	76	54 ng/ml	ND	156
28. HuCC1-14	Colon	P	68	M	ND	24	70	4340 ng CEA/10^6 cells	ND	165
29. COLO 201	Colon ACA	M	70	M	D	ND	72;75	23 ng CEA/10^6 cells	ND	137
30. COLO 205	Colon ACA	M	70	M	D	ND	75	106 ng CEA/10^6 cells	5.5×10^{-4}	137
31. COLO 206	Colon ACA	M	70	M	D	ND	75	12 ng CEA/10^6 cells	5.5×10^{-4}	137
32. COLO 320	Colon	P	55	F	C_2	ND	51	Negative	2.3×10^{-4}	122
33. COLO 321	Colon	P	55	F	C_2	ND	48	Negative	2.6×10^{-4}	122
34. COLO 394	Rectum CA	M	78	M	D	ND	ND	ND	0.015	*d*
35. COLO 395	Rectum CA	M	78	M	D	ND	ND	ND	0.012	*d*
36. COLO 396	Rectum CA	M	78	M	D	ND	ND	ND	0.013	*d*
37. COLO 397	Rectum CA	M	78	M	D	ND	ND	ND	0.013	*d*
38. COLO 399	Rectum CA	M	78	M	D	ND	ND	42 ng CEA/10^6 cells	ND	*d*
39. COLO 405	Rectum CA	M	78	M	D	ND	ND	ND	ND	*d*
40. COLO 416	Rectum CA	M	78	M	D	ND	ND	ND	ND	*d*
41. COLO 463	Rectum CA	M	78	M	D	ND	ND	ND	ND	*d*
42. WiDr	Colon ACA	P	78	F	ND	15	73	12 ng CEA/10^6 cells	0.01	109
43. CCK-81	Colon ACA	M	62	F	ND	45	48	515 ng CEA/10^6 cells	ND	80
44. HCT-15	Colon ACA	P	75	M	C	20	46	Membrane CEA	ND	24
45. DLD-1	Colon ACA	P	45	M	C	19	46;>70	Membrane CEA	ND	25
46. LS123	Colon ACA	P, second	65	F	B	21	57	5.6 ng CEA/10^6 cells[b]	0.004	133
47. HTC 116	Colon ACA	P	ND	ND	ND	21,22	46	76 ng CEA/10^6 cells[e]	ND	9
48. HTC 116a	Colon ACA	P	ND	ND	ND	22,26	46	2360 ng CEA/10^6 cells[e]	ND	9
49. HTC 116b	Colon ACA	P	ND	ND	ND	24,20	46	Negligible[e]	ND	9
50. HCT C	Colon ACA	P	ND	ND	ND	ND	46	24 ng CEA/10^6 cells[e]	ND	10
51. HCT C Col	Colon ACA	P	ND	ND	ND	ND	47	442 ng CEA/10^6 cells[e]	ND	10

(continued)

TABLE I (*Continued*)

Donor tumor tissue					Cell line					
Cell line	Tissue	Tissue origin	Age	Sex	Dukes' class	PDT (hr)	Chromosome mode	Tumor marker CEA	Genetic signature (43)	Reference
52. VACO 1	Colon ACA	M	42	F	D	65	39	63 ng CEA/10^6 cells[f]	ND	97
53. VACO 3	Colon ACA	M	66	M	C2	39	52	40 ng CEA/10^6 cells[f]	ND	[g]
54. VACO 4	Colon ACA	P	59	M	D	30	60	8 ng CEA/10^6 cells[f]	ND	[g]
55. VACO 4L	Colon ACA	P	59	M	D	30	65	17 ng CEA/10^6 cells[f]	ND	[g]
56. VACO 5	Colon ACA	P	78	F	C2	29	47	Negligible[f]	ND	[g]
57. VACO 5A	Colon ACA	P	78	F	C2	42	47	2.5 ng CEA/10^6 cells[f]	ND	[g]
58. VACO 6	Colon ACA	P	63	M	C2	25	90	20 ng CEA/10^6 cells[f]	ND	[g]
59. VACO 8	Colon ACA	P	56	M	D	46	43	ND	ND	[g]
60. VACO 9P	Colon ACA	P	67	M	D	110	42	ND	ND	[g]
61. VACO 9M	Colon ACA	M	67	M	D	80	42	ND	ND	[g]
62. VACO OP	Colon ACA	P	72	F	D	ND	ND	ND	ND	[g]
63. VACO OM	Colon ACA	M	72	F	D	ND	ND	ND	ND	[g]

[a] ACA, Adenocarcinoma; P, primary; M, metastatic origin; F, female; M, male; R, round; E, epithelial; B, bipolar. Dukes' class indicates extent of lesion invasion of the malignant tissue source: A, mucosa and submucosa; B, muscularis; B_1, into muscularis; B_2, muscularis penetrated; C, intestinal wall; C_1, limited to intestinal wall; C_2, extended through all layers with posterior node involvement; D, extent of metastasis includes other organs; ND, no data.

[b] Twenty-one day supernatant.

[c] A. Leibovitz (personal communication).

[d] G. E. Moore (personal communication).

[e] Seventy-two hour supernatant.

[f] Twenty-four hour supernatant.

[g] J. Wilson (personal communication).

cell lines have been characterized and published between 1963 and 1983; most of the lines were published after 1974 (Table I). Within this group 10 sets of multiple cell lines were established from tumor tissues resected from 10 different patients. These sets of cell lines are (1) LS174T and LS180 (Tom *et al.*, 1976); (2) COLO 394, 395, 396, 397, 399, 405, 416, 463 (G. Moore, unpublished data); (3) COLO 205 and 206 (Semple *et al.*, 1978); (4) COLO 320 and 321 (Quinn *et al.*, 1979); (5) HCT 116, HCT 116a, and HCT 116b (Brattain *et al.*, 1981); (6) HCTC and HCTC C Col (Brattain *et al.*, 1983; (7) VACO 4 and VACO 4L (McBain *et al.*, 1983); (8) VACO 5 and VACO 5A (McBain *et al.*, 1983); (9) VACO 9P and VACO 9M (McBain *et al.*, 1983); and (10) VACO O and VACO OM (McBain *et al.*, 1983; Wilson, personal communication). Cell line sets SW480 and SW620 (Leibovitz *et al.*, 1976), VACO 9P and VACO 9M, and VACO O Panel OM (McBain *et al.*, 1983) were established from primary and metastatic lesions resected from the same patient. In the case of the SW and COLO cell lines, genetic signature analysis of allozyme phenotypes identified the origin of multiple lines established from individual patients (Rutzky and Siciliano, 1982). Multiple cell lines from the same tumor tissue that have various properties indicate the heterogeneous nature of many tumors (Fidler, 1978; Heppner *et al.*, 1978).

Analysis of the human colorectal tumor cell lines listed in Table I indicated that 67% were derived from primary colorectal tumors. The cell lines established from tumors of metastatic origin are (1) SK-CO-1 (Fogh and Trempe, 1975); (2) RPMI-4788 (Moore *et al.*, 1963); (3) SW620 (Leibovitz *et al.*, 1976); (4) LoVo (Drewinko *et al.*, 1976); (5) COLO 201, 205, and 206 (Semple *et al.*, 1978); (6) the COLO 394 cell line series (G. Moore, unpublished data); (7) CCK-81 (Koura and Isaka, 1980); and (8) VACO 1, VACO 3, VACO 9M, and VACO OM (McBain *et al.*, 1983). LS123 was established from the second in a series of three primary colorectal tumors resected from one patient (Rutzky *et al.*, 1983). Of the cell lines listed in Table I, 41 and 59% of the cell lines were established from female and male tissue donors, respectively. Of these cell lines the mean age for women tissue donors was 62.3 ± 2.2 (SEM) years and 65.6 ± 2.5 years for men. The cell lines ranged in Dukes' classification of invasiveness from B to D, with D being the most invasive (Dukes, 1932). Most of the cell lines had an epithelial-like morphology, except for some of the COLO cell lines established in the laboratory of Dr. G. E. Moore. These cell lines had rounded and bipolar cells (Semple *et al.*, 1978; Quinn *et al.*, 1979), HCT C cells that grew as grape-like clusters (Brattain *et al.*, 1983), and some of the VACO cell lines (McBain *et al.*, 1983).

B. Genetic Integrity

Genetic identification of cell lines and continuous monitoring of their integrity can be achieved with two methods: (1) allozyme phenotyping and (2) human leukocyte antigen (HLA) identification. Electrophoresis of cell lysates in starch gels followed by visualization of enzyme reaction products has been used to study the expression of gene loci coding for enzymes (isozymes) with more than one allele. Enzymes with frequencies in a population that are greater than 0.01 are considered to be polymorphic (Harris, 1966). When the frequencies of polymorphic isozymes are used in combination, the relationship of an individual to various populations can be determined (Gartler, 1966; Wright, 1980). Enzyme polymorphisms are used in genetic signature analysis to identify phenotypically unique combinations of alleles present in a particular individual and in a cell line (O'Brien *et al.*, 1977). This type of analysis is useful for the detection of HeLa cell cross-contamination (Povey *et al.*, 1976). Wright *et al.* (1981) reported the genetic signatures for 71 human tumor cell lines including two cell lines of colorectal origin. Rutzky and Siciliano (1980, 1982) reported allozyme phenotypes of 13 polymorphic enzyme loci for 24 human colorectal tumor cell lines derived from 18 individuals and for 6 cell lines of nonmalignant origin. Dracopoli and Fogh (1983) studied the gene frequencies of 10 polymorphic isozymes in human tumor cell lines. They reported a loss of heterozygosity for the markers studied.

Human leukocyte antigens (HLA) have been used for phenotyping cell lines (Espmark *et al.*, 1978). A recent study reported the HLA and DR antigen phenotypes for several cell lines (Pollack *et al.*, 1981). The HLA phenotype of the tissue donor of the LS174T and LS180 set of cell lines is known, but it has not been possible to phenotype the cell lines completely because of their apparent resistance to antibody-mediated cytolysis (L. P. Rutzky, unpublished data).

Analysis of the karyotypic modes for 25 cell lines listed in Table I indicates that the mean mode for cell lines with one predominant chromosome mode was 53.93 ± 1.83 (SEM). From this group of cell lines the mean chromosome modes were 56.57 ± 2.78 for 21 cell lines derived from men and 51.29 ± 2.22 for 21 cell lines derived from women. For cell lines with bimodal chromosome distributions, four were derived from male and one was from a female tissue donor. When cell lines with one chromosome mode were analyzed, they ranged from 42 to 90 and 39 to 76 chromosomes for male and female tumor donors, respectively. This analysis shows that many of the published colon tumor cell lines with karyotype data are hyperdiploid with a single chromosome mode, although 14% of the cell lines have a chromosome mode of 46 or less. The cell lines are equally divided between male and female tissue donors. A recent study by Chen *et al.* (1982) reported on the karyotypes of nine human colorectal tumor cell lines banked at the American Type Culture Collection (ATCC). The cell lines belonged to two groups: those characterized by stable chromosomes and others that were karyotypically less stable. Chromosome numbers 7 and 1 had the most structural rearrangements.

C. Tumor Markers

Population doubling times (PDT) of 26 colorectal tumor cell lines listed in Table I range from 12 hours for HT-29 cells to 273 hours for SW742 cells. The mean PDT for all of these cell lines is 54.4 ± 8.24 hours. The amount of carcinoembryonic antigen (CEA), a glycoprotein oncofetal antigen, released by various cell lines into the culture medium ranges from nondetectable amounts to 7500 ng CEA/million SW403 cells (Table II). Leibovitz *et al.* (1976) reported that 11 human colorectal tumor cell lines established in his laboratory were classified into three groups based upon morphology, modal chromosome number, and the ability to synthesize CEA *in vitro*. For example, cell lines belonging to group 1 were characterized by the release of 8–214 ng CEA/million cells and had near normal karyotypes. Group 2 cell lines were characterized by low CEA release (11–21 ng CEA/million cells) and had modal chromosome numbers of 54 and 55. Cell lines in group 3 were distinguished by CEA release between 1200 to 7500 ng CEA/million cells and had chromosome modes greater than 54. When the quantities of reported secreted CEA and modal chromosome numbers were analyzed for 21 other human colorectal tumor cell lines, no correlation was observed (Table II). It must be noted, however, that these data were gathered from studies published by several independent laboratories. Thus the data may not be directly comparable since the cells were grown for varying lengths of time and under different culture conditions (Drewinko *et al.*, 1976). In addition, the analysis of detectable CEA may vary depending upon the method of assay (Fritsche *et al.*, 1978, 1980). The comparison and analysis of these data therefore should be interpreted with caution. Probably the most relevant data are those derived from comparative studies with several cell lines that are analyzed in one laboratory (see Leibovitz *et al.*, 1976) or where similar growth and assay procedures are used (Leibovitz *et al.*, 1976; Rutzky *et al.*, 1979).

TABLE II

RELATIONSHIP OF SECRETED CEA AND CHROMOSOME MODE IN HUMAN COLORECTAL TUMOR CELL LINES

Cell line	Class[a]	ng CEA/10^6 cells	Chromosome mode
VACO 5A	ND[b]	2.5[c]	47
HT-29	ND	3.7[d]	45;60
LS123	ND	5.6[d]	57
SW-48	Group 1	8.0[d]	47
VACO 4	ND	8.0[c]	60
SW-620	Group 2	11[d]	54
COLO-206	ND	12	75
WiDr	ND	12	73
VACO 4L	ND	17[c]	65
VACO 6	ND	20[c]	90
SW-480	Group 2	21[d]	55
COLO 201	ND	23	72;75
HCTC	ND	24[e]	46
SW 707	Group 1	31[d]	47
VACO 3	ND	40[c]	52
LoVo	ND	54	49
VACO 1	ND	63[c]	39
HCT 116	ND	76[e]	46
COLO-205	ND	106	75
SW 802	Group 1	214[c]	47
LS174T	ND	401[d]	45
HCTC Col.	ND	442[e]	47
CCK-81	ND	505	48
LS180	ND	775[d]	45
SW-837	Group 3	1200[d]	42;80
SW-948	Group 3	2000[d]	76
HCT 116a	ND	2360[e]	46
SW-742	Group 3	2500[d]	54
HCCL-14	Group 3	4340[d]	70
SW1116	Group 3	7000[d]	63
SW 403	Group 3	7500[d]	66

[a] From Leibovitz *et al.* (1976).
[b] ND, No data.
[c] Twenty-four hour supernatant.
[d] Twenty-one day supernatant.
[e] Seventy-two hour supernatant.

D. Neoplastic Potential and Differentiation

The neoplastic origin and/or potential of human tumor cell lines is difficult to evaluate. Lacking the availability of appropriate syngeneic hosts for tumor tissue transplantation, tumorigenicity has been evaluated in (1) either immunosuppressed or immunodeficient xenogeneic animal hosts (Giovanella *et al.*, 1972); (2) semisolid agar (Macpherson and Montagnier, 1964); (3) growth on confluent cell monolayers (Tompkins *et al.*, 1974; Tom *et al.*, 1976; Brattain *et al.*, 1981; Rutzky *et al.*, 1983); (4) invasive growth on chick chorioallantoic membranes (Leighton *et al.*, 1972); and (5) in growth on chick embryonic skin (Noguchi *et al.*, 1978; Rutzky *et al.*, 1983). Several human colorectal tumor cell lines have been evaluated for invasiveness in immunodeficient athymic nude mice (Giovanella *et al.*, 1972) or in immunosuppressed hamsters (Goldenberg *et al.*, 1966; Tom *et al.*, 1977) or mice (Semple *et al.*, 1978). In many of these systems, progressively growing tumors develop that express histopathology mimicking the patient's original tumor. Two cell lines, HCT-8 and HRT-18, were grown upon confluent monolayers as a method to evaluate tumorigenicity (Tompkins *et al.*, 1974). Tumor formation in irradiated mice was used to study invasiveness of cell lines (Semple *et al.*, 1978; Quinn *et al.*, 1979). One cell line, LS123, established in this laboratory, may be sensitive to NK cell killing and failed to grow progressively in nude mice but exhibited some invasive growth in chick embryonic skin and grew as colonies upon confluent cell monolayers (Rutzky *et al.*, 1983).

Tumor differentiation to more benign cell types has been observed in human neuroblastoma (Dyke and Mulkey, 1967), mouse melanoma cells treated with 5-bromodeoxyuridine (Silagi and Bruce, 1970), mouse neuroblastoma cells exposed to cylic AMP (Parsad and Hsie, 1971), and mouse teratocarcinoma cells (Pierce and Verney, 1961). Clones of mouse teratocarcinoma cells (Kahan and Ephrussi, 1970) and tumors arising from a single embryonal carcinoma cell (Kleinsmith and Pierce, 1964) indicate the presence of pleuripotential stem cells. Differentiation can be induced *in vitro* following the addition of certain polar solvents, butyric acid, or retinoic acid that produce phenotypic changes in a human promyelocytic leukemia cell line (Collins *et al.*, 1978, 1980).

Most of the cell lines listed in Table I grow as undifferentiated cell monolayers. When some of the cell lines were grown in nude mice, morphological evidence of differentiated growth was observed (Koura and Isaka, 1980; Dexter *et al.*, 1979, 1981). Cell lines DLD-1 and HCT-15 can be induced to differentiate when grown in culture medium containing the polar solvent *N,N*-dimethylformamide or retinoic acid (Dexter *et al.*, 1979, 1981; Hager *et al.*, 1980). In the LS174T colorectal tumor cell system, differentiation was induced *in vitro* without the addition of exogenous chemicals. Glandular growth of LS174T cells and its clones was observed when they were cultured either in a perfused hollow fiber culture system (Rutzky *et al.*, 1980), on floating collagen gels (L. P. Rutzky, unpublished data), or on embryonic chick skin (Ridge and Noguchi, 1983). The glands that formed contained mucin and intracytoplasmic vacuoles lined with microvilli and resembled the patient's original tumor.

III. Biological Properties of Cell Lines LS174T and LS180

A. *Establishment and Cloning*

Models of human colorectal cancer must possess properties associated with the disease and should retain these properties during subculture. Animal tumor cell line models should metastasize to the appropriate secondary organ sites. Recently, a model for metastatic human colon cancer was described in nude mice (Spremulli *et al.*, 1983). Information obtained from characterization of cell line models can be used to study the disease while continuous monitoring can reveal either the stability of cells or the drift of various properties during *in vitro* culture.

A cell line model for human colon cancer, LS174T and LS180, was developed in this laboratory (Tom *et al.*, 1976). Table III outlines some of the biological properties of these cell lines. Serial characterization studies have shown that the properties studied have remained essentially unchanged for almost 10 years. This apparent stability was facilitated by cryopreservation of a large number of early passage cultures. Later studies have been performed on cells that are below 100 serial subcultivations. Whenever low passage cells were retrieved from the liquid nitrogen freezer, a portion of the cell population was expanded in culture and again cryopreserved. This process helps to ensure the future availability of cells with limited subcultivation.

Both the LS174T and LS180 cell lines were established from one primary colon adenocarcinoma that was resected from a 58-year-old Caucasian female in 1974. The tissue was diagnosed as a moderately well-differentiated Dukes' B tumor that had spread to the pericolonic

TABLE III

PROPERTIES OF CELL LINES

Characteristics	LS174T	LS180
Cell morphology	Epithelial	Epithelial
Culture features	Monolayer	Monolayer and multicellular spheroids
Karyotype	45,X	45,X t(Xq→5q)
Genetic signature	2.6×10^{-4}	2.6×10^{-4}
Mucin	+	++
CEA (21 day)	370 ng/10^6 cells	700 ng/10^6 cells
PDT[a]	18–20 hours	72 hours
Tumorigenicity		
Nude mice	+	+
Soft agar	+	+
Stem cell assay	+	+
Monolayers	+	+
Chick embryonic skin	+	ND[b]

[a] PDT, Population doubling time.
[b] ND, No data.

adipose tissue (Fig. 1A). The patient remained disease free until 1981 when a pancreatic carcinoma was diagnosed.

Tumor tissue was treated with several volumes of culture medium containing penicillin, streptomycin, and fungizone. After being minced into small pieces, the tumor was passed through a sterile stainless-steel screen (40 mesh) (Tom *et al.*, 1976). The resulting free cells and small tissue fragments were washed and cultured in 25-cm² culture flasks containing minimal essential medium (MEM) (Eagle, 1959) supplemented with 20% heat-inactivated (56°C, 30 minutes) fetal bovine serum (FBS).

Several of the original flasks had confluent monolayers of epithelial-like tumor cells after 10 months in culture. Half of the flasks were scraped with a rubber policeman and subcultured on MEM containing 10% heat-inactivated FBS. These cells were designated LS180. The remaining flasks were treated with 0.1% trypsin diluted in 0.02% versene, pH 7.4. These cells were subcultured on MEM with 10% heat-inactivated FBS and became the LS174T cell line: "T" represents trypsin. Both cell lines were grown on MEM with either 50 units or 50 μg of penicillin or streptomycin, respectively. The cell lines were grown with antibiotics and tested routinely for mycoplasma, aerobic or anaerobic

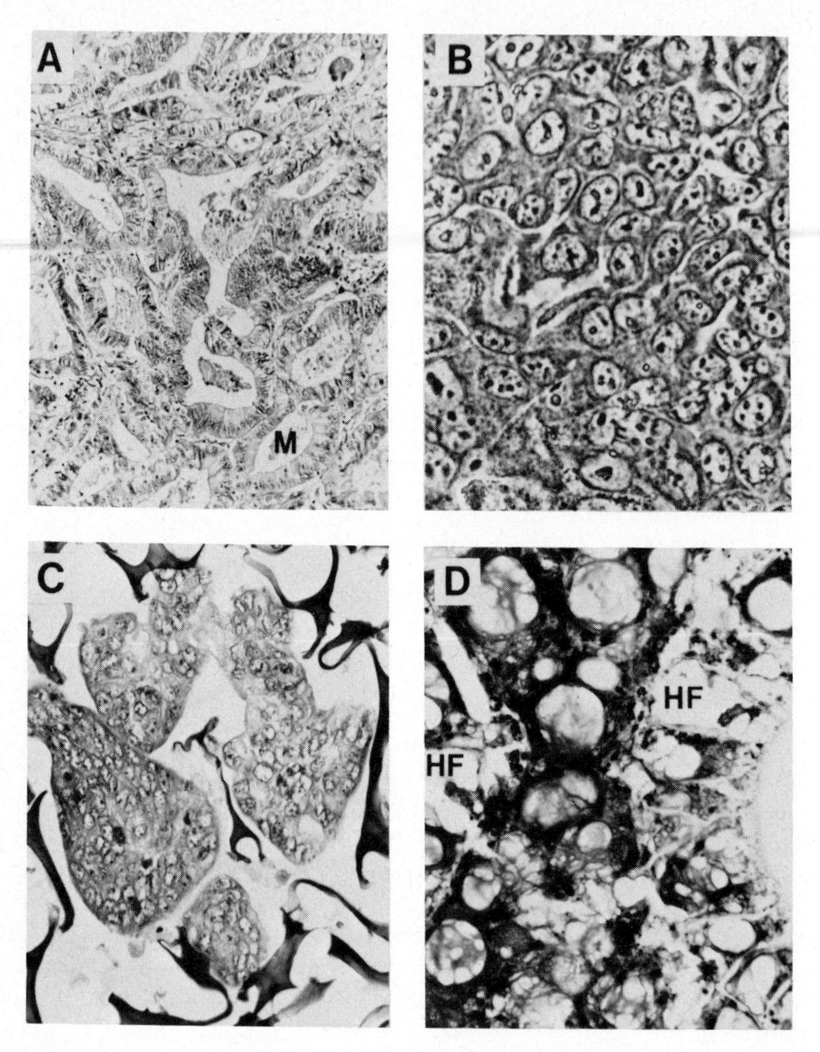

FIG. 1. (A) Cross section of patient's original tumor with histopathology of a moderately differentiated mucinous (M) adenocarcinoma (H and E, ×1000). (B) LS174T cells cultured as a monolayer, exhibiting distinct epithelial-like morphology, a high nucleus to cytoplasm ratio, and multiple nucleoli (phase contrast, ×400). (C) LS174T cells growing in a gelatin sponge matrix as 4 spherules devoid of apparent glandular organization (H and E, ×440). (D) LS174T cells growing between adjacent hollow fibers (HF). Note distribution of darkly staining PAS-reactive carbohydrates within gland lumina and between cells (PAS ×440). Reprinted from Rutzky *et al.* (1979).

bacteria, and virus particles. No evidence of microbial contamination has been detected. Cells from several subcultures preserved in liquid nitrogen have been retrieved successfully.

The LS174T cell line was subcloned at several passage levels. A culture medium was designed to provide maximal growth of low cell numbers. This medium, consisting of an equal mixture of MEM and F-12 (Ham, 1965), is supplemented with 10 μg/ml insulin, 20% heat-inactivated FBS, and 400 m*M* L-glutamine. Several clones were established from single cells that were isolated with glass cloning rings. Clones 10-1, 10-2, 10-3, and 10-4 were isolated from young cells growing at 10 serial population doublings. Clones 3-5 and 6-6 were established by Dr. I. Tribbe (Abbott Laboratories, North Chicago, IL) from cells growing at the fifth and sixth passage levels, respectively. Subclones were isolated from the original clones after inoculation into soft Noble agar (Difco, Detroit, MI) (Macpherson and Montagnier, 1964).

B. Nutritional Requirements

LS174T, LS180, and LS174T clones can be cultured on a variety of media. In addition to MEM, the cells grow on galactose containing L-15 medium (Leibovitz, 1963). Comparison of LS174T cells grown in MEM or L-15 showed that about twofold more CEA is released into the L-15 culture medium. CEA bound to LS180 cells was 15-fold greater on cells grown in L-15; 2.3 times more CEA was released from LS180 cells grown in L-15 medium (Kahan *et al.*, 1979). The kinetics of CEA released was essentially exponential with maximum release in the stationary phase of the growth cycle.

LS174T cells have been cultured in either serum-free or low-serum-containing culture medium. These cells were adapted for serial proliferation in serum-free MEM (Hanks' salts), containing 1% Bactopeptone (Difco), 25 mg/ml dextrose, 2× MEM nonessential amino acids, and 1× BME vitamins (Rutzky, 1981).

In one study serum-free LS174T cells were serially subcultured by trypsin treatment for over 50 population doublings. The growth rate doubled from about 20 hours in serum-containing medium to 45 hours in serum-free medium. Cells growing on the peptone supplement plus 1% FBS had a 25-hour population doubling time. The cells continued to produce mucin and sulfated and nonsulfated acid mucopolysaccharides when grown with or without serum. Cell-associated CEA content was similar for LS174T cells grown in the presence of either 10 and 1% FBS, but a ninefold decrease in CEA shedding was observed in low-serum-containing medium.

Recently, clone 3-5 of LS174T cells were cultured in MEM supplemented with a 10% mixture of growth factors, hormones, attachment factors, transport factors, and serum products (Nu Serum, Collaborative Research, Lexington, MA). In serial subcultivation, cells with this supplement had growth rates similar to those observed with 10% FBS. Nu Serum contains about 1.3 mg protein/ml final concentration. LS174T cells have been subcultured also in a serum-free medium described by Murakami and Masui (1980).

C. Morphological Analysis

LS174T cells and clones are epithelial-like with an average cell diameter of 20 μm. The cells have a large nucleus with at least two prominent nucleoli. LS180 cells appear to have a similar morphology but produce large amounts of mucin which tends to obscure the cell boundaries (Tom *et al.*, 1976). Mucin production can be demonstrated by reacting the cell monolayers with mucicarmine stain. LS174T and LS180 cells produce both sulfated and nonsulfated acid mucopolysaccharides, as illustrated by reacting the cells with high iron diamine and Alcian Blue stains (Spicer and Meyer, 1960; Spicer, 1965). The nonsulfated material is associated with discrete intracellular droplets; the sulfated materials are found in the central portions of the cell colonies (Kahan *et al.*, 1979).

LS174T cells grow as tightly packed islands of cells that coalesce to form an epithelial-like cell sheet (Fig. 1B). All of the clones isolated have a morphology that is similar to LS174T cells except for several of the clones that produce clusters of intracellular vacuoles. LS180 cells are subcultured by mechanical scraping with a rubber policeman. When fragments of the cell sheet are dislodged, they roll up to form spherical structures in the fresh culture medium. When these balls of cells become dense, they settle out of the medium and grow on the culture flask surface. A monolayer forms that has areas of both densely packed flat and elevated cells. Because of the unevenness of the monolayer and the large amounts of released mucin, some of the LS180 cells may appear to be out of focus.

Junctional complexes, interdigitation of adjacent cell membranes, and surface microvilli were found during ultrastructural analysis of LS174T cells (Kahan *et al.*, 1979). The cells contained numerous polysomes, some endoplasmic reticulum, and a well-developed Golgi system (Tom *et al.*, 1976). Scanning electron microscopic examination showed that the cell surface was covered with polymorphic microvilli (Kahan *et al.*, 1979).

D. Genetic Integrity

Isozyme analysis has been used to phenotype the cells. This analysis determines the origin and integrity of human tumor cell lines and ensures maintenance of their unique genetic properties. Isozyme analysis of LS174T, LS180, and LS174T clones was used to (1) prove that the cell lines were derived from the tumor tissue of one individual, and (2) monitor the genetic integrity of the cell lines during serial subcultivation (Rutzky and Siciliano, 1980, 1982). The genetic signatures were 2.6×10^{-4} for LS180, LS174T, and a cell strain isolated from the same patient's bowel mucosa (NB LS174T) (Table IV). The pattern of allozymes present differs from HeLa cells at three loci studied, indicating that LS180, LS174T, and NB (LS174T) are not HeLa contaminated. The identical genetic signature of all of these cell lines is consistent with their derivation from one person. The low frequency (2.6×10^{-4}) of this combination of enzyme phenotypes in the Caucasian population indicates that the probability is negligible that one of these lines is a contaminant of independent origin (Rutzky and Siciliano,

TABLE IV

GENETIC SIGNATURE OF LS174T CELLS[a]

		LS174T	
	HeLa	Parental	Clones
1. Acid phosphatase (AcP)	ab	ab	ab
2. Aconitase (ACONs)	1	1	1
3. Adenosine deaminase (ADA_{rbc})	1	1	1
4. Esterase (EsD)	1	1	1
5. Glucose-6-phosphate dehydrogenase (G_6PDH)	A	B	B
6. α-Glucosidase (α-GLU)	1	1	1
7. Glutamic-oxaloacetic transaminase (GOT_m)	1	1	1
8. Glyoxylase (GLYI)	2	2	2
9. Peptidase			
A (PEP A)	1	1	1
C (PEP C)	1	1	1
D (PEP D)	1	1	1
10. Phosphoglucomutase			
1 (PGM_1)	1	1	1
3 (PGM_3)	1	2	2
11. 6-Phosphogluconate dehydrogenase (6PGD)	A	A	A
12. Malic enzyme (ME_m)	1,2	2	2

[a] Reprinted from Rutzky *et al.* (1980).

1980). Similar analysis of the clones of LS174T revealed that each has a genetic signature identical to the parent LS174T cell line (Rutzky and Siciliano, 1982). Thus, any differences observed among the clones are not due to contamination with other cell lines. The genetic signatures are stable properties in these lines and did not change during serial subculture or when passaged in nude mice and recultured (Rutzky and Siciliano, 1982).

Genetic integrity of cell lines can be verified by HLA phenotyping. Using complement-mediated cytolysis and Trypan Blue dye uptake by damaged cells, it has not been possible to HLA type the LS174T cell line. However, the presence of some components of the patient's HLA phenotype on the tumor cells has been verified indirectly. After *in vitro* production of cytotoxic mouse lymphocytes, polyclonal NIH HLA typing sera were used to block cell-mediated cytolysis (Raphael and Tom, 1982). The HLA phenotype of the patient's lymphocytes is HLA A2, AW30, B13, B35, CW4, CW-, DRW4, DRW7; the patient is ABO blood type O+. Cytolysis was reduced 24.2 and 17.5% when anti-HLA A2 and B13 antisera were added to LS174T cells, respectively. No inhibition of cytolysis was observed when anti-HLA A31 sera was used. Thus, LS174T cells possessed at least some of the patient's HLA antigens (see Fig. 4).

Chromosome analysis of human tumors (Lubs, 1970) has been one approach to understanding cancer cells. Many cultured human colon tumor cell lines have near diploid chromosome complements (Yaniv *et al.*, 1978; Fogh and Trempe, 1975; Drewinko *et al.*, 1976; Rosenthal *et al.*, 1977; Tompkins *et al.*, 1974) whereas others have heteroploid (Leibovitz *et al.*, 1976; Noguchi *et al.*, 1979; Semple *et al.*, 1978; Quinn *et al.*, 1979) or variable karyotypes (Fogh and Trempe, 1975) (Table II). LS174T and LS180 cells are characterized by a stable hypodiploid chromosome complement of 45, X (Rutzky *et al.*, 1980). LS180 cells developed a translocation from the long arm of the remaining X chromosome to the long arm of the number 5 chromosome. The translocation occurred in all LS180 cells examined between the fifth and nineteenth serial subcultivations. The karyotypes were stable through 70 and 34 serial subcultivations for LS174T and LS180 cells, respectively (Rutzky *et al.*, 1980).

Observation of karyotype and polymorphic isozyme stability in the LS174T/LS180 model system provides a good foundation for other studies performed with these cells. Continuous monitoring of traits that are related to genetic integrity emphasizes the need to evaluate the stability of innate properties of the cells with increased *in vitro* age.

Thus, long-term studies with cell lines that are continuously changing may not represent satisfactory model systems.

E. Tumor Markers

Tumor markers are useful in the diagnosis and progression of neoplastic disease. Carcinoembryonic antigen is the only well-established marker for colorectal cancer. Unfortunately, this marker is not unique and is found associated with other neoplastic and nonneoplastic diseases (Moore *et al.,* 1971; LoGerfo *et al.,* 1971). Several cross-reactive CEA-like molecules have been identified. Among them are CEA-S which is similar to CEA (Edington *et al.,* 1975), and nonspecific cross-reacting antigen (NCA) which is not elevated in colon cancer patients (von Kleist *et al.,* 1972). Non-CEA-like antigens include (1) colon-specific antigens (CSA and CSAp) (Pant *et al.,* 1978; Goldenberg *et al.,* 1975, 1976); (2) Thomsen-Friedenreich T antigen (Springer *et al.,* 1975); and (3) the monoclonal antibody detected monosialoganglioside antigen (Koprowski *et al.,* 1979; Herlyn *et al.,* 1979; Koprowski *et al.,* 1981; Magnani *et al.,* 1981). Several recent reviews of human colorectal tumor markers have been published (Zamcheck, 1981; Goldenberg *et al.,* 1981; Sell, 1980).

LS180 and LS174T and its clones have been studied for the presence of tumor markers. These cell lines produce and release CEA under a variety of culture conditions. Early studies indicated that LS180 cells released 900 times more CEA and had 30 times more cell-associated CEA than HT29 cells (Tom *et al.,* 1976). Twice the amount of CEA was released by LS180 cells when compared with LS174T cells (Tom *et al.,* 1977). The apparent difference in CEA release by LS174T and LS180 cells was studied by injecting LS174T cells into immunosuppressed hamsters. The resulting tumor tissue was divided into two portions. Small tumor pieces were grown as explants, and the resulting cell monolayers were either trypsin treated or mechanically scraped. The morphology of the trypsin-treated cells resembled LS174T cells whereas the scraped cells looked like LS180 cells. The scraped cells released 11 times more CEA than the trypsin-treated cells. These findings suggest that the difference in CEA release between LS180 and LS174T cells may reside in the method of subculture. CEA is a peripheral membrane glycoprotein and is loosely associated with the cell membrane. Mechanical scraping of LS180 cells probably preserves more of the glycocalyx-associated CEA than does trypsin treatment. Thus, differences in CEA release may reflect differences in cultivation procedures rather than alterations in innate cell properties (Tom *et al.*, 1977).

The kinetics of CEA release was examined for LS174T cells growing in a perfused hollow fiber culture system (Amicon, Lexington, MA) (Rutzky *et al.*, 1979). After one month of perfusion, organoid growth reminiscent of the patient's tumor tissue was observed (Fig. 1D). Alternate-day determinations of CEA in the intracapillary fluids showed a release rate similar to that observed in identically treated monolayer cultured cells. A significantly accelerated secondary release phase of CEA was seen only for cells cultured on hollow fibers. Maximum CEA release was observed during the final days of incubation for both the perfusion and the stationary phase of monolayer cultures.

Other tumor markers produced by the LS174T cells include CSAp (Pant *et al.*, 1978), CSA (Goldenberg *et al.*, 1975, 1976), the T antigen (Springer *et al.*, 1975), and the monosialoganglioside antigen recognized by the 1116 monoclonal antibody (Koprowski *et al.*, 1979; Herlyn *et al.*, 1979; Koprowski *et al.*, 1981; Magnani *et al.*, 1981; Miller and Tom, 1980). LS174T, LS180, and LS174T clones produce mucins detectable by mucicarmine, high iron diamine–Alcian Blue (Kahan *et al.*, 1979; Spicer and Meyer, 1960; Spicer, 1965) and periodate–Schiff staining reactions.

LS174T and LS180 cells have been examined for ectopic hormones. Tests for specific human chorionic gonadotropin and for gastrin release were close to or below the minumum concentrations that were detectable by radioimmunoassay (Kahan *et al.*, 1979; Halpern *et al.*, 1971). These studies showed that LS174T cells are stimulated to divide in the presence of epidermal growth factor and platelet growth factor.

F. Differentiation Potential

LS180 and LS174T and its clones differentiate in two culture systems in the absence of exogenously added chemical growth modifiers. These cells can be modulated to grow in either differentiated or undifferentiated phenotypes *in vitro*. In one model LS174T and clone 3-5 cells were grown for 21 to 28 days in a perfused hollow fiber culture system (Amicon Corp.). After fixation the fiber bundle was cross-sectioned and prepared for histological examination by light and electron microscopy. Figure 1D illustrates LS174T cells growing between adjacent hollow fibers with a 50,000 MW exclusion. Differentiation was evaluated morphologically by the development of glands with sequestered mucin (Rutzky *et al.*, 1980). No evidence of differentiation

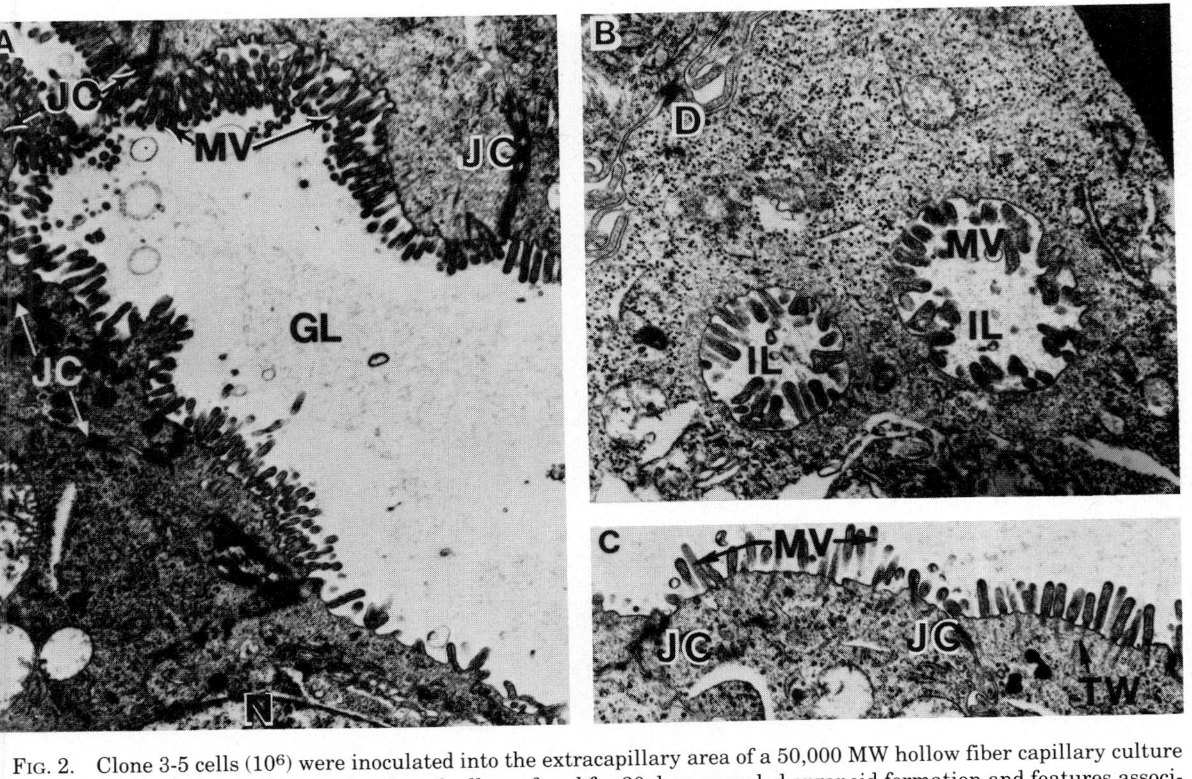

Fig. 2. Clone 3-5 cells (10^6) were inoculated into the extracapillary area of a 50,000 MW hollow fiber capillary culture unit (Amicon). Ultrastructural examination of cells perfused for 30 days revealed organoid formation and features associated with a well-differentiated adenocarcinoma. Organoid growth is not observed in monolayer culture. Therefore, we conclude that clone 3-5 was derived from a pluripotential cell which is capable of differentiation in nude mice and on perfused hollow fibers *in vitro*. (A) A section through clone 3-5 cells lining a gland lumen (GL). The lumen is lined by well-developed microvilli (MV) and several junctional complexes (JC). ×13,800. (B) A cell with 2 intracytoplasmic lumia (IL) and a desmosome (D) at the cell periphery. ×19,200. (C) Brush border-like array of microvilli (MV) and junctional complexes (JC) of cells lining a gland lumen. Note the terminal web (TW) lying below the cell periphery. ×15,000. Reprinted from Mastromarino (1981).

was observed in monolayer cultures that were treated identically or in gelatin sponges inoculated with cells (Fig. 1C).

Ultrastructural examination of clone 3-5 cells grown between 50,000 MW exclusion hollow fibers for 30 days revealed organoid gland formation and features of a well-differentiated adenocarcinoma (Fig. 2). Similar differentiation was observed with clone 3-5 cells inoculated and grown in nude mice. Since LS174T cells can be grown *in vitro* with either differentiated or undifferentiated phenotypes, they are good candidates for studying differentiation-associated antigens.

G. *Tumorigenicity*

Human cell tumorigenicity must be studied in xenogeneic and/or artificial culture systems *in vitro* since ethical reasons prohibit the use of *in vivo* syngeneic or allogeneic test systems. Thus, it is impossible to achieve a true measurement of either tumorigenicity or invasiveness of human tumor cells. The LS180 and LS174T cell lines have been studied using a variety of xenogeneic *in vivo* and *in vitro* procedures. These studies have been used to evaluate the invasive and differentiation potential of the cells.

LS180 and LS174T and its clones grew progressively in chemically immunosuppressed and athymic nude mice (Tom *et al.*, 1977; Rutzky and Siciliano, 1980, 1982). Examination of histologic sections of the tumor tissue revealed organoid gland formation resembling the morphology of the patient's tumor. An inoculum of 10^7 tumor cells formed tumor nodules within 1 week. When these tumors were permitted to grow progressively, the animals died in about 4 weeks. In one study 10 million LS180, LS174T, or clones 3-5, 6-6, 10-2, 10-3, or 10-4 cells were inoculated into nude mice. The tumors that developed were examined histologically, prepared for isozyme analysis, and grown in tissue culture. In every case the tumor cells outgrew the mouse stromal cells, as indicated by the absence of mouse isozymes in the cell cultures (Rutzky and Siciliano, 1980, 1982).

In vitro assessments of tumorigenicity have included growth of tumor cells in soft agar (Macpherson and Montagnier, 1964), the stem cell assay semisolid agar (Hamburger and Salmon, 1977), collagen gels (Yang *et al.*, 1979), and on pieces of chick embryonic skin (Noguchi *et al.*, 1978). LS174T cells have about a 15-day lag period before colonies are visible after inoculation into 0.3% soft agar. This lag period is not observed when the cells are inoculated into the stem cell assay system of Hamburger and Salmon (1977). In contrast, the clones of LS174T do not have a discernible lag phase in 0.3% soft agar (Rutzky *et al.*, in press). When cultured on collagen gels (Vitrogen 100, Flow Laboratories, McLean, VA) by a modification of the procedure of Yang *et al.* (1979) (Rutzky *et al.*, 1983), LS174T cells and clones grew as small colonies. The colonies were glandular and organoid in structure with mucin accumulation within central gland cavities (unpublished data).

Colony formation on confluent cell monolayers is another assay system to measure invasiveness (Tompkins *et al.*, 1974). LS174T cells form colonies on confluent monolayers of human skin fibroblasts, cells isolated from human bowel mucosa, and hamster skin fibroblasts (Tom *et al.*, 1977). Clone 3-5 cells formed colonies under similar conditions on human skin fibroblast cells (unpublished data). When grown upon chick embryonic skin explants, LS174T cells showed an invasive growth pattern and gland function (Ridge and Noguchi, 1983). These studies indicate that LS180 and LS174T and its clones can grow in an invasive manner, which is indirect evidence of tumorigenicity.

H. Gene Expression

Tissue-specific isozymes that are coded by multiple loci were studied for gene expression in LS180, LS174T, 6 LS174T clones, and the NB (LS174T) cell line established from the uninvolved bowel mucosa of the same patient (Rutzky and Siciliano, 1980, 1982). Twelve independently derived SW colorectal tumor cell lines were also studied. These cell lines were designated with SW prefixes 48, 403, 742, 802, 837, 948, 1116, 1222, 1345, 1398, 1417, and 1463. HT-29, an established colon tumor cell line, was provided by Dr. J. Fogh (Sloan Kettering Institute for Cancer Research, Rye, NY). HeLa cells were supplied by Dr. J. Clarkson (The University of Texas System Cancer Center, Smithville, TX). Cell lines LS123 and NB (LS123) (Rutzky *et al.*, 1983) were established in our laboratory from a patient's colon tumor and normal bowel mucosa, respectively. Genetic signature analysis showed that these lines were distinct from the LS174T and LS180 cell lines (Rutzky *et al.*, 1980). Eight colorectal tumor cell lines derived from the colon tumors of 3 individuals were studied and were designated with the COLO prefixes 205, 206, 320, 321, 394, 396, 399, and 463. Control cultures included 4 cell lines from the bowel mucosa of independent origin and 2 cell lines from fetal intestine supplied by Dr. W. Nelson-Rees (University of California Naval Biosciences Laboratory, Berkeley, CA). Six pairs of autochthonous human bowel tumors and normal bowel mucosa surgical specimens were examined for gene expression.

Pellets of the cell culture or tissues were homogenized and centrifuged. Supernatant fluids were subjected simultaneously to starch gel electrophoresis. The isozymes studied included lactate dehydrogenase (LDH), esterase B_2 (ESB2), esterase A_4 (ESA4), creatine kinase brain form (CKB), adenosine deaminase tissue form (ADAt), and adenosine deaminase red blood cell form (ADAR). Strong expressions of ESB2, ESA4, and CKB were stable in all LS174T and LS180 cell materials studied (Table V). Greater expression of LDH A than LDH B was observed. Strong expression of ADAt, but not ADAR, was present. This pattern of isozyme expression was found in bowel mucosa tissues from 6 patients with colorectal cancer, but it was not present in several cell lines established from other bowel mucosa tissues. Among the other colorectal tumor cell lines studied, only SW-742 and SW1345 had the same pattern of isozyme expression as that observed for the LS174T and LS180 cell lines. Gene expression in the 6 adenocarcinoma surgical specimens reflected the various patterns observed in 23 colorectal tumor cell lines studied.

The differences observed in tissue-specific isozyme gene expression

TABLE V

NORMAL COLON MUCOSA AND ADENOCARCINOMA ISOZYME EXPRESSION[a,b]

	ESB2	ESA4	CKB	ADAt	ADAR
Patient no.					
Normal mucosa					
1	+++	+++	+++	++	−
2	+++	+++	+++	++	−
3	+++	+++	+++	++	−
4	+++	+++	+++	++	−
5	+++	+++	+++	++	−
6	+++	+++	+++	++	−
Adenocarcinoma					
1	+++	+++	−	++	++
2	+++	+++	+++	+++	−
3	+++	+++	+++	++	++
4	+	+	+++	−	−
5	−	+++	−	+++	−
6	+	−	+++	+	++
Cell lines					
LS174T	+++	+++	+++	+++	−
LS180	+++	+++	+++	+++	−
NB (LS174T)	−	−	+	−	+++

[a] Paired surgical specimens and cell lines were used, all of which had a greater expression of LDHA than of LDHB.

[b] Reprinted from Rutzky and Siciliano (1982).

were not related to the number of subcultivations of the cell lines nor to the stage of invasiveness that the tumor tissues had achieved at the time of resection. Because the gene expression patterns also differed in the surgical specimens examined, we concluded that the type of cells in the heterogeneous colon mucosa from which the tumor arose may have determined the type of pattern observed. Thus, the cell lines examined in this study reflect isozyme expression present in tumor specimens and therefore represent models for colon cancer. The cell lines derived from nonmalignant colon mucosa did not have isozyme patterns that were observed in mucosa tissue specimens and probably were not appropriate controls. The epithelial cell lines derived from normal human colon mucosa (Danes and Sutano, 1982; Moyer *et al.*, in press) and primary cell cultures established from human premalignant adenomas of the colon (Friedman and Higgins, 1979; Friedman *et al.*, 1981) should be examined to determine if they are more appropriate control materials for these studies.

I. Nuclear Magnetic Resonance Relaxation of Water Protons

Nuclear magnetic resonance (NMR) relaxation times for water protons have been found to be different for tumors and their tissues of origin in numerous studies (Bovee *et al.*, 1974; Damadian, 1971; Damadian *et al.*, 1973; Hazelwood *et al.*, 1972). In normal rat and human gastrointestinal tissues, relaxation times for water protons were lower than those in tumors (Cottam *et al.*, 1972; Eggleston *et al.*, 1975; Goldsmith *et al.*, 1978; Koutcher *et al.*, 1978). Since whole tissues are composed of heterogeneous cell populations and fluids, these results are difficult to interpret. NMR studies on pure populations of cultured tumor cells have eliminated some of the variables in the study of cancer by NMR (Beall *et al.*, 1976a, b, 1978; Ling and Tucker, 1980; Chaughule *et al.*, 1974; Wagh *et al.*, 1977).

Human colorectal tumor cell lines LS180, LS174T, HT-29, SW480, SW1345, and clones 3-5 and 6-6 of LS174T were examined by NMR for water proton spin–lattice relaxation times (T_1) and spin–spin relaxation times (T_2) (Beall *et al.*, 1982). Two cell lines that were established from adult human colon mucosa and two cell lines of human fetal intestines (HS0074 Int. and HS0677 Int.) were studied as controls.

Spin–lattice relaxation time (T_1) for the 5 tumor cell lines and 2 clones studied ranged between 460 and 982 msec. Spin–spin relaxation time (T_2) varied from 83 to 176 msec. Hydration of the colon cancer cells studied ranged from 83 to 90%. Population doubling times were 12 hours for HT-29, 18–22 hours for LS174T and clones, 39 hours for SW480, 51 hours for SW1345, and 76 hours for LS180 cells (L. P. Rutzky, unpublished results; Leibovitz *et al.*, 1976; Tom *et al.*, 1976). The adult and fetal cell lines studied had higher T_1, T_2, and hydration values than the established cell lines. Population doubling times of these cell lines were about 24 hours.

Established human colorectal tumor cell lines maintained water proton relaxation times similar to whole tumor tissues. Clones 3-5 and 6-6 had T_1 and T_2 values similar to the parent LS174T cell line, although clone 3-5 cells were wetter than clone 6-6 cells. The differences observed in water relaxation times among the cell lines were not a function of hydration. Because the adult and fetal human cell lines established from bowel tissues were wetter and had higher T_1 and T_2 values than the tumor cell lines, they were probably inappropriate controls for these studies. The relationship between spin–lattice relaxation time of water protons and cellular hydration was examined. A curvilinear relationship was noted when the reciprocal of T_1 was plotted against the colorectal tumor cell lines and clones. The data indicated that T_1 values are not entirely dependent upon hydration of the cells and are probably related to the types of macromolecules and their conformational states within the cells. The NMR studies on LS174T, LS180, and the clones offer additional support for their use as models for the study of human colon cancer.

IV. Antigenic Analysis of LS174T and LS180 Cells

A. Background

Abnormalities in gene regulation and protein associated with neoplasia are often reflected in altered macromolecular synthesis (Weinhouse, 1982) such as the reappearance of the fetal antigens CEA (Gold, 1971) and α-fetoprotein (Abelev, 1971), the presence of fetal isozymes (Fishman, 1974; Ibsen and Fishman, 1979; Markert and Moller, 1959), and the disruption of normal cell differentiation (Parsad and Hsie, 1971; Pierce, 1970, Pierce *et al.*, 1977; Silagi and Bruce, 1970). Cell surface alterations distinctive of neoplasia are tumor-associated and/or differentiation-associated antigens. The alterations may include spatial rearrangements (Brodsky and Parham, 1982) or changes in surface glycoproteins and glycolipids (Hakomori *et al.*, 1977; Kim *et al.*, 1973; Whitehead *et al.*, 1979).

Hybridoma technology has been useful in detecting and isolating human tumor-associated antigens for colorectal carcinoma (Koprowski *et al.*, 1979; Herlyn *et al.*, 1979), melanoma (Steplewski *et al.*, 1979; Herlyn *et al.*, 1980; Loop *et al.*, 1981), lung tumors (Mazauric *et al.*, 1982), and breast carcinoma (Thompson *et al.*, 1983). Hybridoma technology has exploited the immune response by selecting and amplifying the single antibody product from one B cell (Kohler and Milstein, 1975; Koprowski, *et al.*, 1977; Olson and Kaplan, 1980; Croce *et al.*, 1980). This approach has been successful because the hybridoma clone produces an antibody population with a unique specificity to a tumor-associated antigenic epitope. Animal immunization produces a polyclonal response with multiple antibody specificities that usually mask the presence of antibodies to rare antigenic epitopes on the cell surface.

The development of monoclonal antibodies that recognize determinants on an antigen that are only expressed during differentiated growth of cultured cells is an approach being used with the LS174T/LS180 cell set. Isolation of the moieties that are present only on the differentiated phenotype may help to clarify the role of the differentiation-associated antigens in colorectal tumor cell growth and their relationships to oncofetal antigens. Such monoclonal antibodies may be useful diagnostically to identify changes on mucosal cells prior to the appearance of overt carcinoma. Oncofetal antigens are expressed during normal embryogenesis, and they may play a role in cell–cell interaction during development. The reappearance of fetal gene products on tumor cells may be due to genetic changes in the cells leading to blocked differentiation (Potter, 1980). The interaction of monoclonal antibodies with colon tumor cells may alter organoid growth as shown with dome formation and monoclonal antibodies against mouse Thy-1 antigen (Dulbecco *et al.*, 1979).

Differentiation and development are believed to be directed by the orderly generation of and response to specific factors produced among neighboring cells (Moscona, 1974, 1975; Marchase, 1976). Cell adhesion has been postulated to occur through the interaction of complementary surface moieties (Weiss, 1947) by macromolecular multivalent ligands binding surface receptors (Moscona, 1974; Rutishauser *et al.*, 1976) or via glycosyltransferases (Roth *et al.*, 1971). Changes in glycoproteins and glycolipids have been found in several diseases of the gastrointestinal tract. Various galactosyltransferase alterations have been reported (Kim, 1977), and high levels of these enzymes are secreted by two human colon tumor cell lines (Whitehead *et al.*, 1979). The cell lines had unique isozyme profiles that were found in colon tumor patients (Weiser *et al.*, 1976).

The approach to the present studies has been to produce monoclonal antibodies to clones of LS174T cells that are growing either as an undifferentiated monolayer or the glandular differentiated phenotype. To date, the first part of the approach has been accomplished. One monoclonal antibody has been produced to phenotypically undifferentiated monolayer grown clone 3-5 cells. This antibody appears to be colon tumor specific when tested for binding on a panel of cell lines from various human tumors.

B. Polyclonal Xenoantisera

In studies published from this laboratory, LS174T cells were examined with antisera prepared by immunization of rabbits with 3 *M* KCl extracts suspended in Freund's adjuvant (Kahan *et al.*, 1979). Extensive absorptions with other cell lines and fresh tissues were used to decrease nonspecific reactivities. Using Ouchterlony double diffusion analysis of absorbed antisera with isoelectrically focused fractions of 3 *M* KCl extracts of LS174T cells, antigens were detected at pIs 3, 4, 5, 6.5-7.5, and 8-8.5. Reactivities to both fetal bovine serum and to normal bowel mucosa cells indicate broad antibody specificities. Recently, colon tumor cell (LS174T) membranes were used either within or upon liposomes as an immunogen in rabbits (Tom *et al.*, 1982). Although the efficiency of antibody production increased by this method of presentation, the specificity of the antisera remained broad.

C. Monoclonal Xenoantisera

Monoclonal antibodies produced by fusion of immune mouse spleen cells to a mouse myeloma cell line (P3-63-Ag-8.653) (Kearney *et al.*, 1979) have diminished most of the broad binding specificity of polyclonal antisera. Such monoclonal antibodies can be used to probe the presumably subtle antigenic structures of colorectal tumor cells.

Recently, a hybridoma, 5E113 of the IgG^{2a} subclass, was produced in our laboratory. It has binding specificity for some human colon cancer cell lines (Rutzky *et al.*, 1984). Spleen cells from BALB/c mice immunized with clone 3-5 cells were hybridized with the nonsecretor mouse myeloma P3-63-Ag-8.653. Specificity was evaluated by an indirect radioimmunoassay (Fig. 3). Figure 3A–D shows target cell lines for clone 3-5 (black bars), SW780 neoplastic bladder, MDA-MB436 neoplastic breast, SW1083 neoplastic colon, FI-81 fetal intestine, clone 10-1, LS174T, LS180, and LS123, another neoplastic colon tumor cell line

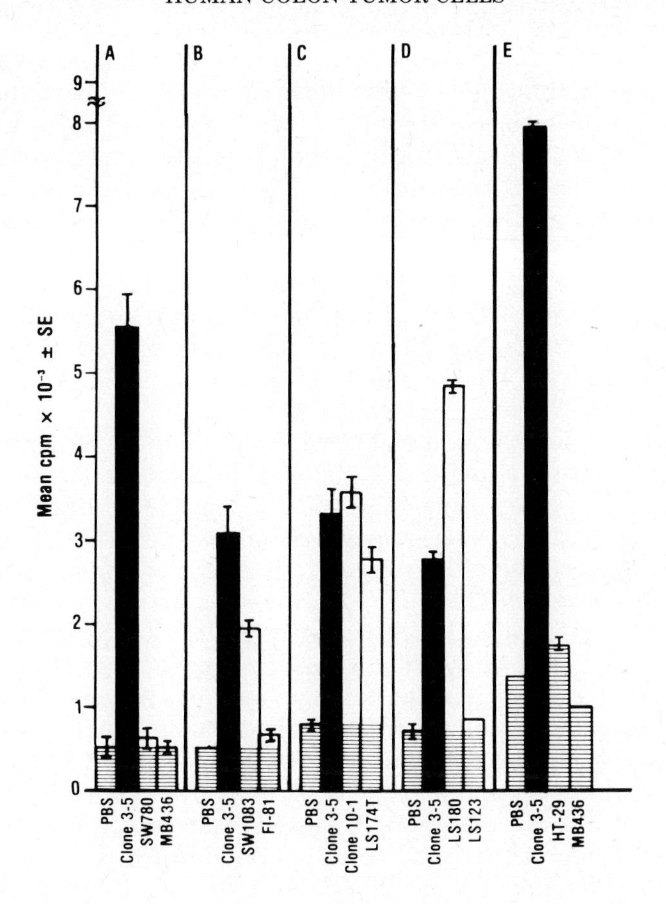

FIG. 3. Specificity screen of hybridoma 5E113 evaluated by indirect radioimmunoassay using trypsin-treated cells. The target cells include the immunizing cell clone 3-5, SW780 (bladder tumor), MB436 (breast tumor), SW1083 (colon tumor), FI-81 (fetal intestine), clone 10-1, LS174T, LS180, LS123, and HT-29 (colon tumor). The background binding of ^{125}I-labled goat antimouse antibody is indicated by PBS crosshatched. The binding of PBS and parent myeloma culture supernatant fluids was about the same. A–E indicate separate experiments; the data are presented as ±SE. Reprinted from Rutzky (1984).

from our laboratory. Controls include PBS and 653 parent myeloma culture supernatant fluids to test the background binding of the radiolabeled second antibody. Monoclonal antibody 5E113 had a low level of binding with the neoplastic breast and bladder, fetal intestine, and LS123 cells. The antibody displayed little binding with colon tumor cell line HT-29 (Fig. 3E). About 50% of the level of binding to clone 3-5 cells was observed on SW1083 cells. Approximately equal binding was observed with LS174T, clone 3-5, and clone 10-1 cells. Additional studies indicated that mAB 5E113 does not recognize antigenic determinants on CEA (J. Tomita, North Chicago, IL), CSAp (D. Goldenberg, Lexington, KY), or HLA antigens presented by a panel of 68 HLA typing cells (R. Kerman, Houston, TX).

The relationship between an antigenic determinant recognized by a monoclonal antibody and that recognized by immune lymphoid cells has been investigated using a method developed to generate cytotoxic mouse T cells *in vitro*. A series of polyclonal and monoclonal antibodies was used to block the cell-mediated lysis of LS174T colon tumor cells (Fig. 4). The target colon cancer cells were first treated with test antibodies followed by addition of immunocytotoxic lymphocytes (CTL). The polyclonal antibodies include the NIH typing reagents anti-HLA, A2, B13, A31, and the ABO blood group antigens A and B. The monoclonal reagents' antiframework DR antibodies were kindly provided by Dr. D. Capra. Monoclonal antibodies 480, 1116, and 1083 (Herlyn *et al.*, 1979), which were prepared against human colon cancer cell lines, were the gift of Dr. H. Koprowski. The HLA phenotype of the lymphocytes from the tumor donor of the LS174T and LS180 cell lines is HLA A2, AW30, B13, B35, CW4, CW, DRW4, DRW7. The anti-A, -B blood group antibodies were used to test whether the patient's original O blood type had been altered on the tumor cells to either A or B, as previously demonstrated by Hakomori *et al.* (1977).

The data show that the anti-HLA-A2 and -B13 antibodies can significantly inhibit cytolysis by 24.8 and 17.5%, respectively. Monoclonal ABs 480, 1083, and mAB 5E113 inhibited cytolysis by 45.2, 32.3, and 19.8%, respectively. Monoclonal antibody 480 has been shown by Herlyn *et al.* (1979) to have broad specificity; mAB 1116 binds to glycolipid, and mAB 1083 recognizes a glycoprotein moiety. No additional blocking of cytolysis was observed when mAB 1083 was combined with either anti-HLA-A2 or -B13 antibodies. Thus, the blocking observed was not additive, suggesting that the antibodies recognize closely grouped antigenic sites. No inhibition of cytolysis was observed with anti-A and anti-B blood group antibodies, indicating that a shift to either A or B antigens had not occurred.

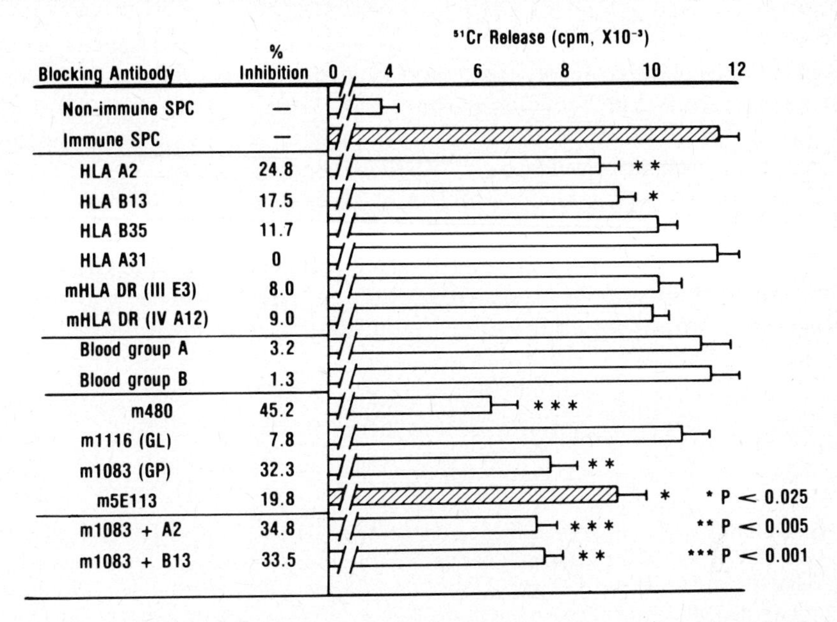

FIG. 4. Lymphocytes primed *in vitro* with liposomes bearing membrane vesicles from LS174T cells. The antibody panel includes several monoclonal antibodies (mABs) to colorectal cancer antigens [M480, M1116, M1083 (from Dr. H. Koprowski) and 5E113], 2 mABs to HLA-DR framework antigens (IIIE3 and IV A12 from Dr. D. Capra), absorbed NIH anti-HLA typing sera A2, B13, B35, and A31, blood group A and B antisera. The ability of antibodies to block cytotoxicity was measured by inhibition of ^{51}Cr release. The HLA of the tumor tissue donor of LS174T cells is HLA A2,AW30; B13,BW35; CW4, CW-; DRW4, DRW7. The patient's blood group is O+. Reprinted from Rutzky (1984).

V. SUMMARY AND CONCLUSIONS

Colon cancer, although aggressively treated by various combinations of surgery, chemotherapy, and radiotherapy, claims the lives of thousands of patients each year. *In vitro* models have been developed to study the biology of colorectal tumor cells. Only cell systems that retain and display properties of the disease are useful as models. When studied at less than 100 population doublings, the LS174T/LS180 set of cell lines has the following stable properties: (1) 45,X hypodiploid karyotype; (2) carcinoembryonic antigen (CEA) and mucin production; (3) organoid growth in nude mice, in hollow fiber culture, and on collagen gels; (4) the presence of colon-specific antigen (CSAp); (5) a pattern of tissue-specific isozymes that is found in some patient colorectal tumors; and (6) nuclear magnetic resonance values for the relaxation of water protons that are similar to those found in tumor tissues. The LS174T/LS180 set of cell lines and clones provides a useful model system to study human colon cancer.

The LS174T/LS180 cell set has provided a source of antigens to probe the composition of the cell surface. Monoclonal antibodies have been prepared against some of these cells. Some of these reagents appear to be colon tumor specific when screened against various human tumor and fetal cell lines. Such studies may provide the basis for the development of useful diagnostic agents to detect the early stages of colon cancer and to identify unique antigens for immunotherapy.

Acknowledgments

I would like to acknowledge the support of Dr. Barry D. Kahan during these studies. I would like to thank the following people for their excellent technical assistance at various times: Mr. Thomas Goodwin, Ms. Diane Dunlap, Mrs. Diana Chan, and Mrs. Miriam Calanoff. I would like to thank Dr. F. J. Attermeier for help in editing and the Department of Surgery Word Processing Center in the preparation of this manuscript. This work was supported by the following grants: CA 22370 from the National Cancer Institute through the National Large Bowel Cancer Project, CA 23871, and CA 27124. Finally, I would like to thank my colleague, Dr. Baldwin H. Tom, for his encouragement and helpful discussions throughout these studies.

References

1. Abelev, G. I. (1971). *Adv. Cancer Res.* **14,** 295–358.
2. Beall, P. T., Hazelwood, C. F., and Rao, P. N. (1976a). *Science* **192,** 904–907.
3. Beall, P. T., Hazelwood, C. F., and Rao, P. N. (1976b). *Science* **194,** 214.
4. Beall, P. T., Chang, D. C., and Hazelwood, C. F. (1978). *In* "Biomolecular Structure and Function" (Paul Agris, ed.), pp. 233–237. Academic Press, New York.
5. Beall, P. T., Hazelwood, C. F., and Rutzky, L. P. (1982). *Cancer Biochem. Biophys.* **6,** 7–12.
6. Berman, L., and Stulberg, C. S. (1956). *Proc. Soc. Exp. Biol. Med.* **92,** 730–735.
7. Bovee, W., Huisman, P., and Smidt, J. (1974). *J. Natl. Cancer Inst.* **52,** 595–597.
8. Bowen, J. M., Cailleau, R., Giovanella, B., Siciliano, M., and Pathak, S. (1980). *In Vitro* **16,** 208.
9. Brattain, M. G., Brattain, D. E., Fine, W. D., Khaled, F. M., Marks, M. E., Kimball, P. M., Arcolano, L. A., and Danbury, B. H. (1981). *Oncodev. Biol. Med.* **2,** 355–366.
10. Brattain, M. G., Marks, M. E., McComb, J., Finely, W., and Brattain, D. E. (1983). *Br. J. Cancer* **47,** 373–381.
11. Breborowicz, G., Easty, G. C., Bribeck, M., Robertson, D., Nery, R., and Neville, A. M. (1975). *Br. J. Cancer* **31,** 559–569.

12. Brodsky, F. M., and Parham, P. (1982). *J. Immunol.* **128,** 129–135.
13. Carrel, S., Sordat, B., and Merenda, C. (1976). *Cancer Res.* **36,** 3978–3984.
14. Chang, R. S. (1954). *Proc. Soc. Exp. Biol. Med.* **87,** 440–443.
15. Chaughule, R. S., Kasturi, S. R., Vijayaraghavan, R., and Ranade, S. S. (1974). *Indian J. Biochem. Biophys.* **11,** 256–261.
16. Chen, T. R., Hay, R. J., and Macy, M. L. (1982). *Cancer Gen. Cytogen.* **6,** 93–117.
17. Collins, S. J., Ruscetti, F. W., Gallagher, R. E., and Gallo, R. C. (1978). *Proc. Natl. Acad. Sci. U.S.A.* **75,** 2458–2462.
18. Collins, S. J., Bodner, A., Tinge, R., and Gallo, R. C. (1980). *Int. J. Cancer* **25,** 213–218.
19. Cottam, G. L., Vasek, A., and Lusted D. (1972). *Res. Commun. Chem. Pathol. Pharmacol.* **4,** 495–502.
20. Croce, C. M., Linnenbach, A., Hall, W., Steplewski, Z., and Koprowski, H. (1980). *Nature (London)* **288,** 488–489.
21. Damadian, R. (1971). *Science* **171,** 1151–1153.
22. Damadian, R., Zaner, K., and Hor, D. (1973). *Physiol. Chem. Phys.* **5,** 381–402.
23. Danes, B. S., and Sutano, E. (1982). *J. Natl. Cancer Inst.* **69,** 1271–1276.
24. Dexter, D. L., Barbosa, J. A., and Calabresi, P. N. (1979). *Cancer Res.* **39,** 1020–1025.
25. Dexter, D. L., Spremulli, E. N., Fligiel, Z., Barbosa, J. A., Vogel, R., Van Voorhees, A., and Calabresi, P. (1981). *Am. J. Med.* **71,** 949–956.
26. Dracopoli, N. C., and Fogh, J. (1983). *J. Natl. Cancer Inst.* **70,** 83–87.
27. Drewinko, B., Romsdahl, M. M., Yang, L. Y., Ahearn, M. J., and Trujillo, J. M. (1976). *Cancer Res.* **36,** 467–475.
28. Dukes, C. E. (1932). *J. Pathol. Bacterol.* **35,** 323–332.
29. Dulbecco, R., Bologna, M., and Unger, M., (1979). *Proc. Natl. Acad. Sci. U.S.A.* **76,** 1848–1852.
30. Dyke, P. C., and Mulkey D. A. (1967). *Cancer* **20,** 1343–1349.
31. Eagle, H. (1955). *Proc. Soc. Exp. Biol. Med.* **89,** 362–364.
32. Eagle, H. (1959). *Science* **130,** 432.
33. Edington, T. S., Astarita, R. W., and Plow, E. F. (1975). *N. Engl. J. Med.* **293,** 103–107.
34. Eggleston, J., Saryan, L. A., and Hollis, D. P. (1975). *Cancer Res.* **35,** 1326–1332.
35. Epstein, A. L., and Kaplan, H. S. (1979). *Cancer Res.* **39,** 1748–1759.
36. Espmark, J. A., Ahlquist-Roth, L., Sarne, L., and Persson, A. (1978). *Tissue Antigens* **11,** 279–286.
37. Fidler, I. J. (1978). *Cancer Res.* **38,** 2651–2656.
38. Fishman, W. E. (1974). *Am. J. Med.* **56,** 617–650.
39. Fjelde, A. (1955). *Cancer* **8,** 845–851.
40. Fogh, J., and Trempe, G. (1975). *In* "Human Tumor Cells *in Vitro*" (J. Fogh, ed.), pp. 115–154. Plenum, New York.
41. Friedman, E. A., Higgins, P. J., Lipkin, M., Shinya, H., and Gelb, A. M. (1981). *In Vitro* **17,** 632–644.
42. Friedman, E. A., and Higgins, P. J. (1979). *J. Cell Biol.* **83,** 102.
43. Fritsche, H. A., Tashima, C. K., Romsdahl, M. M., Holoye, P. Y., and Geitner, A. (1978). *Am. J. Clin. Pathol.* **69,** 140–146.
44. Fritsche, H. A., Tashima, C. K., Collinsworth, W. L., Geitner, A., and Van Oort, J. (1980). *J. Immunol. Meth.* **35,** 115–128.
45. Gartler, S. M. (1966). *Natl. Cancer Inst. Monogr.* **26,** 167–195.
46. Gey, G. O., Coffman, W. D., and Kubicek, M. T. (1952). *Cancer Res.* **12,** 264–265.

47. Giard, D. J., Aaronson, S. A., Todaro, G. J., Arnstein, P., Kersey, J. H., Dosik, H., and Parks, W. P. (1973). *J. Natl. Cancer Inst.* **51,** 1417–1423.
48. Giovanella, B. C., Yim, S. O., Stehlin, J. S., and Williams, L. J., Jr. (1972). *J. Natl. Cancer Inst.* **48,** 1531–1533.
49. Gold, P. (1971). *Prog. Exp. Tumor Res.* **14,** 43–58.
50. Goldenberg, D. M., Witte, S., and Elster, K. (1966). *Transplantation* **4,** 760–763.
51. Goldenberg, D. M., Bhan, R. D., and Pavia, R. A. (1971). *Cancer Res.* **31,** 1148–1152.
52. Goldenberg, D. M., Pavia, R. A., Hansen, H. J., and Vanderoorde, J. P. (1972). *Nature (London) New Biol.* **239,** 189–190.
53. Goldenberg, D. M., Pegram, C. A., and Vazquez, J. J. (1975). *J. Immunol.* **114,** 1008–1013.
54. Goldenberg, D. M., Pant, K. D., and Dahlman, H. L. (1976). *Cancer Res.* **36,** 3455–3463.
55. Goldenberg, D. M., Neville, M., Carter, A. C., Go, V. W., Holykoe, E. D., Isselbacher, K. J., Shein, P. S., and Schwartz, M. (1981). *Ann. Intern. Med.* **94,** 407–409.
56. Goldenberg, D. M., and Pavia, R. A. (1981). *Science* **212,** 65–67.
57. Goldsmith, M., Koutcher, J., and Damadian, R. (1978). *Cancer* **41,** 183–191.
58. Hager, J. C., Gold, D. V., Barbosa, J. A., Fligiel, Z., Miller, F., and Dexter, D. L. (1980). *J. Natl. Cancer Inst.* **64,** 439–446.
59. Hakomori, S., Wang, S. M., and Young, W. W., Jr. (1977). *Proc. Natl. Acad. Sci. U.S.A.* **74,** 3023–3027.
60. Halpern, B., Eckman, T. R., and Dolkart, R. E. (1971). *Am. J. Obstet. Gynecol.* **110,** 412–414.
61. Ham, R. G. (1965). *Proc. Natl. Acad. Sci. U.S.A.* **53,** 288.
62. Hamburger, A. W., and Salmon, S. E. (1977). *J. Clin. Invest.* **60,** 846–854.
63. Harris, H. (1966). *Proc. R. Soc. Lond. (Biol.)* **164,** 298–310.
64. Harris, N. L., Gang, D. L., Quay, S. C., Poppema, S., Zamecnik, P. C., Nelson-Rees, W. A., and O'Brien, S. J. (1981). *Nature (London)* **289,** 228–230.
65. Hazelwood, C. F., Chang, D. C., Medina, D., Cleveland, G., and Nichols, B. L. (1972). *Proc. Natl. Acad. Sci. U.S.A.* **69,** 1478–1480.
66. Heppner, G. H., Dexter, D. L., DeNucci, T., Miller, F. R., and Calabresi, P. (1978). *Cancer Res.* **38,** 3758–3763.
67. Herlyn, M., Steplewski, Z., Herlyn, D., and Koprowski, H. (1979). *Proc. Natl. Acad. Sci. U.S.A.* **76,** 1438–1442.
68. Herlyn, M., Clark, W. H., Mastrangelo, M. J., DuPont, G. IV., Elder, D. E., LaRossa, D., Hamilton, R., Bondi, E., Tuthill, R., Steplewski, Z., and Koprowski, H. (1980). *Cancer Res.* **40,** 3602–3609.
69. Ibsen, K. H., and Fishman, W. H. (1979). *Biochem. Biophys. Acta* **560,** 243–280.
70. Kahan, B. W., and Ephrussi, B. (1970). *J. Natl. Cancer Inst.* **44,** 1015–1036.
71. Kahan, B. D., Rutzky, L. P., LeGrue, S. J., and Tom, B. H. (1979). *Meth. Cancer Res.* **18,** 197–275.
72. Kearney, J. F., Radbruch, A., Liesengang, B., and Rajewsky, K. (1979). *J. Immunol.* **123,** 1548–1550.
73. Kim, Y. S., Isaacs, R., and Perdomo, J. M. (1973). *Proc. Natl. Acad. Sci, U.S.A.* **71,** 4869–4873.
74. Kim, Y. S. (1977). *In* "Pathophysiology of Carcinogenesis in Digestive Organs" (E. Farberg, ed.), pp. 233–245. Univ. Park Press, Baltimore, Maryland.
75. Kleinsmith, L. J., and Pierce, G. B. (1964). *Cancer Res.* **24,** 1544–1551.
76. Kohler, G., and Milstein, C. (1975). *Nature (London)* **256,** 495–497.

77. Koprowski, H., Gerhard, W., and Croce, C. M. (1977). *Proc. Natl. Acad. Sci. U.S.A.* **7,** 2985–2988.
78. Koprowski, H., Steplewski, Z., Mitchell, K., Herlyn, M., and Fuhrer, P. (1979). *Som. Cell Genetics* **5,** 957–972.
79. Koprowski, H., Herlyn, M., Steplewski, Z., and Sears, H. F. (1981). *Science* **212,** 53–54.
80. Koura, M., and Isaka, H. (1980). *Gann.* **71,** 313–318.
81. Koutcher, J. A., Goldsmith, M., and Damadian, R. (1978). *Cancer* **41,** 174–182.
82. Lavappa, K. S., Macy, M. L., and Shannon, J. E. (1976). *Nature (London)* **259,** 211–213.
83. Leibovitz, A. (1963). *Am. J. Hyg.* **78,** 173–180.
84. Leibovitz, A., Stinson, J. C., McCombs, W. B., III, McCoy, C. E., Mazur, K. C., and Mabry, N. D. (1976). *Cancer Res.* **36,** 4562–4569.
85. Leibovitz, A., Wright, W. C., Pathak, S., Siciliano, M. J., Daniels, W. P., Fogh, H., and Fogh, J. (1979). *J. Natl. Cancer Inst.* **63,** 635–645.
86. Leighton, J., Mansukkhani, S., and Fetterman, G. H. (1972). *Eur. J. Cancer* **8,** 281–285.
87. Ling, G. N., and Tucker, M. (1980). *J. Natl. Cancer Inst.* **64,** 1199–1207.
88. Lo Gerfo, P. J., Krupey, J., and Hansen, H. J. (1971). *N. Engl. J. Med.* **285,** 138–141.
89. Loop, M., Nishiyama, K., Hellstrom, I., Woodbury, R. G., Brown, J. P., and Hellstrom, K. E. (1981). *Int. J. Cancer* **27,** 775–781.
90. Lubs, H. A. (1970). *In* "Carcinoma of the Colon and Antecedent Epithelium" (W. J. Burdette, ed.), pp. 319–332. Charles C Thomas, Springfield, Illinois.
91. Macpherson, I., and Montagnier, L. (1964). *Virology* **23,** 291–294.
92. Magnani, J. L., Brockhaus, M., Smith, D. F., Ginsburg, V., Blaszczyk, V., Blazczyk, M., Mitchell, K. F., Steplewski, Z., and Koprowski, H. (1981). *Science* **212,** 55–56.
93. Marchase, R. B., Vosbeck, K., and Roth S. (1976). *Biochem. Biophys. Acta* **457,** 385–416.
94. Markert, C. L., and Moller, F. (1959). *Proc. Natl. Acad. Sci. U.S.A.* **45,** 753–763.
95. Mastromarino, A. J. (1981). *Cancer Bull.* **33,** 156–179.
96. Mazauric, T., Mitchell, K. F., Letchworth, G. J. III, Koprowski, H., and Steplewski, Z. (1982). *Cancer Res.* **42,** 150–154.
97. McBain, J., Meisner, L., Weese, J., Wolberg, W., and Wilson, J. (1983). *Proc. Am. Assoc. Cancer Res.* **24,** 4.
98. Miller, A., and Tom, B. H. (1980). *Proc. Am. Assoc. Cancer Res.* **21,** 221.
99. Moore, A. E., Sabachewsky, L., and Toolan, H. W. (1955). *Cancer Res.* **15,** 598–602.
100. Moore, G. E., Mount, D., Tara, G., and Schwartz, N. (1963). *Cancer Res.* **23,** 1735–1741.
101. Moore, T. L., Kupchick, H. Z., Marcon, N., and Zamcheck, N. (1971). *Am. J. Digest. Dis.* **16,** 1–7.
102. Moscona, A. A. (1974). *In* "Cell Surface in Development" (A. H. Moscona, ed.), pp. 67–100. Wiley, New York.
103. Moscona, A. A. (1975). *In* "Extracellular Matrix Influence on Gene Expression" (H. S. Slavin and R. C. Greulich, eds.), pp. 57–67. Academic Press, New York.
104. Moyer, M. P., Page, C. P., and Moyer, R. C. (in press). *In* "*In Vitro* Models of Human Disease" (M. Webber and L. Sekely, eds.). CRC Press, Boca Raton, Florida.
105. Murakami, H., and Masui, H. (1980). *Proc. Natl. Acad. Sci. U.S.A.* **77,** 3464–3466.
106. Nelson-Rees, W. A., and Flandermeyer, R. R. (1976). *Science* **191,** 96.
107. Nelson-Rees, W. A., Hunter, L., Darlington, G. J., and O'Brien, S. J. (1980). *Cytogenet. Cell Genet.* **27,** 216–231.

108. Noguchi, P. D., Johnson, J. B., O'Donnell, R., and Petricciani, J. C. (1978). *Science* **199,** 980–983.
109. Noguchi, P., Wallace, R., Johnson, J., Earley, E. M., O'Brien, S., Ferrone, S., Pellegrino, Milstein, J., Needy, C., Browne, W., and Petricciani, J. (1979). *In Vitro* **15,** 401–408.
110. O'Brien, S. J., Kleiner, G., Olson, R., and Shannon, J. E. (1977). *Science* **195,** 1345–1348.
111. Olson, L., and Kaplan, H. S. (1980). *Proc. Natl. Acad. Sci. U.S.A.* **77,** 5427–5431.
112. Orfeo, T., Tiso, J., Sharkey, F. E., Fogh, J. M., and Daniels, W. P. (1980). *Exp. Cell Biol.* **48,** 229–239.
113. Pant, K. D., Dahlman, H. L., and Goldenberg, D. M. (1978). *Cancer* **42,** 1662–1635.
114. Parsad, K. N., and Hsie, A. W. (1971). *Nature (London) New Biol.* **233,** 141–142.
115. Pattillo, R. A., and Gey, G. O. (1968). *Cancer Res.* **28,** 1231–1236.
116. Pierce, G. B., and Verney, E. L. (1961). *Cancer* **14,** 1017–1029.
117. Pierce, G. B. (1970). *Fed. Proc.* **29,** 1248–1254.
118. Pierce, G. B., Nakane, P. K., Martinez-Hernandez, A., and Ward, J. M. (1977). *J. Natl. Cancer Inst.* **58,** 1329–1345.
119. Pollack, M. S., Heagney, S. D., Livingston, P. O., and Fogh, J. (1981). *J. Natl. Cancer Inst.* **66,** 1003–1012.
120. Potter, V. R. (1980). *Yale J. Biol. Med.* **53,** 367–388.
121. Povey, S., Hopkinson, D. A., Harris, H., and Franks, L. M. (1976). *Nature (London)* **264,** 60–63.
122. Quinn, L. A., Moore, G. E., Morgan, R. T., and Woods, L. K. (1979). *Cancer Res.* **39,** 4914–4924.
123. Raphael, L., and Tom, B. H. (1982). *Cell Immunol.* **71,** 224–240.
124. Ridge, J., and Noguchi, P. D. (1983). *In Vitro* **19,** 274.
125. Rosenthal, K. L., Tompkins, W. A. F., Frank, G. L., McCulloch, P., and Rawls, W. E. (1977). *Cancer Res.* **37,** 4024–4030.
126. Roth, S., McGuire, E. J., and Roseman, S. (1971). *J. Cell Biol.* **51,** 536–547.
127. Rutishauser, U., Thiery, J. P., Brackenburg, R., Sela, B. A., and Edelman, G. M. (1976). *Proc. Natl. Acad. Sci. U.S.A.* **73,** 577–581.
128. Rutzky, L. P., Tomita, J. T., Calenoff, M. A., and Kahan, B. D. (1979). *J. Natl. Cancer Inst.* **63,** 893–902.
129. Rutzky, L. P., and Siciliano, M. J. (1980). *Cancer Bull.* **32,** 52–54.
130. Rutzky, L. P., Kaye, C. I., Siciliano, M. J., Chao, M., and Kahan, B. D. (1980). *Cancer Res.* **40,** 1443–1448.
131. Rutzky, L. P. (1981). *In* "The Growth Requirements of Vertebrate Cells *in Vitro*" (C. Waymouth, R. G. Ham, and P. J. Chapple, eds.), pp. 277–292. Cambridge Univ. Press, Cambridge.
132. Rutzky, L. P., and Siciliano, M. J. (1982). *J. Natl. Cancer Inst.* **68,** 81–88.
133. Rutzky, L. P., Giovanella, B. C., Tom, B. H., Kaye, C. I., Noguchi, P. D., and Kahan, B. D. (1983). *In Vitro* **19,** 99–107.
134. Rutzky, L. P., Tom, B. H., and Kahan, B. D. (1984). *In* "Markers of Colonic Cell Differentiation" (S. Wolman and T. Mastromarino, eds.), pp. 135–145. Raven, New York.
135. Rutzky, L. P. (1984). *In* "*In Vitro* Model for Human Disease" (M. Webber and L. Sekely, eds.), Vol. I. CRC Press, Boca Raton, Florida (in press).
136. Sell, S. (1980). "Cancer Markers: Diagnostic and Developmental Significance." Humana Press, Clifton, New Jersey.
137. Semple, T. U., Quinn, L. A., Woods, L. K., and Moore, G. E. (1978). *Cancer Res.* **38,** 1345–1355.

138. Silagi, S., and Bruce, S. A. (1970). *Proc. Natl. Acad. Sci. U.S.A.* **66,** 72–78.
139. Smith, H. S., Lan, S., Ceriani, R., Hackett, A. J., and Stampfer, M. R. (1981). *Cancer Res.* **41,** 4637–4643.
140. Spicer, S. S., and Meyer, D. B. (1960). *Am. J. Clin. Path.* **33, 453–459.**
141. Spicer, S. S. (1965). *J. Histochem. Cytochem.* **13,** 211–233.
142. Spremulli, E. N., Dexter, D. L., Young, P., Cambell, D., and Calabresi, P. (1983). *Proc. Am. Assoc. Cancer Res.* **24,** 28.
143. Springer, G. F., Desai, P. R., and Banatwala, I. (1975). *J. Natl. Cancer Inst.* **54,** 335–339.
144. Steplewski, Z., Herlyn, M., Herlyn, D., Clark, W. H., and Koprowski, H. (1979). *Eur. J. Immunol.* **9,** 94–99.
145. Tom, B. H., Rutzky, L. P., Jakstys, M. A., Oyasu, R., Kaye, C. I., and Kahan, B. D. (1976). *In Vitro* **12,** 180–191.
146. Tom, B. H., Rutzky, L. P., Oyasu, R., Tomita, J. T., Goldenberg, D. M., and Kahan, B. D. (1977). *J. Natl. Cancer Inst.* **58,** 1507–1512.
147. Tom, B. H., Goodwin, T. J., Sengupta, J., Kahan, B. D., and Rutzky, L. P. (1982). *Immunol. Comm.* **11,** 315–323.
148. Tompkins, W. A. F., Watrach, A. M., Schmale, J. D., Schultz, R. M., and Harris, J. A. (1974). *J. Natl. Cancer Inst.* **52,** 1101–1110.
149. Thompson, C. H., Jones, S. L., Whitehead, R. H., and McKenzie, I. F. C. (1983). *J. Natl. Cancer Inst.* **70,** 409–419.
150. Toolan, H. W. (1953). *Cancer Res.* **13,** 389–394.
151. Toolan, H. W. (1954). *Cancer Res.* **14,** 660–666.
152. Toolan, W. H. (1957). *Cancer Res.* **17,** 418–420.
153. van der Bosch, J., Masui, H., and Sato, G. (1981). *Cancer Res.* **41,** 611–618.
154. von Kleist, S., Chavanel, G., and Burtin, P. (1972). *Proc. Natl. Acad. Sci. U.S.A.* **69,** 2492–2494.
155. Wagh, U. V., Kasturi, S. R., Chaughule, R. S., Shah, S. S., and Ranade, S. S. (1977). *Physiol. Chem. Phys.* **9,** 167–174.
156. Watkins, J. F., and Sanger, C. (1977). *Br. J. Cancer* **35,** 785–794.
157. Weinhouse, S. (1982). *J. Natl. Cancer Inst.* **68,** 343–349.
158. Weiser, M. M., Podolsky, D. K., and Iselbacher, K. J. (1976). *Proc. Natl. Acad. Sci. U.S.A.* **73,** 1319–1322.
159. Weiss, P. (1947). *Yale J. Biol. Med.* **19,** 236–278.
160. Whitehead, J. S., Fearney, F. J., and Kim, Y. S. (1979). *Cancer Res.* **39,** 1259–1263.
161. Wolff, E., and Wolff, E. (1975). *In* "Human Tumor Cells *in Vitro*" (J. Fogh, ed.), pp. 207–237. Plenum, New York.
162. Wright, W. C., Daniels, W. P., and Fogh, J. (1981). *J. Natl. Cancer Inst.* **66,** 239–247.
163. Wright, W. C. (1980). *In Vitro* **16,** 875–883.
164. Yang, J., Richards, J., Bowman, P., Guzman, R., Enami, J., McCormick: K., Hamamoto, S., Pitelka, D., and Nandi, S. (1979). *Proc. Natl. Acad. Sci. U.S.A.* **76,** 3401–3405.
165. Yaniv, A., Altboum, Z., Gazit, A., Block-Schtacher, N., and Eylan, E. (1978). *Exp. Cell Biol* **46,** 220–230.
166. Zamcheck, N. (1981). *Cancer Bull.* **33,** 141–151.

ADVANCES IN CELL CULTURE, VOL. 4

CELL SHAPE AND GROWTH CONTROL

C. A. Heckman

Department of Biological Sciences
Bowling Green State University
Bowling Green, Ohio

I. INTRODUCTION

A. *Geometric Configuration and Growth*

A long history of experimentation has suggested that cells comprising a tissue exercise a refined degree of control over their own density (reviewed by Folkman and Greenspan, 1975). It can be anticipated that this regulatory feat is accomplished through a variety of different mechanisms. One of these may be a mechanism whereby modulating the shape of cells directly influences their growth rate. This idea has been the central focus of research projects carried out in my laboratory since 1978. In reviewing the literature on cellular configuration and growth control at that time, we found a close circumstantial relationship between these parameters. In Rous sarcoma virus-transformed chick embryo cells (Temin and Rubin, 1958) and spontaneously transformed mouse embryo cells (Sanford *et al.*, 1967), changes could be detected at either a gross or microscopic level. At a gross level, virally transformed foci could be detected by the darker staining of the rounded and multilayered cells. Differences were found between spontaneously transformed cells and their normal counterparts in the area occupied on substrata (Cherny *et al.*, 1975), in cytoplasmic basophilia,

ISBN 0-12-007904-6

in dry weight, and in the ratio of nuclear and cytoplasmic areas (Fox *et al.*, 1976; Handleman *et al.*, 1977). Due to features of the experiments, however, increased basophilia and nuclear–cytoplasmic ratios could result directly from a reduction in lamellar spreading. As the extent of spreading of the peripheral cytoplasm decreased, an increase in height might be expected. The transformed cells would have a thicker optical cross section, thus appearing more densely stained. We think it probable that changes in cell spreading underlie the observed cytological changes in nuclear–cytoplasmic ratio as well. Regardless of the quantitative extent of spreading, the nucleus would assume a shape more nearly spherical than the cell as a whole and would thus change less drastically with a decrease in the extent of spreading.

Not only was it apparent that different cytological expressions of altered growth control were attributable to changes in cell spreading, but the latter changes were almost invariably present in transformed cells. The prevalence of reduced lamellar spreading *in vitro* can be gauged by the number of reports in which "rounding up" has been described. The original reports on commonly used model systems for transformation are summarized in Table I. While most of these reports mentioned shape in a descriptive context, there are few, if any, model systems in which shape changes have not been noticed. Shape change is one of the earliest manifestations of the "pleiotropic response" in temperature-sensitive transformation of chick embryo cells by Rous sarcoma virus (Hanafusa, 1977). In the commonly used mutants, Buchanan's group and Bader have observed rapid morphological changes in the cells after shift to a temperature permissive for *src* gene expression (Ambros *et al.*, 1975; Bader, 1972). These changes may be closely coupled to the activity of the *src* gene cAMP-independent protein kinase (Collett and Erikson, 1978; Levinson *et al.*, 1978; Collett *et al.*, 1980). Although a single gene product is responsible for transformation in *ts*68 (Vogt, 1977), it has recently become clear that the product is an enzyme which not only has broad specificity for substrates but which may also be altered in its subcellular distribution within transformed cells.

Studies of the acute transforming RNA tumor viruses suggest that their mechanism of action involves the cell-derived genes which encode tyrosine-specific protein (TSP) kinase activities. Twelve replication-defective retroviruses, some closely related to one another, have been identified as producers of TSP kinase activities. The first such activity to be discovered was the pp60^{src} gene product from cells transformed by Rous sarcoma virus (Hunter and Sefton, 1980). It was found to have an unusual specificity for the tyrosine residues of polypeptide

TABLE I

MORPHOLOGICAL CORRELATES OF TRANSFORMATION

Cell type	Species	(Strain)	Transforming agent	Morphological changes	Reference
Mesenchymal					
Embryonic	Chicken	—	Rous sarcoma virus	More rounded cell shape	Kawai and Hanafusa (1971)
	Mouse	(C3H)	SV40, 20-methylcholanthrene	Reduced lamellar spreading	Cherny *et al.* (1975)
	Mouse	(C3Hf/HeN)	Spontaneous	Reduced spreading, spherical shape at metaphase	Fox *et al.* (1976)
	Mouse	(C3Hf/HeN)	Spontaneous	More pleiomorphic cell shape	Wetzel *et al.* (1977)
	Mouse	BALB/c	SV40	More rounded or spindle-shaped	Porter *et al.* (1973)
	Hamster	(Syrian)	Herpes simplex virus	More rounded cell shape, loss of actin cables	Goldman *et al.* (1974)
	Hamster	(golden)	X irradiation	More rounded cell shape, more complex surface features	Borek and Fenoglio (1976)
Adult	Rat	(various)	Various	More rounded cell shape at confluency	Cloyd and Bigner (1977)
Epithelial, adult					
Liver	Rat	(Wistar)	4-Dimethylaminoazobenzene, spontaneous	More domed cell shape, higher microvillar density	Allen *et al.* (1976)
	Rat	(Wistar)	4-Dimethylaminoazobenzene, acetoxyacetylaminofluorine	Reduced cytoplasmic spreading, more complex surface features	Karasaki *et al.* (1977)
	Rat	(BDVI, BUF)	Dimethylnitrosamine, *N*-methyl-*N*′-nitro-*N*-nitrosoguanidine	Increased cytoplasmic basophilia, increased nuclear–cytoplasmic ratio, variation in cell size and shape	Montesano *et al.* (1977)
Lung	Rat	(F-344)	7,12-Dimethylbenz[*a*]anthracene	Spindle-shaped cells within colonies	Heckman and Olson (1979)

substrates, as opposed to the more commonly phosphorylated residues, serine and threonine. The kinase, moreover, was itself phosphorylated on a tyrosine residue (Smart *et al.*, 1981).

The extensive studies summarized in Table II indicate that six families of genes encode TSP kinase activities. Evidence of such an activity has also been sought for transforming gene products of avian acute leukemia viruses which also induce sarcomas, namely, the MC29 virus and the avian erythroblastosis virus (AEV). No such activity has been found associated with the AEV protein (Anderson and Hanafusa, 1982). In the case of the MC29 protein, it has been found to have a kinase activity by some investigators (Bister *et al.*, 1980) but not by others (Levinson *et al.*, 1978).

Of the transforming gene products known to lack protein kinase activity altogether, there are only two whose cellular functions have been discovered to date. One gene, *sis,* of the Simian sarcoma virus, presumably originated from the cellular sequences of the woolly monkey. Its product, $p28^{sis}$, is 87% homologous to the human platelet-derived growth factor, PDGF-2 (Doolittle *et al.*, 1983). The homology suggests that the gene product exerts its effects by producing a qualitatively or quantitatively inappropriate PDGF-like growth factor. Furthermore, it has recently been found that the binding of human PDGF to mouse 3T3 cells can cause a two- to threefold increase in phosphotyrosine content (Cooper *et al.*, 1982). Thus, this molecular lesion may be a change occurring "upstream" from that induced by the TSP kinase-type retroviruses. Interestingly, the *sis* gene product has been identified in cultured cells derived from cancers of human connective tissues, but not of epithelial tissues (Marx, 1984). More recently, the product of a second gene, *erb* B, which is the avian erythroblastosis virus, has been found to share partial homology with the polypeptide composing the EGF receptor (Downward *et al.*, 1984).

Like the transforming retroviruses, polyoma virus also has a single gene which induces malignant transformation in its host cells. However, polyoma viruses are double-stranded DNA viruses differing from the retroviruses in host specificity and life cycle. In spite of these differences, the middle T protein encoded by the polyoma transforming gene has been found to carry out its transforming function through a TSP kinase activity (Eckhart *et al.*, 1979). The TSP kinase thus emerges as a major mechanism of virally mediated transformation. TSP kinases may, however, be only one of several possible means of implementing an altered program of growth control. Other viral transforming genes possess or bind kinase activities which phosphorylate amino acids other than tyrosine. A kinase activity showing specificity

TABLE II

RELATIONSHIP AMONG TRANSFORMING GENES OF RETROVIRUSES

Species origin	Putative proto-oncogene	Class	Virus	Transforming gene product	TSP kinase activity	Reference
Chicken	*src*	Avian sarcoma virus I	Rous sarcoma virus	P60	+	Hunter and Sefton (1980)
	fps	Avian sarcoma virus II	Fujinami sarcoma virus	P130	+	Feldman *et al.* (1980)
			UR1 Yamaguchi	P150	+	Wang *et al.* (1981)
			PRCII	P105	+	Neil *et al.* (1981)
			PRCIIp	P170	+	Breitman *et al.* (1981)
	yes	Avian sarcoma virus III	Y73	P90	+	Kawai *et al.* (1980)
			Esh sarcoma virus	P80	+	Ghysdael *et al.* (1981)
	ros	Avian sarcoma virus IV	UR2	P68	+	Feldman *et al.* (1982)
Mouse	*abl*	Murine leukemia virus	Abelson	P120	+	Witte *et al.* (1980)
Cat	*fes*	Feline sarcoma virus	Snyder-Theilen	STP110	+	Barbacid *et al.* (1980)
				STP85	+	
			Gardner-Arnstein	GAP115	+	Reynolds *et al.* (1980)
			McDonough	SMP180	+	Barbacid and Lauver (1981)
				SMP120		

for serine and threonine residues was associated with the gene product of hamster adenovirus type 5, expressed in human cells (Branton *et al.*, 1981). Additionally, the p21 protein encoded by the transforming gene of Harvey sarcoma virus occurred in both phosphorylated and nonphorphorylated forms, but was phosphorylated only at a threonine residue. Although the p21 protein bound GTP, no enzymatic activity has yet been associated with it (Shih *et al.*, 1980). While these examples of virally mediated transformation suggest that non-TSP kinase mechanisms are also implicated in growth regulation, they emphasize the importance of understanding the role of protein phosphorylation in these regulatory processes.

The kinases which are best characterized in terms of their interactions with substrates are those encoded by the avian retroviral genes, particularly *src*, *fps*, and *yes*. The apparent molecular weights of substrates heavily phosphorylated by these kinases are identified in Table III. Identities have been proposed for several of these proteins, of 130K, 46K, 39K, and 35K molecular weight, respectively. The 130K protein is vinculin, which is an integral component of the adhesion plaques of fibroblasts, occupying an intermediate position between actin cables and the cell surface (Sefton *et al.*, 1981). Phosphorylation of even a fraction of the vinculin residues could change their binding properties in such a way as to account for the redistribution of vimentin (David-Pfeuty and Singer, 1980) and the reduced frequency of adhesion plaques in transformed fibroblasts (Rohrschneider, 1980).

Of the cytosolic proteins, 35K has been identified as malic dehydrogenase, the metabolic branch point between the glycolytic pathway and amino acid anabolic pathways (Rubsamen *et al.*, 1982). Compared to the unphosphorylated form of MDH, the phosphorylated form is more active and is no longer regulated by fructose 1,6-bisphosphate (Cassman and Vetterlein, 1974). Thus, glutamate may be converted to pyruvic acid and then to lactic acid at an unregulated rate. The 46K and 28K proteins may also correspond to glycolytic enzymes, as they are now thought to be enolase and phosphoglycerate mutase, respectively (Cooper and Hunter, 1983). A cytosolic protein of 90K appears to be bound to $pp60^{src}$ in immunoprecipitates but not to be phosphorylated. It is identical with the ubiquitous 90K heat shock protein (Oppermann *et al.*, 1981b). The 36/39K protein is found at the inner surface of the plasma membrane and may have a cytoskeletal function (Nigg *et al.*, 1983; Radke *et al.*, 1983). The remaining substrates (81K and 43/42K) are thought to be localized in membranes or in the cytoskeleton, but their functions are unknown. Because the $pp60^{src}$ protein takes up a location in the plasma membrane, the identity of its

TABLE III

SUBSTRATES AND PROTEINS BOUND TO RETROVIRUS GENE PRODUCTS

Putative proto-oncogene	Avian virus	Transforming gene product	Target protein(s)										Reference
src	Rous sarcoma virus	P60					50K		80K	90K[a]			Hunter and Sefton (1980)
							50K			90K[a]			Brugge *et al.* (1981)
								58K[b]			130K	250K[b]	Sefton *et al.* (1981)
				36K									Radke *et al.* (1980)
				36K			50K				130K		Amini and Kaji (1983)
			28K		39K	46K							Cooper and Hunter (1983)
fps	Fujinami sarcoma virus	P130	28K		39K	46K							Cooper and Hunter (1983)
	PRCII	P105					50K			90K[a]			Adkins *et al.* (1982)
yes	Y73	P90	28K		30K	46K							Cooper and Hunter (1983)
											130K		Sefton *et al.* (1981)

[a] Coprecipitated with antibody to transforming protein but apparently not a substrate for the enzyme activity.
[b] Found in trace amounts.

membrane-bound substrates is likely to be of interest. The cytoplasmic form, which is synthesized on free polyribosomes, becomes complexed with two cellular proteins, of 50K and 90K, respectively (Brugge *et al.*, 1981; Oppermann *et al.*, 1981a). In this form, it appears to exhibit little tyrosine-specific activity (Brugge *et al.*, 1981; Courtneidge and Bishop, 1982), but rapidly translocates to the plasma membrane (Levinson *et al.*, 1980). The temperature-sensitive product encoded by the NY*ts*68 mutant of RSV appeared markedly less able to intercalate into membranes at temperatures that were nonpermissive for transformation (Courtneidge and Bishop, 1982). A severalfold decrease in enzyme activity was correlated with this change in locus (Hunter and Sefton, 1980).

Interestingly, mutations causing alterations in the 8K sequence at the amino-terminal end of the polypeptide also decreased the ability of the protein to enter the membrane. Such mutant viral strains showed decreased tumorigenicity when tested in animals. Complete regression of the tumors was found with one of these mutant isolates, indicating that the membrane location of the protein is important to its role in growth control (Krueger *et al.*, 1982). Sefton *et al.* (1982) discovered that a [^{3}H]palmitate-labeled lipid moiety could be detected bound to the pp60^{src} gene product. It has been postulated that the role of the bound lipid(s) is to anchor the protein within the plasma membrane.

The data clearly indicate that the membrane location is a crucial factor underlying the activity of pp60^{src}. Although the relationship of normal growth-permissive and nonpermissive configurations to the behavioral anomalies and the uncontrolled growth shown by cancer cells is still unclear, it can now be suggested that the structure and composition of the plasma membrane are probably altered in microdomains or overall as a consequence of the inappropriate intercalation of the viral gene product.

Some of the proteins also known to interact with the *src* gene kinase serve as substrates for the products encoded by the *fps* and *yes* genes (Table III). Clearly, further investigations of these proteins will lead to an understanding of the mechanisms mediating the characteristically increased glycolytic metabolism and decreased adhesion of retrovirus-induced transformants. These findings fall short of specifically demonstrating a relationship between the growth-permissive and nonpermissive configuration and the type of uncontrolled growth shown by cancer cells. However, the discovery of membrane involvement in the process of retrovirus-mediated transformation may suggest a central event in the expression of these genes. Considering the structure of anchorage-dependent cells, the intercalation of the gene product into the membrane would be able to affect sites of extracellular adhesion, the lipid bilayer itself, and/or the internal macromolecular cytoskeletal constituents which integrate with the inner aspect of the plasma membrane. There are good reasons for postulating (see Section I, C) that cytoskeletal and/or adhesive mechanisms may be critically related to growth control. An additional finding which makes the idea of a cytoskeletal target an attractive one is that certain functions of cytoskeletal proteins are known to be regulated by kinase-mediated phosphorylation, for example, the Mg^{2+}-ATPase activity of the actomyosin complex in both muscle and nonmuscle cells (Chacko *et al.*, 1977; Daniel and Adelstein, 1976; Scordilis *et al.*, 1977). In terms of major reorganization of cytoskeletal components in cells transformed by *src*, it is clear that both the normal actin cable network and the microtubule-intermediate filament network are perturbed (Edelman and Yahara, 1976; Ball and Singer, 1981). However, in the strict sense, we do not know whether cytoskeletal changes have a mechanistic role in transformation, even for Rous sarcoma virus transformants.

It also remains to be proved whether increases in glucose uptake, which accompany the expression of the transformed phenotype in the temperature-sensitive Rous sarcoma virus transformants, are related to growth control. In fact, there is reason to question whether the cytoskeletal and glycolytic expressions of the genotypic alteration are metabolically independent. It is clear that cell shape in some model systems can be changed by altering the levels of glucose in the media (Kalckar *et al.*, 1973; Amos *et al.*, 1976). This suggests that alterations in actin cable and microtubule structure in Rous sarcoma virus transformants could be secondary consequences of changes in the levels of intracellular glycolytic intermediates. It would be advantageous to be able to test the degree of interdependence among these factors experimentally. A major obstacle to this approach has been the difficulty of quantifying the configuration of any of the cytoskeletal networks or of cell shape. The lack of a technical approach to cytoskeletal problems has meant that perturbations short of a complete structural reorganization were difficult to characterize. Thus, even limited goals, such as learning which network is altered first and the relative rates of reorganization in each of the networks, have not been attained.

Transformation of avian embryo cells by introducing a retroviral genome is undoubtedly the altered growth program best understood at the cellular level. To what extent does it serve as a model for any other kind of growth modulation? It must be admitted that the relevance of transformation by *src* to other processes of growth or regeneration is obscure. However, there are reasons for thinking that the viral transformation model at least offers some interesting and provocative parallels to the processes taking place more generally in neoplasia.

B. Transformation in Epithelial Model Systems

Our interests over the past several years have centered on model systems representative of epithelial tissues. These are of practical importance due to the fact that 80–90% of all clinically diagnosed neoplasms arise in the epithelia. Unfortunately, progress in these model systems has lagged behind that in many other areas of *in vitro* research. Recently, significant advances have been made by investigators who have developed special methods for culturing the epithelia (Michalopoulos and Pitot, 1975; Rheinwald and Green, 1975; Richman *et al.*, 1976; Wigley and Franks, 1976; Stoner *et al.*, 1980; Hennings *et al.*, 1980; Marceau *et al.*, 1982). While long-term growth of normal epithelial cells in culture is clearly feasible (Heckman *et al.*, 1978; Heckman, 1983), the ease of culturing these cells, even under the most favorable conditions, is only now approaching that of culturing normal mesenchymal cell strains. The latter, of course, have been used in innumerable laboratories over the past 30 years.

In any discussion of *in vitro* tissue models, it is important to recognize that the requirements of epithelial cells for substrates, nutrients, and hormones differ from those of mesenchymal cells and may even show specificity at the organ and tissue level. The impact of these requirements on the choice of conditions for culturing the epithelia can scarcely be overestimated. Whereas a large fraction of the cells in populations of human skin fibroblasts appeared capable of dividing after enzyme-mediated dissociation from a tissue (Schneider and Mitsui, 1976; Martin *et al.*, 1981), the vast majority of epithelial cells from rat liver and respiratory tract failed to divide following dissociation (Richman *et al.*, 1976; Heckman *et al.*, 1978). These studies clearly implied that epithelial cell lines were produced infrequently under routine culture conditions and, when produced, were probably unrepresentative of the normal population. The ability of dissociated cells from human epidermis and bovine kidney to proliferate when cultured on mesenchymal cells or their extracellular matrices suggested that a substrate requirement was the major barrier to proliferation (Rheinwald and Green, 1975; Gospodarowicz, 1983). Recent work, however, has shown that the epithelia from a few organs of the rat, namely, trachea, esophagus, salivary gland, epidermis, and prostate, grew well in primary culture in the absence of mesenchymal cells or their matrices (Heckman, 1983).

The characteristic pattern of epithelial growth from cultured tracheal explants is illustrated in Fig. 1. The cells migrated onto plastic culture substrates while undergoing division, forming a circular sheet of epithelium around the explant. If fibroblasts were present around the edge of the cultures, these cells appeared to cause the exfoliation of the epithelial sheet. This effect may account for the problems encountered in attempting to coculture the two cell types. Examination of outgrowth cultures by transmission electron microscopy demonstrated the typical features of noncornifying epithelial populations (Fig. 2). Moderate dedifferentiation of the epithelium occurred in cultures derived from the upper airway of rats (Heckman *et al.*, 1978). The growth characteristics of the tracheal epithelium were remarkably uniform in replicate experiments. Figure 3 shows the plots of outgrowth area vs time for populations derived from the inbred F344 rat strain in experiments conducted approximately 18 months apart. Since the growth rates were unusually high for epithelial cells, it was advisable to rule out the possibility of fibroblast overgrowth. When cultures established for several weeks were examined by scanning electron microscopy, we found that they had a unique organization. They consisted of upper and lower strata of cells and resembled a squamous epithelium. The lower cells had a flattened cuboidal shape and unusual surface features (Fig. 4). Thus, they could not be fibroblast cultures.

The rapidity and uniformity of growth in these cultures has contributed to the productivity of recent studies on chemical transformation of rat airway cells (Marchok *et al.*, 1978; Steele *et al.*, 1978, 1979), and on the morphological (Heckman and Olson, 1979; Manger and Heckman, 1982), behavioral (Manger and Heckman, 1984), and biochemical (Scott *et al.*, 1979; see also page 106) characteristics of the resulting epithelial populations. *In vitro* models of "transformation" in other epithelia have also been developed in the form of cell lines from rat salivary gland (Wigley and Franks, 1976) and liver (Karasaki *et al.*, 1977; Montesano *et al.*, 1975), and from the mouse skin (Yuspa *et al.*, 1980) and mammary gland (Butel *et al.*, 1977). While the characterization of these models at the cellular and molecular level has only recently been initiated, it is already clear that some epithelial lines undergo a process of oncogenic transformation while being grown *in vitro* (Montesano *et al.*, 1975; Karasaki *et al.*, 1977; Marchok *et al.*, 1978). For these models, the alterations undergone by the cells resemble in key respects the prototype established by Rous sarcoma virus- and spontaneously transformed mesenchymal cells. Montesano *et al.*

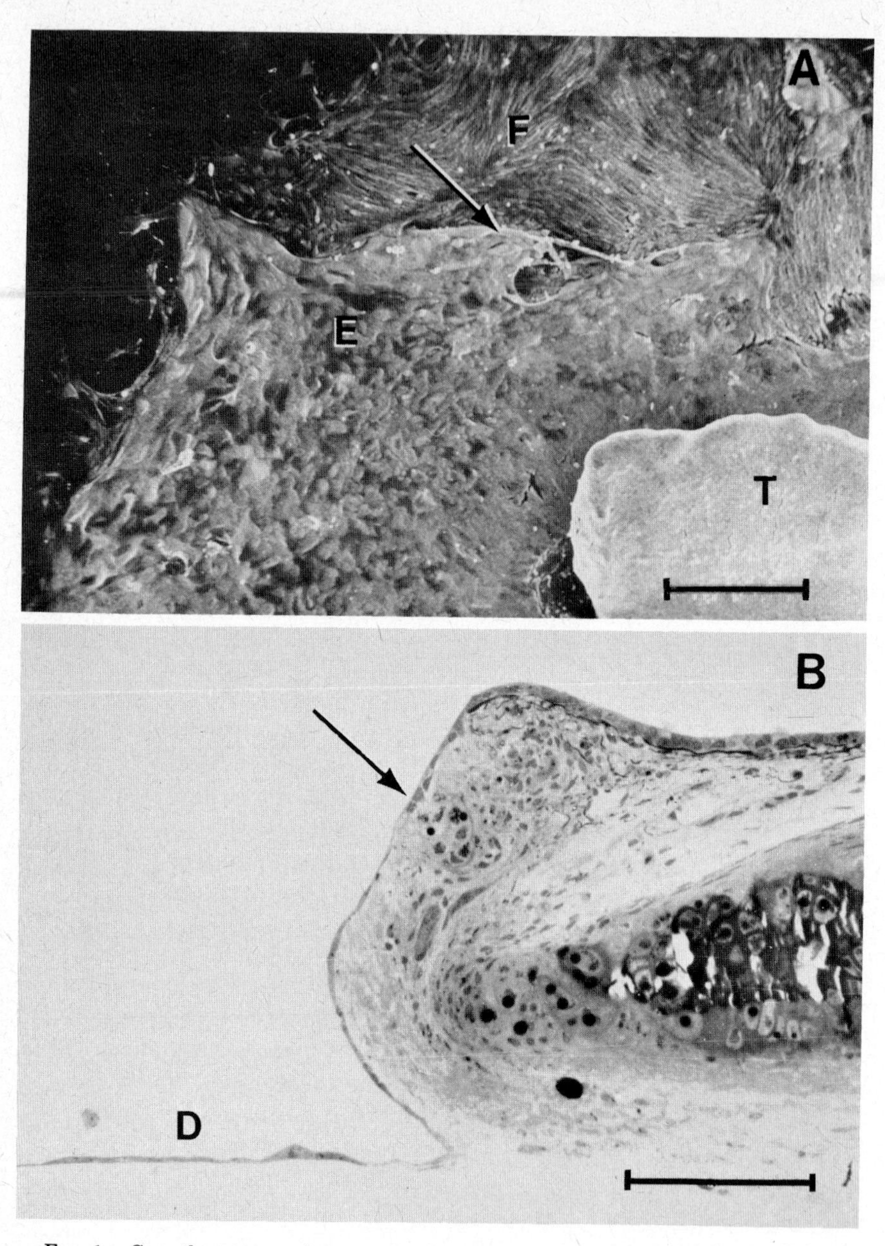

FIG. 1. Growth pattern of rat tracheal epithelium *in vitro.* (A) scanning electron micrograph showing part of a rat tracheal explant after 7 days in culture. Sheets of epithelial cells (E) and fibroblasts (F) have migrated out from the tissue (T). At the interface between the two cell types, the fibroblasts appear underneath the advancing edge of the epithelium (arrow). ×36. Bar = 0.5 mm. (B) Section through part of a tracheal explant and outgrowth, after 7 days of maintenance *in vitro.* The epithelium has grown over the cut edge of the explant (arrow) and out onto the surface of the culture dish (D). ×50. Bar = 0.5 mm.

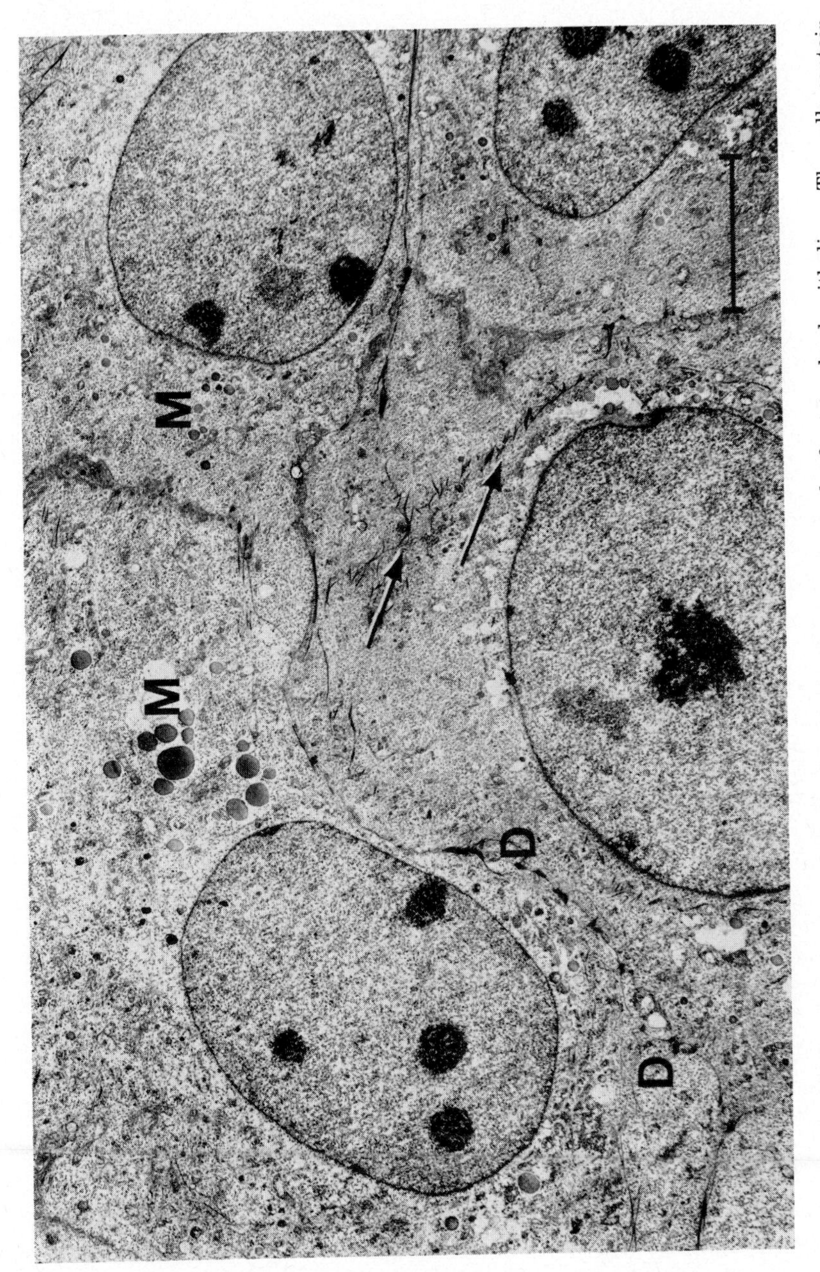

FIG. 2. Transmission electron micrograph of basal cells in an outgrowth of rat tracheal epithelium. The cells contain prominent keratin tonofilaments (arrows) and desmosomes (D), as well as mucin granules (M). These features are characteristic of the tracheal mucociliary epithelium. ×1900. Bar = 10 μm.

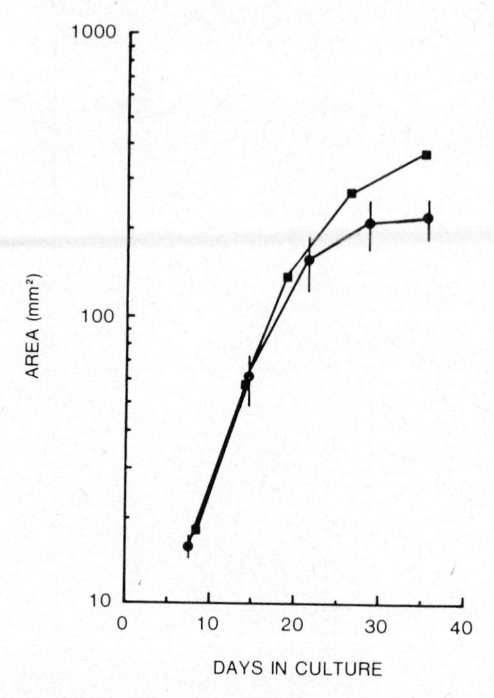

FIG. 3. Curves showing the rates of expansion of rat tracheal outgrowths in culture. The experiments were conducted in 1979 (circles) and 1978 (squares). The explants were removed 7 and 10 days after initiation of the cultures, respectively.

(1977), comparing tumorigenic and nontumorigenic liver cell lines, found greater cytoplasmic basophilia, a higher nuclear–cytoplasmic ratio, and greater variations in size and shape in the cells of colonies from tumorigenic lines. In separately derived liver lines which became tumorigenic over a prolonged period of *in vitro* culture, cells viewed by scanning electron microscopy showed gradual changes in shape (Karasaki *et al.*, 1977).

We have studied morphological and behavioral changes in cells from the rat respiratory tract over the period during which they become oncogenically transformed *in vitro* (Heckman and Olson, 1979). The morphological changes in the epithelial lines were more subtle than in Rous sarcoma virus-transformed and spontaneously transformed embryo cells, where they were obvious upon light microscopic examination. In the lung cell lines, we initially perceived morphological differences by looking at the shapes of cells in colonies. Early passages of the progressing line, 1000 W, were compared with early passages of a subline derived from a tumor obtained after injection of 1000 W cells into a host animal. After development of the latter cell population into a cell line (1000 WT), comparisons were performed at a similar passage level as in the 1000 W line so as to negate the influence of the immediate *in vitro* history of the cell populations.

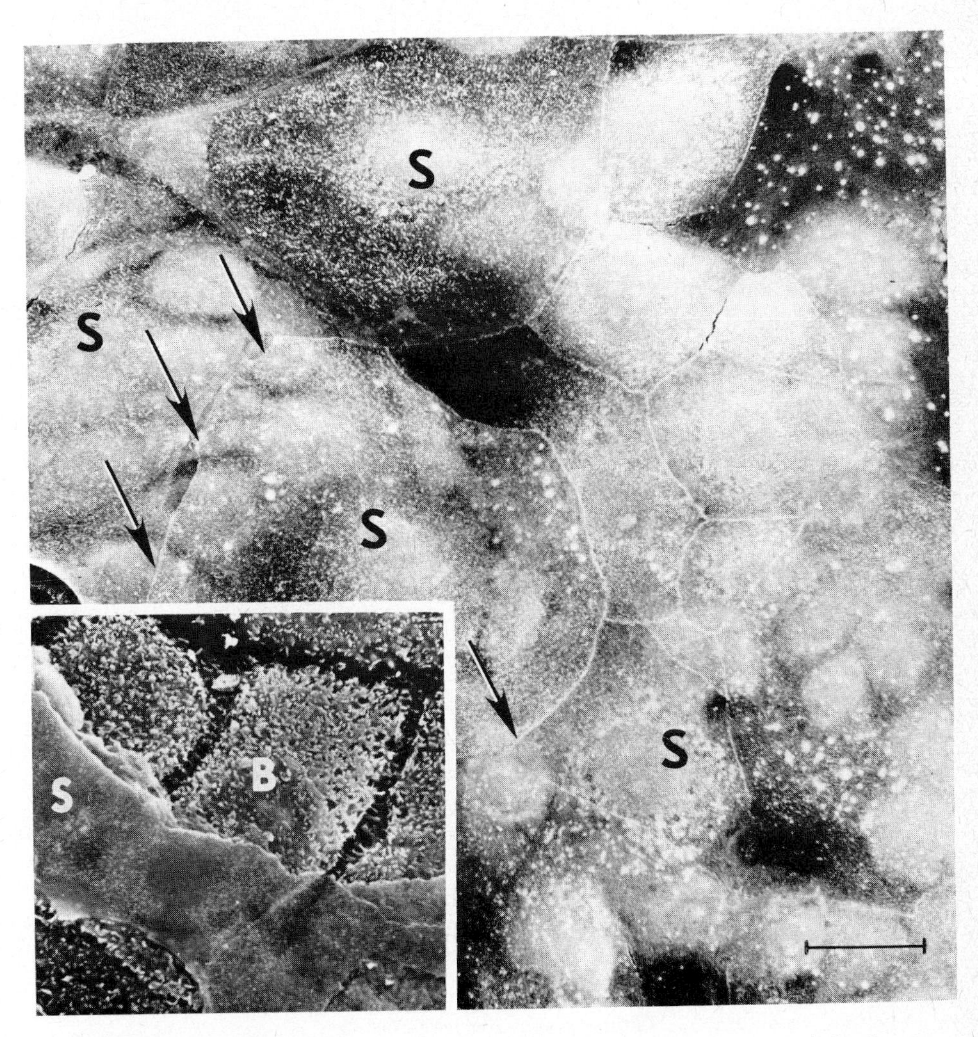

FIG. 4. Scanning electron micrograph showing the surface of an epithelial culture derived from rat trachea. While large squamous cells (S) are apparent at the surface of the culture, their borders sometimes traverse the outlines of smaller cells underneath (arrows). ×580. Bar = 30 μm. Inset: The basal layer of these cultures contains small cells (B) with highly villous surfaces. This layer can be exposed by stripping off the squamous cells (S). ×2050. Bar = 10 μm.

When these epithelial populations were studied by scanning electron microscopy, the cells in colonies from the earliest passage interval studied appeared flat and quasi-circular in profile. Even in relatively elongated cells near the edges of the colonies, the ratio of long and short axes rarely exceeded 2:1 (Fig. 5). Cells from the 1000 WT subline typically appeared more rounded than those from early passages of the 1000 W line. They were more extended at sites of intercellular contact and, at some locations, the cells were separated by conspicuous spaces (Fig. 5). Moreover, cells at the edge of the colonies were frequently quite elongated, with a ratio of long and short axes exceeding 2:1. Additionally, some cells within colonies from the 1000 WT subline (as well as from late passages of the 1000 W line) showed extreme deviations from epithelioid morphologies. We concluded that changes in cell shape and cell–cell interactions were suggestive of increased deformability and that they could only be reversed to a minor extent by manipulating *in vitro* conditions (Heckman and Olson, 1979).

Neither in the epithelial model systems discussed above nor in comparisons of tumorigenic and nontumorigenic lines from the mouse mammary gland (Voyles and McGrath, 1976; Butel *et al.*, 1977) did investigators see consistent differences in morphology *in single cells* by standard light microscopic methods. Moreover, changes in cytoskeletal constituents implicated in the maintenance of epithelial cell architecture have been debatable in many instances. Asch *et al.* (1979) found no difference in the organization of microtubules or actin-containing filaments in normal and neoplastic mouse mammary gland cells *in vitro*. Subsequently, in a detailed study of human mammary carcinoma cell lines, Brinkley *et al.* (1980) were unable to find consistent changes in microtubule organization. Some of the cell lines exhibited a structural phenotype similar to normal cells. In one of the most thoroughly characterized of the epithelial model systems, the mouse keratinocyte, cells from one line were found to acquire vimentin filaments upon transformation (Franke *et al.*, 1979), but a later study comparing normal cells with 12 tumorigenic lines did not confirm the appearance of vimentin biochemically. However, a bladder epithelial cell line, on-

---

FIG. 5. Scanning electron micrographs of colonies from early passages of the 1000 W line and from the tumor-derived cell line, 1000 WT. (A) Cells from the 1000 W line appeared flat and rigid in shape and were frequently demarcated by linear borders (arrows). (B) Cells from the 1000 WT line appeared less rigid and more deformable than those from the early passage of 1000 W. The nuclear region was protruberant and the intercellular contacts appeared highly irregular (arrows) in comparison to those observed in cells from the original line. ×420. Bar = 100 μm.

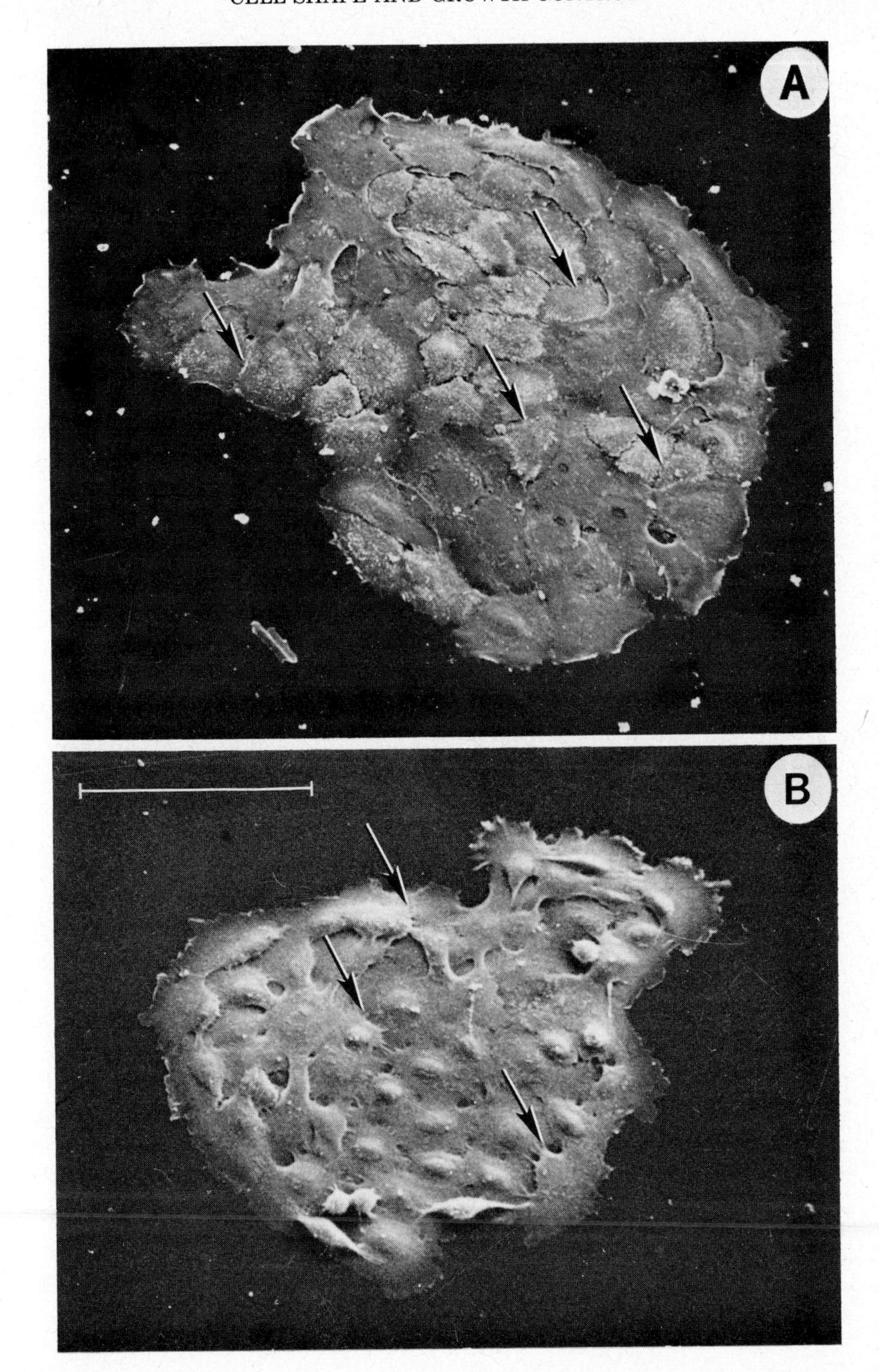
A
B

cogenically transformed *in vitro* by benzo[*a*]pyrene treatment, acquired vimentin filaments (Summerhayes *et al.*, 1981).

Some of the most promising cellular models in terms of cytoskeletal alteration during transformation have been those derived from the liver. Miettinen and Virtanen (1978) found that the actin microfilaments of cultured hepatoma cells appeared less prominent than in primary hepatocytes. More recently, cells from tumorigenic and nontumorigenic liver lines have been compared with respect to the organization of microtubules and actin fibrils (Bannikov *et al.*, 1982). In these studies, isolated cells from nontumorigenic lines had a ring-shaped, marginal bundle of microfilaments. This distribution tended to be discontinuous in cells from tumorigenic lines, so that only part of the cell edge was delineated by actin immunofluorescence. Three of the five tumorigenic lines exhibited long, cable-like structures which were nearly parallel to one another across the long axis of the cell. A fourth line exhibited no microfilament bundles whatever. Although all of the cells had a matrix-type distribution of microtubules, the cells from nontumorigenic lines appeared to have more prominent marginal bands of microtubules. The changes in internal organization were thought to relate to the reduced lamellar spreading of the tumorigenic cells, which was readily noted by scanning electron microscopy. The tendency of isolated, nontumorigenic cells to have a discoid shape also appeared reduced in tumorigenic cells (Bannikov *et al.*, 1982).

The results of the studies on epithelial cells are so varied as to provide little insight into the question of whether there is a single cytoskeletal lesion common to the different models of transformed epithelia. Even when we attain more complete information on cytostructural organization in these cellular models, significant difficulties in interpretation will likely persist due to two complicating factors. First, studies on the interactions of cytostructural components have shown that a network composed of one constituent can be disrupted by perturbations specific for a different constituent (Ball and Singer, 1981). Second, as demonstrated by our studies on the respiratory tract epithelium, changes in cell conformation can be missed in one set of experiments while being detected in another. The latter problem may be due to the subtlety of architectural changes in cells and can be partially overcome by conducting experiments on "progressing" cell lines, i.e., those becoming tumorigenic over a period of *in vitro* cultivation. With these experimental models, the observations can be conducted on populations of cells which are relatively homogeneous. Thus, it may be significant that, in all the instances in which progressing epithelial lines have been studied with methods of sufficient resolution, changes in cell shape have been seen (Karasaki *et al.*, 1977; Montesano *et al.*, 1977; Heckman and Olson, 1979; Bannikov *et al.*, 1982).

The question of whether cytoarchitectural changes could be linked to growth control modulation in epithelial cells was an intriguing one in light of the early timing and extensive nature of cytoskeletal changes following temperature shift of Rous sarcoma virus-transformed mesenchymal cells. However, the structure of mesenchymal and epithelial cells differed in fundamental ways. For example, it was clear that the predominant species of cytoskeletal proteins were dissimilar. Although actin was the major constituent of the cytoskeleton in embryonic fibroblasts, it was a minor component in respiratory tract epithelial cells. The structural changes could hardly be identical at a molecular level unless they involved a constituent of the cytoskeleton which was common to both cell types. The results of various studies on epithelial model systems indicated that radical structural reorganization, such as seen in the mesenchymal models discussed above, was rare. This may have been an integral feature of epithelial cells, perhaps related to the stable properties of the keratin cytoskeleton which constituted the main structural framework of the cells.

An experiment was designed to answer the question of whether profound structural alterations were ever likely to occur. We conducted comparisons of the keratin cytoskeletons in cells from three lines of marked malignancy (transformed) and three of marginal malignancy (nontransformed). Although the progressing epithelial lines increased in oncogenic potential over the course of *in vitro* culture, it was clear that they only attained a moderate degree of malignancy (Marchok *et al.*, 1978; Steele *et al.*, 1979; Manger and Heckman, 1982). Even in very late passages of these lines, large populations of cells, e.g., 10^4–10^6 cells, were required in order to produce tumors in host animals. Thus, it was essential to obtain highly malignant cell lines from transplantable tumors (Heckman and Olson, 1979; Scott *et al.*, 1979; Manger and Heckman, 1982). Although several such lines were tested, they were only considered highly malignant if at least 50% of the host animals developed tumors following injection with doses of 100 cultured cells. The TD_{50} (tumorigenic dose) of the transformed lines was therefore 10^2 or lower. A summary of the tumorigenicity studies is presented in Table IV.

In subsequent examination of the cells directly by immunofluorescent staining for keratins, we could be sure that the populations we saw would contain *bona fide* malignant stem cells. These comparisons also allowed us to identify some of the presumed expressions of the transformed phenotype. Comparisons among the six lines were carried

TABLE IV

TUMORIGENICITY OF CELL LINES TESTED IN IMMUNE-DEFICIENT HOSTS

Cell line	Number of tumor-bearing animals after injection of various cell lines[a]				
	1×10^6	1×10^5	1×10^4	1×10^3	1×10^2
B2-1	nd[b]	nd	nd	5/5	5/10
BP 3-0	nd	nd	nd	5/5	5/5
MCA_7	nd	nd	nd	10/10	9/10
2C1	0/6	nd	nd	nd	nd
4C9	3/7	nd	nd	nd	nd
165S	1/8	nd	nd	nd	nd

[a] Abstracted from Manger and Heckman (1982).
[b] nd, not determined.

out by having human observers unfamiliar with the immunofluorescence images categorize encoded micrographs. The micrographs were sorted into groups and the groups subsequently combined into two major classes (Manger and Heckman, 1982). The descriptions of the observers indicated that cells from all three of the transformed lines had a nonuniform distribution in the field of view. Cells from two of the lines also showed nonuniform fluorescence at the periphery, indicative of keratin-poor cytoplasmic regions. Nearly all observers recognized a dissimilarity between the MCA_7/BP3-0 pair and the 2C1/165S pair, which represented the extremes of oncogenic potential. When the classification patterns of all observers were considered, they provided a ranking of cell lines that was precisely in parallel with their TD_{50} values.

The data could not be subjected to statistical analysis because they were based on the individual perceptions of different observers. Therefore it was hard to determine how much confidence we should properly have in the existence of these differences. For this reason, we entertained the possibility that the anomalous structures characteristic of transformed cells could also be found at random in nontransformed lines. We examined cells from five additional, marginally tumorigenic lines of respiratory tract epithelium (7C5, 10C9, 1000 W, BP4, and M(F) clones A and B). Although images from thousands of these cells were recorded, little evidence of structural anomalies was seen. Cells from one of these lines, 1000 W, are shown along with those from the transformed line, B2-1, in Fig. 6. It was clear that cells from the two lines were of approximately the same size, but were dissimilar in other

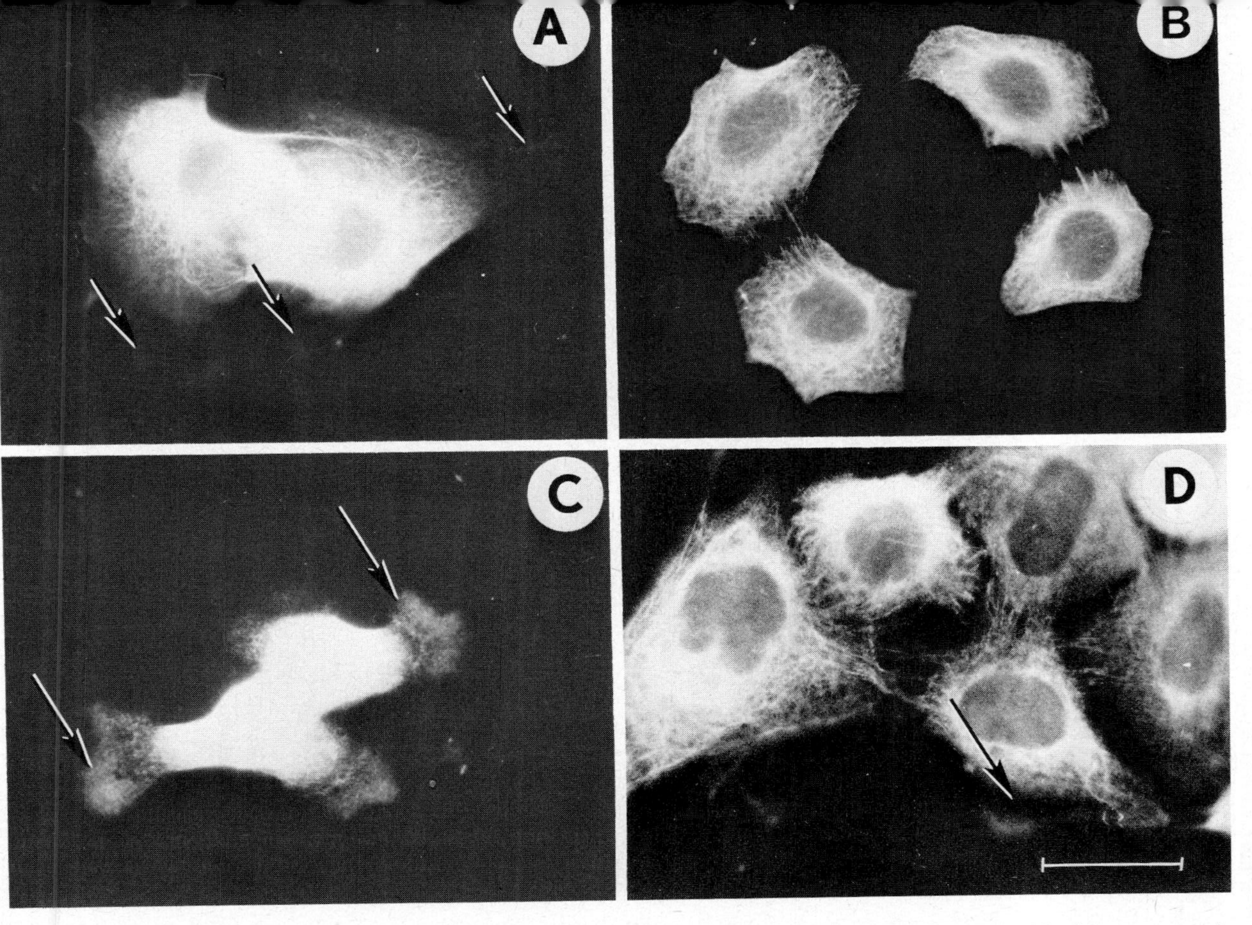

FIG. 6. Immunofluorescent localization of rabbit antiserum to mouse K1 keratin. Nontumorigenic and highly tumorigenic lines were compared. The nontumorigenic cells had exceptionally thin, keratin-poor peripheries (A, arrows). Highly malignant cells were unusual in that masses of keratins were visualized in the cell periphery (arrows). Nontumorigenic lines: (A) 4C9 cells. (B) 1000 W cells. Highly tumorigenic lines: (C) BP3-0 cells. (D) B2-1 cells. ×1100. Bar = 20 μm.

respects. Examples of cells from the transformed BP3-0 and nontransformed 4C9 lines are also shown (Fig. 6). These results could then be interpreted unequivocally as an indication that alterations in epithelial structure, namely, extreme irregularities in cell shape, nonuniform distribution of cells in the field, and keratin-poor peripheral regions within the cells, were unique to highly malignant cells.

The criteria used by those observers who misclassified the intermediate lines, B2-1 and 4C9, were of great interest because they allowed us to identify features that were unrelated to the oncogenic potential of the lines. These included the size of the cells, the extent of contact among them, and the size and intensity of staining of the keratin filaments. Changes in the organization of the filaments appeared of greater importance than their actual structure in relation to the potential malignancy of the lines. The findings suggested that alterations in the molecular composition of major keratin species, while possible, would necessarily have limited effects on intrafilament associations. Some insight into the mechanism underlying cytostructural changes could be gained by identifying and quantifying proteins of the cytoskeleton after extraction of the cytosol with a nonionic detergent. We have initiated investigations along these lines, using our models of highly malignant and nonmalignant cells. Although the majority of proteins and their isoelectric variants resembled those characteristically found in the normal parental cell type, the content of basic keratins focusing above pH 5.52 appeared reduced in samples from the tumorigenic cell lines. The differences appeared related to the differentiated state of the cells, since the line showing the greatest reduction gave rise to relatively undifferentiated squamous cell carcinomas. Of the acidic proteins, three constituents of M_r 41,000 focusing at pHs of 5.10–5.48 were more prevalent in samples from the highly tumorigenic lines. The variety and quantity of polypeptides with molecular weights of 42,000 and 43,000 focusing in a range of pHs between 5.43 and 6.15 also seemed increased in highly tumorigenic lines. These species appeared to increase quantitatively at the expense of the major keratin species (Heckman *et al.*, 1985).

The identification of structural anomalies in transformed cells raised a question about whether these features were related to behavioral changes. While it is difficult to imagine that the discontinuous distribution of keratins was directly related to invasiveness, for example, it is easy to imagine that altered motile or adhesive properties were related. Using the combined techniques of phase microscopy and video recording, we studied the movements of cells from the transformed and nontransformed lines described above. As a group, the transformed cells tended to move approximately 150% faster than the nontransformed cells. Although the difference was statistically significant, the motility rates varied widely among the cell lines (Manger and Heckman, 1984). Examples of the motility patterns seen in these epithelial cells are shown in Fig. 7. Two of the cells shown, from the MCA_7 line, traveled rapidly and tended to reverse their direction of travel frequently. The other two cell "tracks" illustrate the alternative pattern. These cells moved slowly and, except for brief changes or reversals of direction, tended to persist in their original trajectory. Although the rates of travel illustrated here were characteristic for the individual cell lines, cells from all of the lines exhibited a tendency to change their direction of travel frequently.

The video recordings also revealed an unexpected difference in the behavior of transformed and nontransformed cells—one which could also have accounted for the differences in motility rates. The ad-

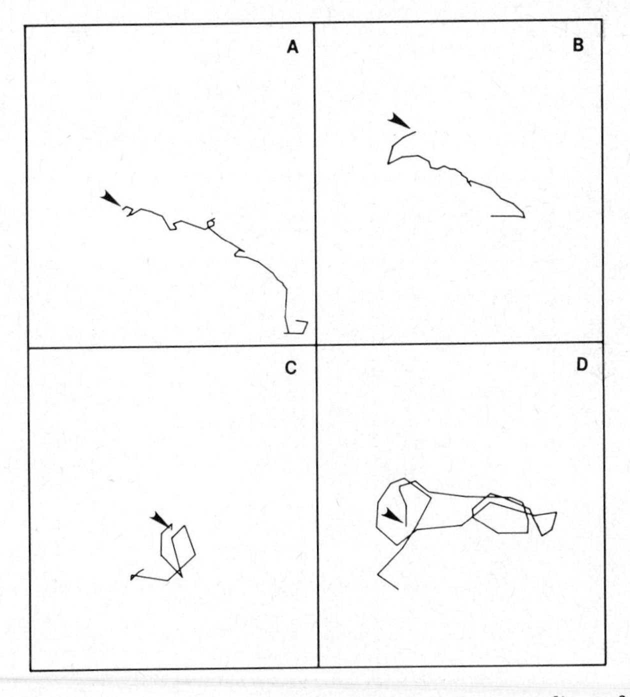

FIG. 7. Trajectories taken by two different lines, based on recordings of cell position every 15 minutes. The start of recording is marked (arrows). (A,B) Cells from the BP3-0 line travel relatively slowly so that sequential points along their trajectory tend to be closely spaced. (C,D) Cells from the MCA_7 line had a rapid rate of movement. The distance between sequential points along the trajectory of these cells appeared longer.

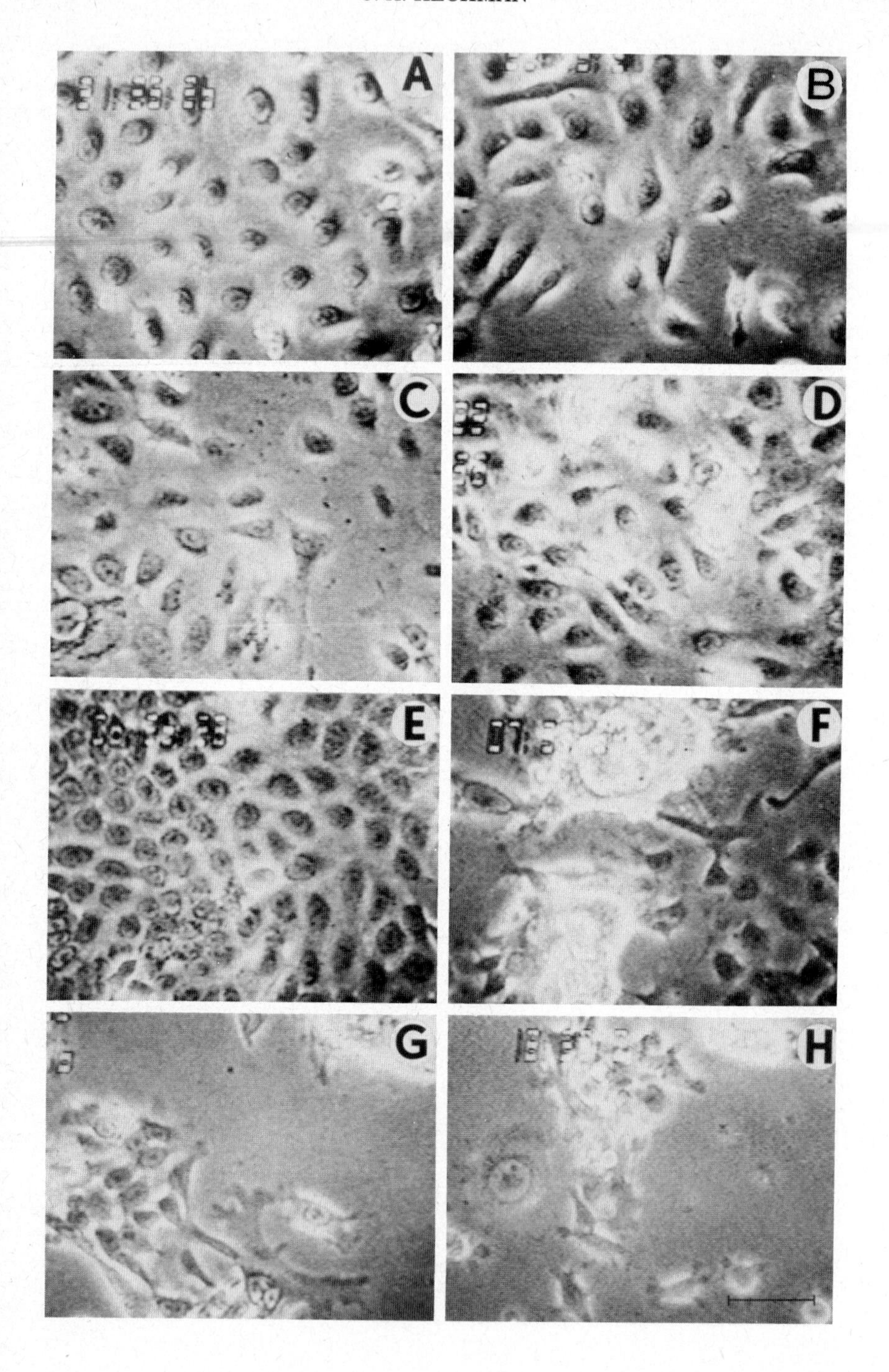
A
B
C
D
E
F
G
H

hesivity of the transformed cells was markedly reduced when we plated them on glass substrates. We initially chose to use glass substrates for the video recordings because of their optical superiority to plastic. However, the behavior of the cells was so profoundly altered that single cells whose movements could be tracked could rarely be found in the recordings. While the transformed cells adhered closely to one another, forming multicellular spheroids, nontransformed cells showed little tendency to discriminate between glass and plastic substrates (Manger and Heckman, 1984). Examples of these diverse behavioral patterns are shown in Fig. 8.

C. Evidence of a Configurational Requirement for Growth

A few studies conducted in recent years have had a revolutionary impact on the field of growth control. Each demonstrated a configurational requirement for cell growth under *in vitro* conditions. Folkman and Moscona (1978) showed that human fibroblasts and endothelial cells could be grown on substrates modified by adsorption of poly(2-hydroxyethylmethacrylate). The tendency of the cells to spread out on the substrate was progressively modified as higher concentrations of poly(HEMA) were adsorbed. As the cells underwent a reduction in spreading, their rate of entry into S-phase was correspondingly decreased. The minimum estimated thickness of adsorbed polymer, 35 Å, scarcely inhibited [^{3}H]thymidine incorporation. However, interposing a polymer layer of approximately 350 nm thickness between the cell and the plastic substrate inhibited cell cycling to a marked degree.

Westermark (1978) and Ponten and Stolt (1980) obtained evidence of cells' growth potential being restricted by conditions which affected their configuration. Westermark's work suggested that growth was limited by cell density even when only 2–3 cells were present on a small substrate. By developing elegant, new methods for culturing single cells *in vitro,* Ponten and Stolt (1980) demonstrated that cell–cell interactions were not involved in the configuration-sensitive process. Agarose films containing palladium-coated "islands" were manufactured by a photolithographic technique similar to that used for the

Fig. 8. Video frames showing cells grown on plastic (A,C,E,G) and glass (B,D,F,H) substrata. When plated on glass surfaces, the nontumorigenic cells, 2C1 and 165S, were affected by reduced adhesion, but the disturbances of adhesion for the highly malignant cells, BP3-0 and MCA_7, were severe. These cells formed clusters to such an extent that it was difficult to trace their time-lapse movements. (A,B) 2C1; (C,D) 165S; (E,F) BP3-0; (G,H) MCA_7. ×480. Bar = 50 μm.

fabrication of integrated circuits. When human fibroblasts and glia-like cells were placed on these films, they attached specifically to the islands and grew at a normal rate. However, cells settling on islands with a surface area below 10,000 μm^2 did not sustain a normal rate of division. While the cells were able to attach to the smallest surfaces manufactured, 2000 μ^2, 70% of them were unable to divide. For human fibroblasts, 2000 μ^2 approximates the surface area of one fully spread cell, so that the data may indicate a mechanism by which cells can examine their environs to determine if there are spatial restrictions to spreading. The fulfillment of a specific spatial criterion would then confer a growth-permissive status on the cells.

In addition to the effects of cell configuration on division rate discussed above, two groups of investigators have shown that differentiation of normal epithelial cells could also be modulated by configurational change. One of these groups demonstrated that certain differentiated functions, namely, synthesis of casein and triglycerides, were not ordinarily expressed by epithelial cells cultured from the mouse mammary gland, but could be induced by conversion of the cells from a flat to a cuboidal shape (Emerman and Pitelka, 1977; Emerman *et al.*, 1977, 1979). The cells were attached to a collagen layer which was then floated, resulting in shrinkage of the collagen and alteration of cell shape. The expression of certain differentiated functions was independent of cellular configuration; for example, tight junctions were found among the cells even in the flat configuration (Pickett *et al.*, 1975). Cultured hepatocytes, used to develop the floating gel model, were thought to acquire differentiated functions when floating gels were used as substrates (Michalopoulos and Pitot, 1975). However, neither hepatocytes cultured routinely on plastic substrata nor floated hepatocytes appeared to be able to progress through mitosis (Richman *et al.*, 1976).

It was not clear from the early studies on collagen gels whether the effects of flotation were mediated by the increased access of hormones and nutrients in the culture medium to the basolateral surfaces of the cells. Shannon and Pitelka (1981) conducted experiments designed to explore this possibility. By cross-linking the collagen gel with glutaraldehyde, these investigators produced an inflexible floating substrate. Kraehenbuhl and co-workers (1982) used a variation of this procedure, namely attaching the collagen to a Millipore filter, to prevent retraction of the gel upon flotation. The cellular geometry resembled that on attached gels, and protein synthetic activity was low, demonstrating that the expression of differentiated functions did not occur under these conditions. The results suggested that if the cells were prohibited from assuming a cuboidal shape, they also failed to increase casein production (Shannon and Pitelka, 1981). The flexibility of the collagen substrate appeared important in allowing the cell to assume a preferred shape.

Kraehenbuhl's group (1982) also carried out experiments in which the cell number was monitored in addition to casein, α-lactalbumin, and transferrin production. When mammary cell aggregates were plated on or embedded within a collagen gel, one doubling ensued. When the gels were floated, the cell number actually declined, presumably due to overcrowding of cells on the shrunken gel. The only situation that appeared to favor cell proliferation was the embedment of aggregates in the collagen matrix. The aggregates formed tubuloalveolar structures, thus establishing a tissue architecture similar to that found in the gland *in vivo*. After a 9-day lag time, the embedded cells began to grow with a doubling time of about 3 days. These results, however, can only be considered definitive when evidence is presented confirming the epithelial nature of the dividing cells.

Separate studies on corneal and lens epithelial cells, carried out by Gospodarowicz and co-workers (1978), offered further evidence about the interrelationship between cell shape and growth control. These cells showed a mitogenic response to epidermal growth factor (EGF) only when they were maintained in a columnar form, although they were able to bind EGF when in flattened form. These cells also responded to fibroblast growth factor as a mitogen, but only when in a flattened configuration. However, the role of the collagenous substrate, which was used to promote the columnar shape, complicated the interpretation of the studies. It seemed likely that the effects on growth control were partly attributable to inductive effects in addition to changes in the steric interactions among cells. The development of experimental approaches to separate the steric and inductive effects of substrates would be of value for future studies.

The accumulating evidence seems to indicate that there are configurational requirements for the growth of anchorage-dependent cells. However, important distinctions must be drawn between the shape-responsive behavior of mesenchymal cells, including those of mesodermal origin, and that of epithelial cells. The experiments discussed above have shown that spatial restrictions can be imposed on normal mesenchymal cells simply by interfering with normal spreading behavior. Although the *effect* of interference is equivalent to "contact inhibition" of growth, the elegant studies of Ponten and co-workers suggest that intercellular contact is not required to implement a "density-dependent"-like response. One may reasonably conclude that contact inhibition may occur independently of mechanisms for cell–cell communication. Epithelial cells appear to behave differently in several respects. When grown from tissue explants without having first undergone dissociation and extensive spreading on substrata, epithelial cells from a number of organs become considerably flatter than normal but still maintain high growth rates *in vitro*. Even under identical nutritional conditions, however, the epithelia from most organs of rodent species failed to grow (Heckman, 1983). Moreover, cells from several organs are unresponsive to growth stimuli when isolated in a flat configuration on plastic substrata (Pickett *et al.*, 1975; Richman *et al.*, 1976; Heckman *et al.*, 1978). As has been suggested for bovine kidney epithelial cells, which could be grown on the extracellular matrix secreted by PF HR-9 teratocarcinoma cells, many epithelia may require contact with extracellular matrix constituents (Gospodarowicz, 1983).

Even at this early stage in our understanding of growth regulatory mechanisms, it can be concluded that the principles governing the growth-permissive state of epithelial cells differ from those previously elucidated for mesenchymal cells. Although there is evidence that regulation of epithelial growth is subject to configurational constraints (Gospodarowicz *et al.*, 1978; Heckman *et al.*, 1978), the exact constraints that apply to the two major classes of cells may be dissimilar. Key contributions to our current understanding of regulatory mechanisms can be expected from future studies of the biochemical and spatial prerequisites for epithelial growth.

II. Methods of Cell Shape Analysis

Considering the ubiquity of changes in cell shape accompanying transformation, shape appeared to be an intrinsically interesting aspect of this process. By the mid-1970s, classification schemes had been developed that were appropriate for considering shape as a global feature of cells. However, there was a need to develop techniques for imaging and measuring shape as well as to develop variables to encompass the many geometric aspects of shape. The most comprehensive technical approach would involve the embedment and sectioning of cultured cells, the collection of electron micrographs from serial sections, and the subsequent reconstruction of a three-dimensional shape from serial images. The resolution of the technique would be limited by the section thickness to a value of approximately 70 nm. Although this technical approach would be adequate to detect even fine variations in cell structure, our previous studies conducted by thin-sectioning techniques had disclosed difficulties in making quantitative comparisons among cells with this approach. Transects through different cells differed in length, so that it was not possible to make point-for-point comparisons of the thickness profile, and relative measures had to be used (Heckman *et al.*, 1977). This had the disadvantage that information related to the actual dimensions of cytoskeletal elements could be lost.

An alternative possibility for analyzing cell shape took advantage of a technique used previously to study the interactions of cells with proteins adsorbed onto culture substrata (Heckman *et al.*, 1977). In this optical method, termed anodic oxide interferometry, selective interference is conducted in the oxidized surface of a highly reflective metal. A precise pattern of destructive interference in certain wavelengths is established due to the high refractive index of the oxide and the opportunity for light to pass through the oxide layer (Fig. 9). Since the metallic surface was viewed by reflected light, the light path included a minimum of at least two passes through the oxide layer. This optical technique had unique advantages for imaging cultured cells. First, the pattern of interference was perturbed by the adsorption of even thin layers of biological material. It proved possible to resolve a thickness of adsorbed material of less than 3 nm (Heckman *et al.*, 1977). Thus, the thin margins of cells were resolved more effectively than by conventional techniques, and even the finest cytoplasmic projections at the cell periphery were visible. A second favorable feature was that the repetition of interference orders in the image provided some information about the third dimension of cell shape, namely,

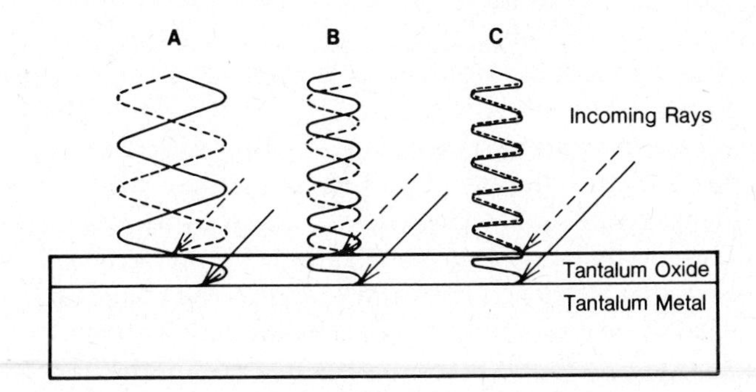

FIG. 9. The anodic oxide interferometer, composed of tantalum metal and its oxide. The variety of wavelengths represented in white light create optical interactions representing complete destructive (A), partially destructive (B), and constructive (C) interference effects.

height. This feature, for example, provided us with an immediate appreciation of the differences in height between fibroblasts and epithelial cells, since cells of most epithelial lines exhibited far more contours than those from fibroblast lines.

Cells cultured on the interferometers grew normally and could be fixed and viewed as geometric objects. Due to the serial interference contours, they resembled topographical maps (Fig. 10). The contrast of these images could be modified by adjusting the thickness of the oxide layer. After optimizing the contrast, we found these images far easier to interpret than the images of stained cells viewed by transmitted light. Further experimentation with these images disclosed that the contours could be viewed at high magnification and traced via a projection device. These images could then be converted into digital form by entering the coordinates of their perimeters into a computer through a digitizing tablet. A comparison of the "paper" and "binary" images showed that the original features of the contours were retained with high fidelity (Fig. 11). Details of the procedures were presented in an earlier publication (Olson *et al.,* 1980). While we also explored the possibility of automating the data collection step, the resolution of images viewed by standard video techniques was not sufficient to preserve information at the level attained by the digitizing tablet.

By 1978, my co-worker, Ann Olson, had developed a number of new variables that described geometrical aspects of cell shape. Since the optical technique provided high-resolution information about the fine structure of the cell edge, a number of these variables evaluated edge features. For example, the perimeter could be "unfolded" and the points along the contour plotted according to their distance from the centroid. The number of projections on the perimeter could be counted by determining the number of "peaks" on such a plot (Fig. 12). Alternatively, the points could be plotted according to their distance from the foci to negate the effect of eccentricity in the shape of the contour. The projections or evaginations along the contour could be considered as triangles and the dimensions of their median, altitude, and base could be measured. Finally, the angles formed by sequential segments of the contour could be measured. The high values recorded at the tips of the projections would elevate the mean of the curvature measurements. These descriptors, along with those representing three-dimensional aspects of cell shape, were unique features of the interferometric methods of analysis (Olson *et al.,* 1980).

Before using the descriptor set in a practical way to assay for cell shape, we wanted to test its efficiency and effectiveness. To see whether there was any formal redundancy among the descriptors, we

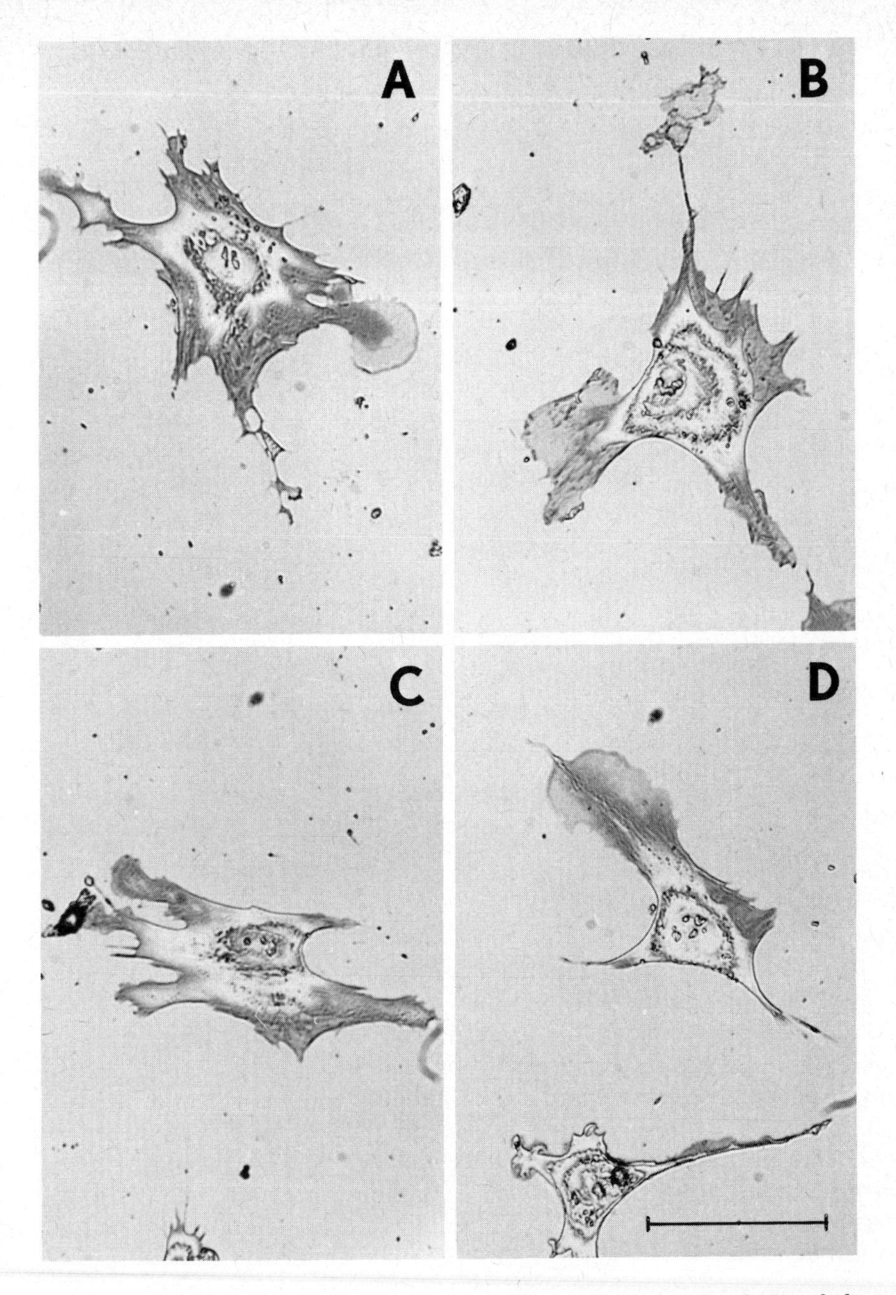

FIG. 10. Images of cells grown on an anodic oxide interferometer. Resolution of edge details was optimized by positioning the first dark fringe at the cell margin. This interference color was dark due to the subtraction of certain wavelengths from the spectrum of reflected light. The appearance of a topographic map was due to the alteration of dark and light colors within the interference order. ×600. Bar = 40 μm.

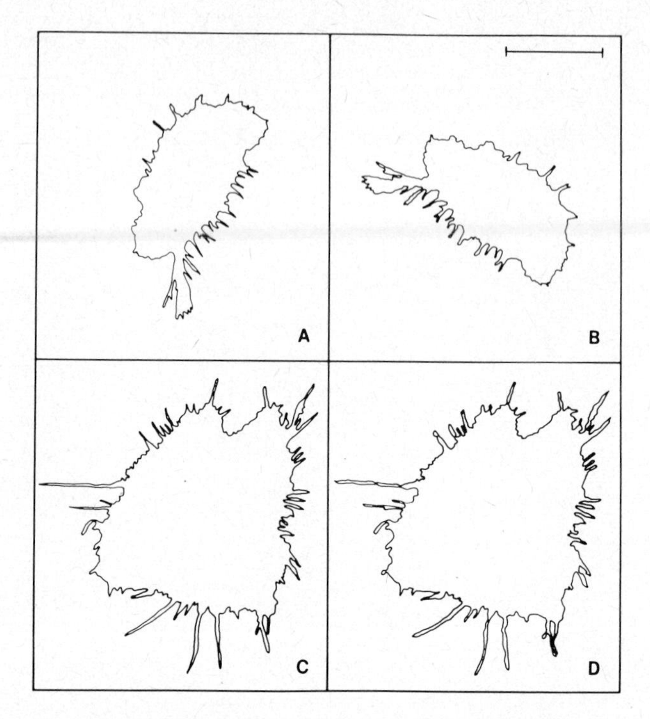

FIG. 11. Comparison of hand-drawn images (A,C) with those reconstructed from points in virtual array storage (B,D). The resolution, approximately 0.5 μm, is close to the limit of light optical resolution. We have previously described the solution to the problem of retaining good resolution while avoiding digitization error (Olson *et al.*, 1980). (A,B) 165S; (C,D) 2C1. ×750. Bar = 40 μm.

determined the extent to which various descriptor values were correlated with one another. The values of the descriptors for a sample of 350 cells were subjected to hierarchical cluster analysis (Fig. 13). This is one of several pattern recognition methods which are well established for mathematical and statistical uses and which can be applied to any large sets of quantitative data. When clustering distances among the descriptor values were plotted, only a few descriptors showed negligible distance values indicating they were redundant. By eliminating the redundant descriptors, we achieved a final set of 34 dimensionless descriptors per contour to be applied to two or three of the contours extracted from the interference image.

A second question of importance concerned the effectiveness of the descriptor set in discriminating among different populations of cells. We tested the power of the descriptor set in a preliminary classification trial conducted with two cell lines which could be distinguished from one another readily on the basis of phase microscopy. By simple bivariate plots of some of the descriptors, we could resolve the objects into two populations with misclassifications occurring in fewer than 5 out of 100 instances. In addition to the examples shown in Fig. 14, there were several other combinations of parameters which could be used to discriminate between the populations with this level of certainty. However, the populations representing normal and abnormal mechanisms of growth control could be identical in origin so that the differences between them were expected to be more subtle. To see whether these

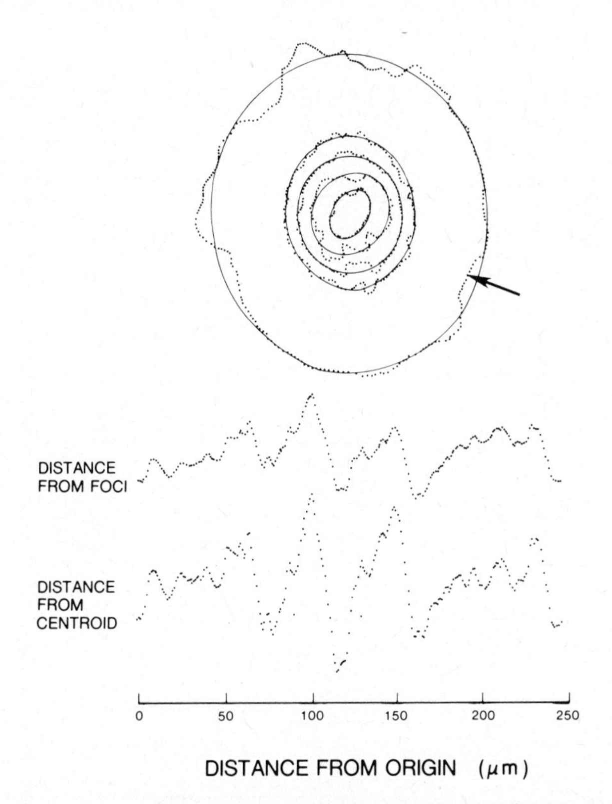

FIG. 12. Enumeration of points collected from contours of a 1000 W cell. The original contours are represented by points. The ellipse representing the model figure, the ellipse of concentration, is shown as a continuous line. The origin of the tracing is indicated in the first contour (arrow). The center of and foci of the first ellipse have been used to "unfold" the contour. This is achieved by calculating the distance between these internal points and each point on the contour. The distance is calculated based on the foci (above) and centroid values (below) and plotted as a function of position on the contour. The means of these distances, FOCI and CENT, respectively, provide information about the degree of deviation of the contour from circularity or ellipticity.

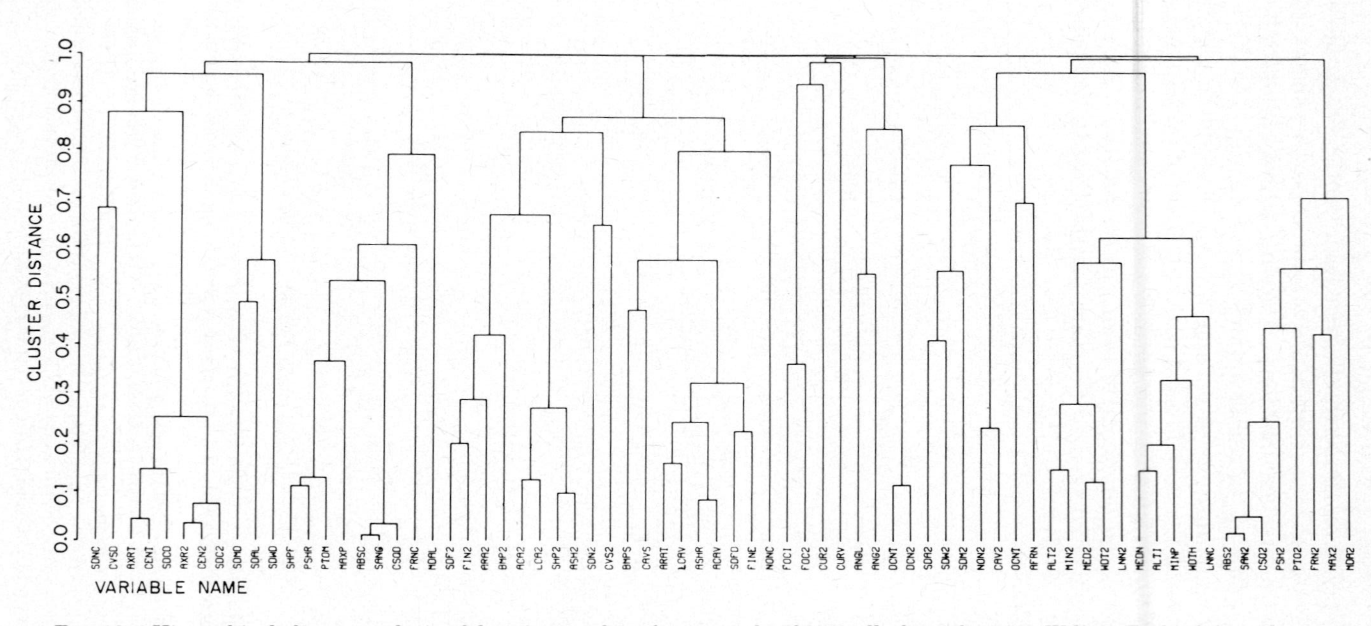

FIG. 13. Hierarchical cluster analysis of descriptor values for a sample of 350 cells from the 1000 W line. Each of the values was autoscaled to reduce its mean to 0 and its standard deviation to 1. Clusters are formed first from the most highly correlated values.

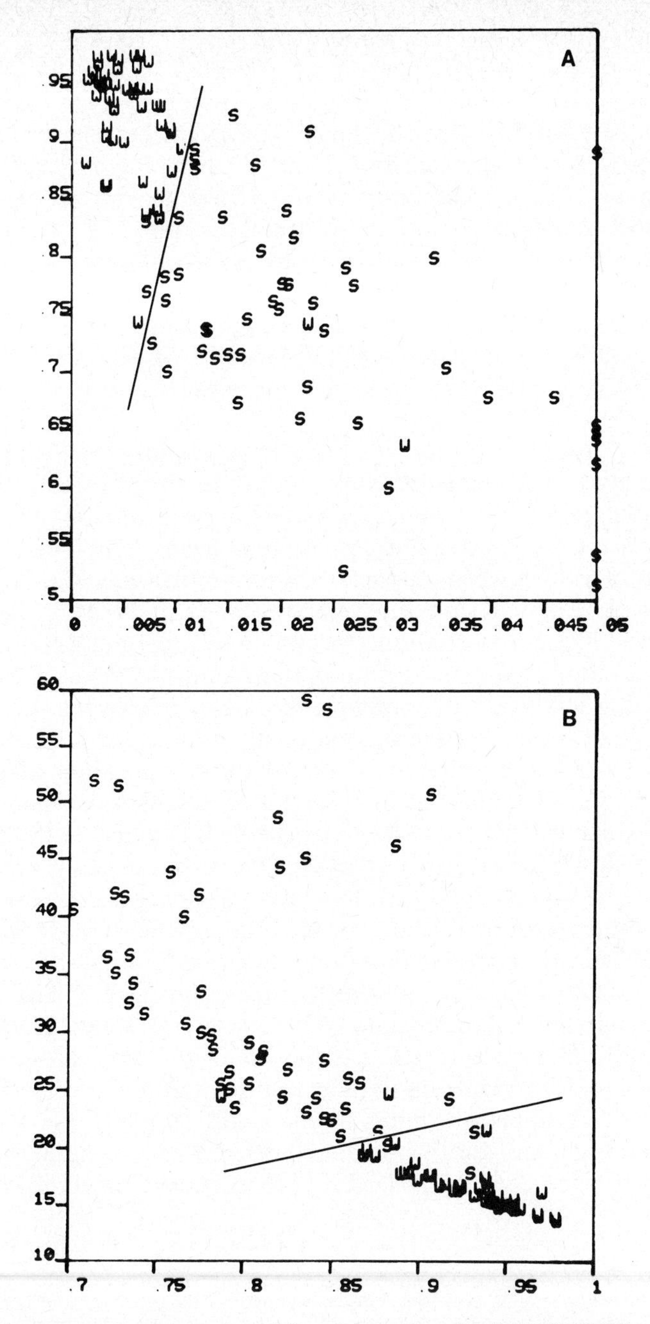

FIG. 14. Bivariate plots of selected descriptor values for the 1000 W (W) and 165S (S) cell lines, showing the decision boundaries. (A) A plot of CENT (ordinate) against ACAV (abscissa) allows the 100-cell sample to be resolved into the correct groups with 4 misclassifications. (B) A plot of SHPF (ordinate) against PSHR (abscissa) allows the cells to be resolved into the correct groups with 5 misclassifications.

differences were detectable, we conducted comparisons of various descriptor values similar to those shown in Fig. 14, but using early and late passages of a single cell line, 1000 W. Even the most efficacious combinations of descriptors only resulted in a separation of approximately 80% of these cells into discrete classes (Fig. 15). In addition to the combinations of descriptors illustrated, any combination plotted for PSHR, PTOM, and MAXP was found to resolve the populations at this level. Certain other combinations of descriptors, especially plots of the most efficacious descriptors (MAXP, PSHR, PTOM, and SDFD) against ASHR, LCAV, or SHPF, also provided correct classifications of 75% of the cells.

The above results suggested that any two samples obtained from the same line might show considerable overlap in the values of most descriptors. There were obvious advantages to be gained by evaluating the statistical differences among such populations. Thus, tests for the equality of variance were conducted on different kinds of cell populations. We anticipated that lines derived separately from the same tissue would exhibit more numerous or more profound differences in the distribution of descriptor values than sublines derived from the same cell line. A three-way comparison of descriptor values generated from cell lines and sublines was conducted, with the results shown in Table V. The data suggested some interesting relationships among the populations. For example, parameters measuring actual dimensions of the cells tended to differ in variance regardless of whether the populations were closely or distantly related. The values of other measures, such as spatial and model figure comparisons, shape factors, and measures of minor irregularities on the contour, tended to have identical variance in closely related populations. The results suggested that the variables, and therefore the cellular features, which were least malleable were related to the complexity of the perimeter. Possibly, this was because the dimensions of the smallest features were based on the organization of the cytoplasm in marginal regions of the cells. We also visualized the numerical values of the features measured for these populations to ensure that their distribution was consistent with the statistical measures. Comparisons like those shown on the BP3-0 and

---

FIG. 15. Bivariate plots of selected descriptor values for early (1) and late (2) passages of the 1000 W line, showing the decision boundaries. (A) A plot of PSHR (ordinate) against MAXP (abscissa) allowed the majority of the cells to be resolved but resulted in 20 misclassifications. (B) A plot of PTOM (ordinate) against SDFD (abscissa) allowed the majority of the cells to be resolved but resulted in 22 misclassifications.

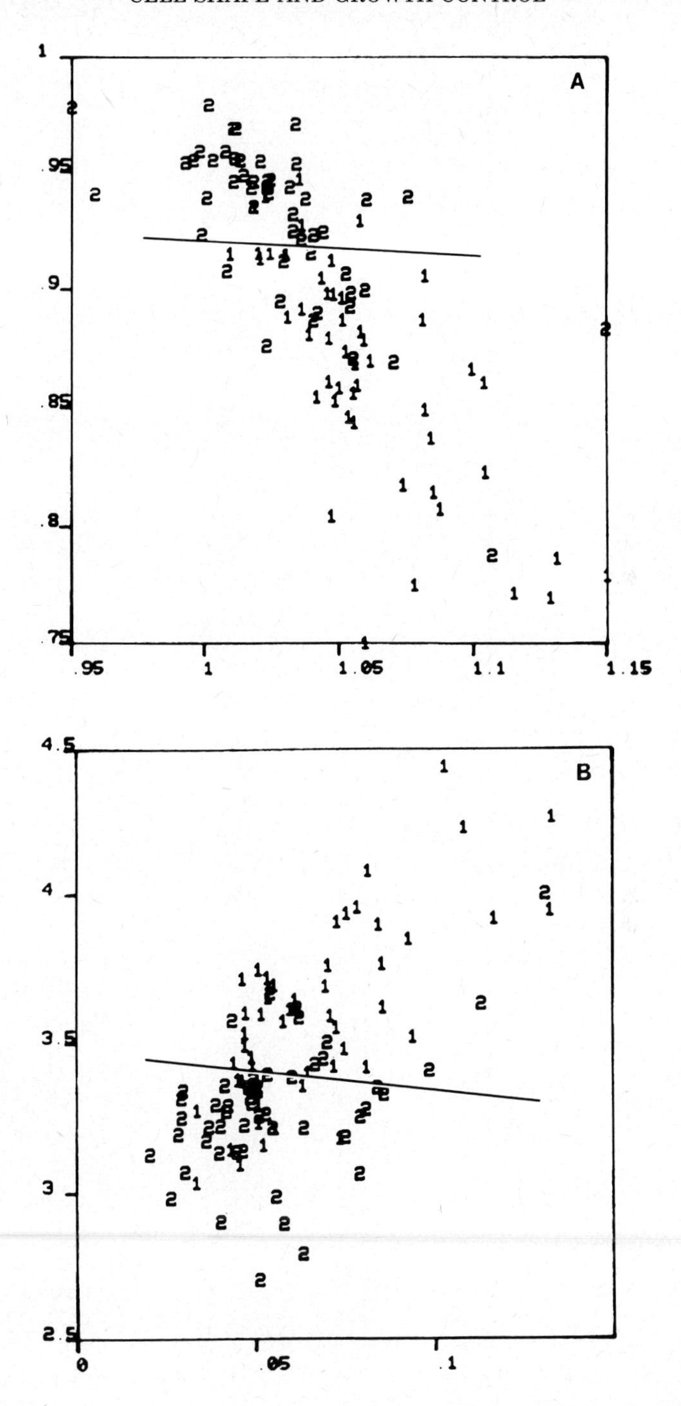
A
1
.95
.9
.85
.8
.75
.95
1
1.05
1.1
1.15
B
4.5
4
3.5
3
2.5
0
.05
.1

TABLE V

THREE-WAY COMPARISON OF THE VARIANCE OF DESCRIPTOR VALUES BY THE CHI-SQUARE TEST[a]

	Descriptor	Class	Critical levels for		
			100 BP_3 vs 100 WT clone 9	100 BP_3 vs 100 WT clone 10	100 WT clone 9 vs 100 WT clone 10
1. OCNT	Number of contours	Dimensioned parameters	0.001	—[b]	0.001
2. AREA	Area, A		0.001	0.001	0.001
3. PERI	Perimeter, P		0.05	—[d]	—
4. MJAX	Major axis, M		—	—	0.05
5. SHPF	P^2/A	Shape factors	0.001	0.001	—[c]
6. PTOM	P/2M		0.025	0.001	—
7. AXRT	M/m, ratio of major and minor axes	Spatial comparisons	0.001	0.001	—
8. AFRN	A/AO, area of contour/area of largest contour		0.005	0.005	—
9. DCNT	Distance from top of cell to centroid/M		—	—	—
10. ANGL	Angle between M and line joining DCNT		—	—	—
11. CURV	Curvature, Σc	Measures related to curvature	—	—[e]	—[e]
12. CSQD	Curvature squared, or "bending energy," Σc^2		—	—	0.025
13. SANG	Absolute angle, $\Sigma\|\alpha\|$		—	0.001	0.025
14. NONC	Number of regions with negative curvature	Measures related to nega-	—	—[e]	0.01
15. FRNC	Fraction of perimeter in these regions		—	—	0.025

16. LNNC	Mean length of regions with negative curvature/M	tive	—	—	—
17. SDNC	Coefficient of variation of length	curvature	0.01	0.005	—
18. BMPS	Number of "bumps"	Measures relat-	—	—	—[e]
19. MEDN	Mean median/M	ed to minor	0.025	—[b]	—
20. SDMD	Coefficient of variation of median	irregularities	—	0.005	—[b]
21. ALTI	Mean altitude/M	of the contour	—	—	—
22. SDAL	Coefficient of variation of altitude		0.05	0.005	—[c]
23. WDTH	Mean width/M		—	—	—
24. SDWD	Coefficient of variation of width		0.005	0.005	—[d]
25. MDAL	Median/altitude (skewness)		—	—[d]	—[e]
26. ARAT	Area of ellipse/area of contour	Measures relat-	0.001	0.001	—[b]
27. CENT	Mean distance from centroid/M	ing to devia-	0.001	0.001	—
28. SDCD	Coefficient of variation of distances from centroid	tion from	0.001	0.001	0.05
29. FOCI	Mean distance from foci/M	circularity	—	0.025	—[b]
30. SDFD	Coefficient of variation of distance	and ellip-	0.005	0.001	0.025
31. FINE	Fraction of contour area contained in ellipse	ticity	0.001	0.001	0.001
32. MAXP	Area of polygon formed by joining maxima/A	Model figure	0.001	0.001	—
33. MINP	Area of polygon formed by joining minima/A	comparisons	0.001	0.001	—

[a] A dash indicates that the populations were not significantly different below $p = 0.05$.
[b] Means significantly different in t test at $p < 0.005$.
[c] Means significantly different in t test at $p < 0.010$.
[d] Means significantly different in t test at $p < 0.025$.
[e] Means significantly different in t test at $p < 0.050$.

WT clone 9 lines (Fig. 16) indicated obviously altered distributions for the feature values found to differ by the chi-square test (Table V).

Additional insight into the statistical basis of the differences between samples was gained by comparing the mean descriptor values for closely and distantly related populations. Comparisons of closely related lines revealed that there were numerous descriptors with equivalent variance that differed with respect to their means. Interestingly, nearly all of these descriptors fell into the classes of shape factors, measures relating to the deviation from circularity (SHPF, ARAT, FOCI), or measures of minor irregularities in the contour (BMPS, SDMD, SDAL, SDWD, and MDAL). Some of the latter features were also found to have different means when distantly related populations were compared (Table V). In this case, however, only a fraction of the total number of descriptors could be examined due to the requirement that the variance of their values be equivalent. Thus, the

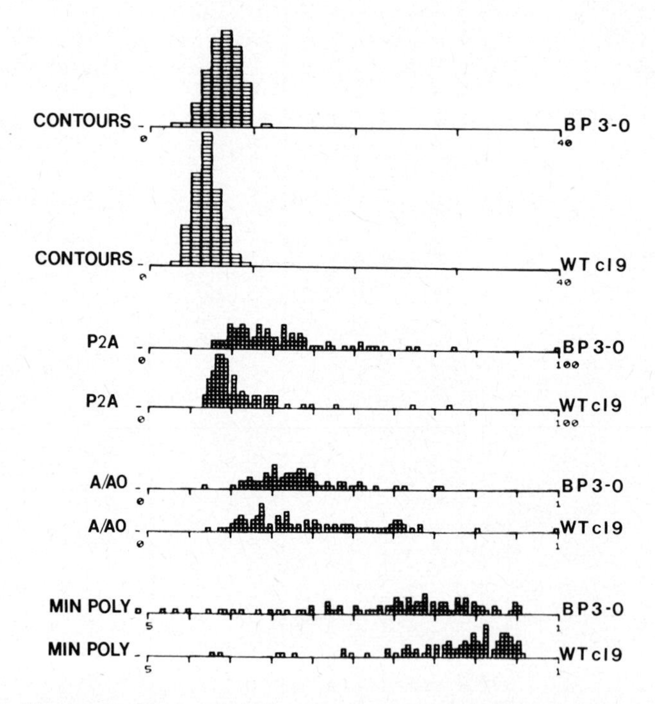

FIG. 16. Frequency histograms illustrating the distribution of selected descriptor values for the cell lines BP3-0 and WT clone 9. For the respective populations, the descriptors and their means are OCNT, 7.7, 6.1; SHPF, 31.4, 21.3; AFRN, 0.364, 0.395; MINP, 0.802, 0.876. All of these values differed in distribution for the two populations (see Table V).

values of the descriptors could be compared more readily among similar populations. The analysis mainly served to indicate whether most of the descriptors could be expected to be useful for discriminating among populations. Except for a few spatial comparisons (DCNT, ANGL) and measures of minor irregularities (ALTI, WDTH) and a single negative curvature descriptor (LNNC), all of the descriptors showed a significant difference in either variance or mean for at least one of the pairwise comparisons. The results suggested that most of the descriptors had a high information content and should be retained in the final set. Similar statistical comparisons carried out for the same lines as initially studied on bivariate plots, 1000 W and 165S, showed that only six descriptors had equivalent variances (MEDN, SDMD, SDAL, MDAL, MINP, CAVS). Thus, these lines appeared remarkably dissimilar.

In early attempts to classify cells from dissimilar populations (Fig. 14), it became clear that the classes were not completely separable by a single decision surface on a bivariate plot. Evidence both from these trials and from the univariate statistical analysis (Table V) suggested that data from a number of descriptors should be employed to optimize the classifications. Many multivariate pattern classification methods, however, were based on the assumption that the populations being sampled were normally distributed in feature space. Some required the additional assumption that the population variances were equal. A multivariate normal distribution would entail the samples falling into a single cluster having a center defined by the mean vector of the descriptor values and a shape defined by the covariance matrix. To visualize the distribution of populations in multidimensional feature space, we obtained descriptor values for cells from five different cell lines. These multivariate data were then transformed by principal component analysis to yield composite variables having the same variance. As viewed on plots of the four principal components, the great majority of cells sampled from a line appeared normally distributed (Fig. 17). Although most samples contained a few cells which occupied outlying positions on the plots, there was little evidence that the lines contained recognizable subsets, consistent with a bimodal distribution. In the one case where a distinct subset was resolved, the population was derived from a primary culture of normal tracheal epithelium and probably included a minor fraction of mesenchymal cells. Thus, the indications from these studies were consistent with an assumption that the *established* epithelial cell lines had a unimodal distribution.

Certain pattern recognition methods allowed us to study the distribution of pattern classes without making any assumptions about the

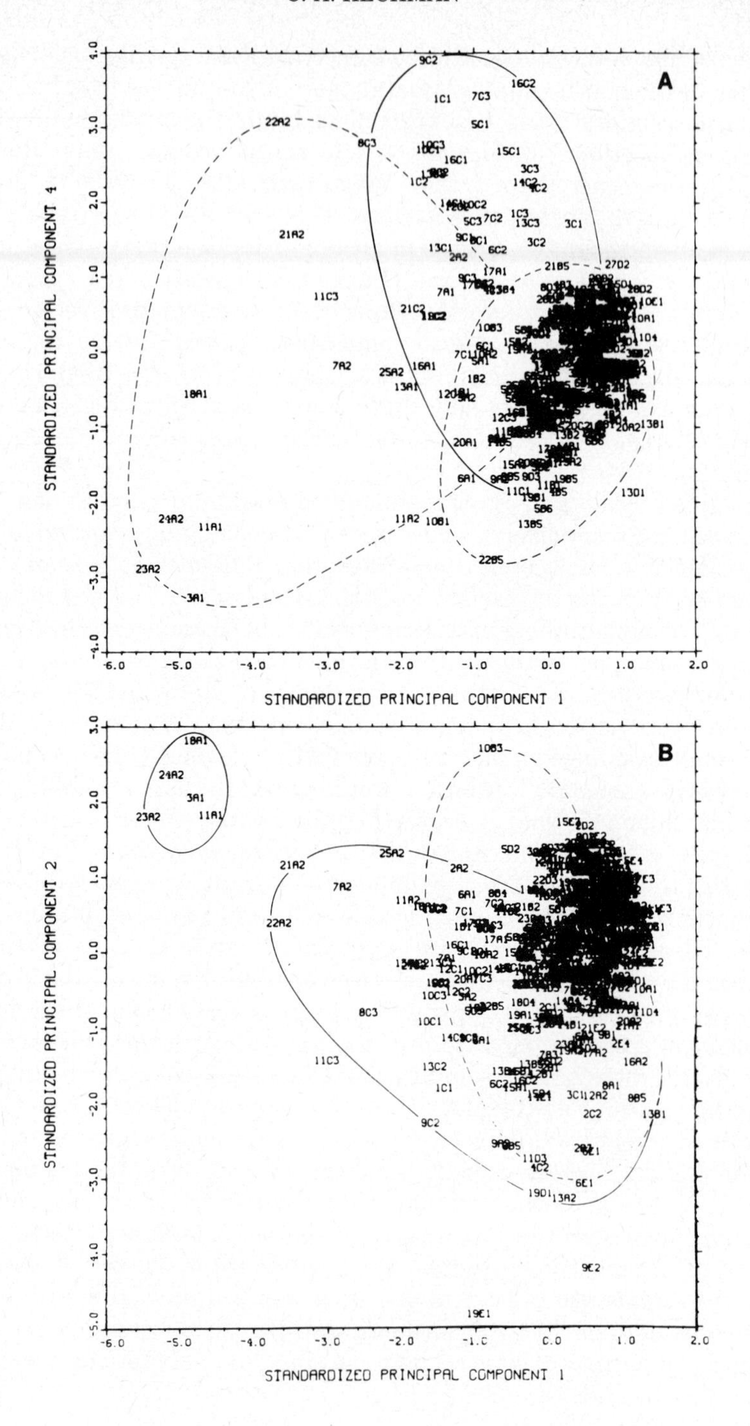
A
STANDARDIZED PRINCIPAL COMPONENT 4
STANDARDIZED PRINCIPAL COMPONENT 1
B
STANDARDIZED PRINCIPAL COMPONENT 2
STANDARDIZED PRINCIPAL COMPONENT 1

data structure of the populations being sampled. Minimum distance classifiers, for example, were based on measurements of Euclidean distance among objects of any class membership, identified by their coordinates in multivariate feature space. These classifiers were advantageous at early stages of our research because we lacked a priori knowledge about the shape characteristics of cultured cells. We investigated the "native" distribution of shape phenotypes using four cell lines that were also of interest in the context of oncogenic transformation. Two of these lines, 1000 W and 165S, became tumorigenic (Marchok *et al.*, 1978) and had been studied previously by myself and co-workers (Heckman and Olson, 1979; Olson *et al.*, 1980; Manger and Heckman, 1982). The third line was nontumorigenic in early passages and closely resembled the 165S line, while the fourth, 8-10-5, was dissimilar from the other lines both in origin and morphology (Fig. 18). When samples of 50 cells from each of these lines were classified by means of a hierarchical clustering routine, 49 of the 8-10-5 cells formed a single cluster (Fig. 19), indicating that they were more similar to one another than to cells from any other line. The fiftieth cell was included in a cluster made up predominantly of 165S cells. A third cluster was comprised of 47 of the 1000 W cells. While these three lines could obviously be discriminated based on cell shape, the results were altered considerably when the 2C5 line was substituted for the 8-10-5 line. While 47 of the 1000 W cells again formed a single cluster, the 2C5 cells were organized into one smaller cluster of 23 cells, with the remainder found in two additional groups intermixed with 165S cells.

The results implied that the 2C5 and 165S populations were widely interspersed in n space. Since the number of cells in the 165S outlier group varied and 1000 W cells could also be included, depending on the decision criterion used, these outliers appeared to have an area of interface with the 1000 W group. The 1000 W and 8-10-5 cells, however, tended to be discretely distributed in n-dimensional space relative to the other two lines, as the maximum inclusion in other clusters was 6%. To see whether the cells which fell outside the major cluster were

---

FIG. 17. Plots showing the distribution of principal component values for 50-cell samples taken from five different epithelial populations. (A) All of the populations sampled are projected as elliptical distributions on a plot of principal components 1 and 4. The distributions of the A (dashed line, large ellipse), B (dashed line, small ellipse), and C (solid line) populations are indicated. (B) Most of the populations sampled are projected as elliptical distributions on a plot of principal components 1 and 2. The distributions of A (solid line) and B (dashed lines) are indicated. A few cells from Group A, however, are projected into an outlying location separate from the main group (upper left).

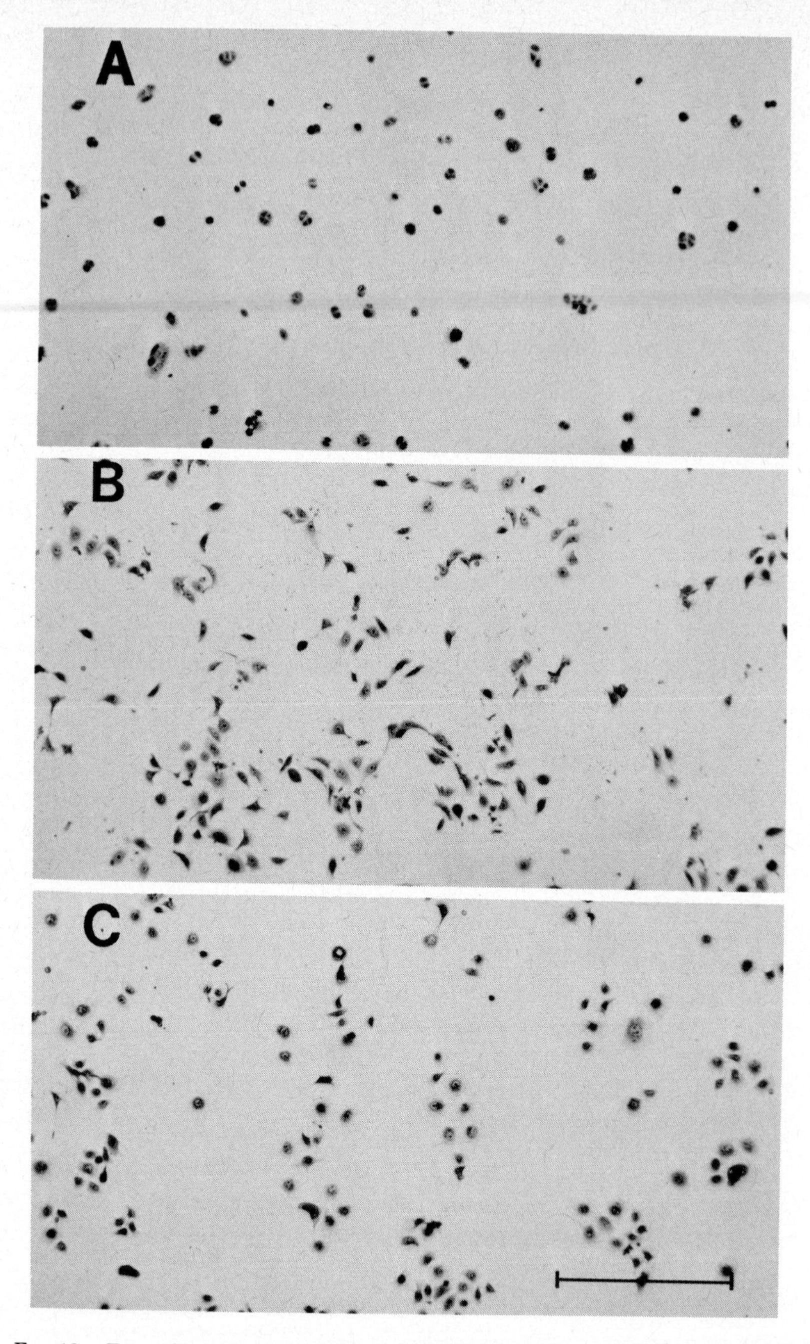

FIG. 18. Examples of three respiratory airway epithelial cell lines, following growth in culture and Giemsa staining. (A) 1000 W and (B) 165S, derived from tissue treated *in vivo* with 7,12-dimethylbenz[*a*]anthracene. (C) 8-10-5, derived from tissue treated *in vitro* with *N*-methyl-*N'*-nitrosoguanidine. ×50. Bar = 0.5 mm.

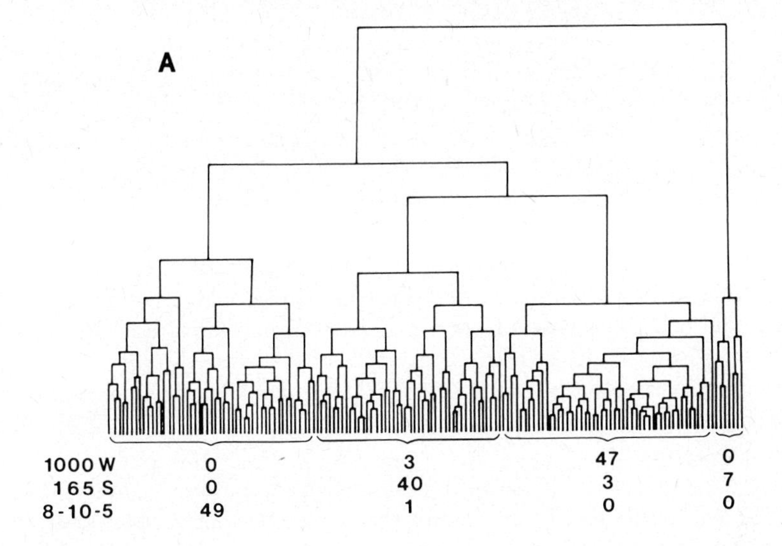

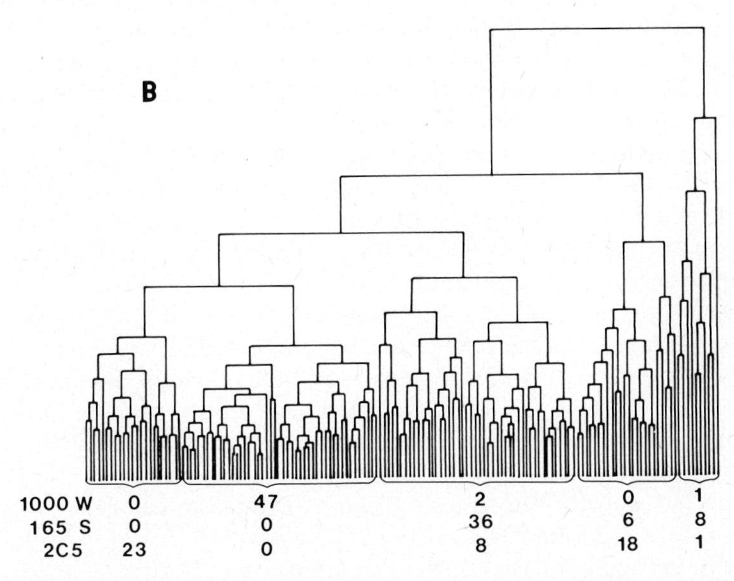

FIG. 19. Hierarchical cluster diagrams representing classifications of cell lines, based on the criterion of minimum centroid-to-centroid distances. (A) An analysis of three dissimilar lines shows that three major groups are formed, each composed of the majority of cells from one cell line. There is also an outlier group containing 165S cells. (B) An analysis of three cell lines, two of which appeared similar by light microscopy, shows that the majority of the 1000 W cells are represented in a single group, as in (A). The majority of the 2C5 cells, however, are grouped with 165S cells, implying that these two cell types were not readily segregated, based on cell shape criteria.

remarkable in any way, we classified populations of 150 cells each from the 1000 W and 165S lines using an algorithm that operated to minimize the within-cluster standard deviation (data not shown). Three of the 1000 W cells, or 2%, were scattered in a cluster containing a majority of 165S cells. On the other hand, 11 of the 165S cells were distributed in various subregions of the 1000 W cluster, while a separate group contained 3 additional 165S outliers. The results suggested that the 165S line was morphologically more variable than the 1000 W line, but confirmed that the distribution of phenotypes in the populations was characteristic of each cell line. Since all shape descriptors were weighted equally in these analyses, however, the results did not have any particular bearing on the question of whether tumorigenic properties on the part of cell lines entailed any common cell shape characteristics.

The cluster analysis raised an interesting question, namely, do dissimilar cell lines have population characteristics such that the membership of the two classes can be separated completely? To test the separability of the populations examined previously by cluster analysis, we used an algorithm which has the unusual property that it indicates at termination whether or not the classes are separable. With this algorithm, least-mean-square error (LMSE), a vector representing the summation of the squared errors is minimized in successive iterations. A function of this vector can be defined, the final values of which are zero if the classes are separable and nonpositive if not. Thus, the algorithm offers a test of separability (Tou and Gonzalez, 1977). Other iterative training schemes for pattern classification (such as the perceptron algorithm) will continue to oscillate ad infinitum if the pattern classes are not separable. In any specific case, it is difficult to tell if the solution is long because it is unattainable or merely because a large but finite number of iterations are required. When we performed the LMSE analysis using a linear decision surface, the results indicated that the 1000 W and 165S populations were nonseparable (Fig. 20). The results were consistent with those from the hierarchical cluster analysis, as expected due to the similarity in axes and decision criteria. For example, we found 95% of the cells classified correctly by LMSE analysis, using all 72 variables calculated from the first two contours and by hierarchical cluster analysis using the eight most efficacious of these descriptors (Olson *et al.*, 1980). However, this did not imply that the populations were inseparable by any criteria, since only linear separations have been attempted.

In the course of developing methods for measuring cell shape, we have revealed new information about the shape characteristics of

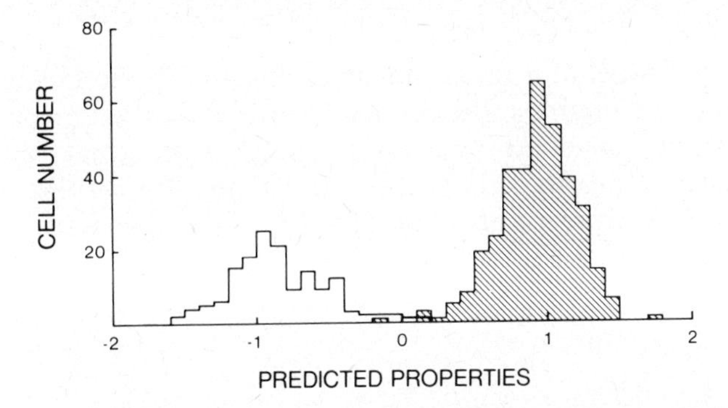

FIG. 20. Least-mean-square error analysis performed to classify 150 165S cells and 350 1000 W cells. At the termination of the iterative algorithm, one of each cell type was classified within the group containing the majority of the other cells. The results suggested the populations could not be separated completely on the basis of linear decision criteria.

cultured cells. First, we found that a large number of different geometrical characteristics appeared useful in defining cell shape. Second, it became clear that certain populations were so dissimilar that they could be distinguished by a variety of different combinations of descriptors. Although the combinations that appeared maximally effective for making comparisons between cell lines varied, they included descriptors from within each class (Table V). The descriptors that appeared to be especially information laden were the shape factors, model figure comparisons, measures relating to the deviation from circularity, and measures of concavities on the perimeter. It is noteworthy that out of these categories, it might be possible to compute all the measures except those relating to the deviation from circularity, based on low resolution data, such as obtained from ordinary optical images.

Further results confirmed that cell lines which appeared dissimilar by eye were readily resolved into the appropriate classes by image analysis and pattern recognition methods. Conversely, lines that appeared similar were readily combined into the same clusters by minimum distance classifiers. It should be emphasized that the unique advantage of these analytical methods was not that we could detect differences among populations, but that we obtained precise quantitative data about the extent of the differences. A fourth major finding was that cells sampled from within a line showed great diversity with respect to shape. In analyzing the 1000 W and 165S cell lines, we found that they differed so radically that 95% of the individual cells sampled were linearly separable on the basis of just two decision criteria. Even when decisions were based on multivariate pattern recognition analyses, such as LMSE and hierarchical clustering, the frequency of misclassifications remained the same. Classification attempts such as these implied that some extreme phenotypes from each population bore a resemblance to cells of the other type. The results, including those from principal component analysis, however, were consistent with the assumption that the populations were normally distributed in *n* space.

III. Shape Changes in Models for Epithelial Transformation

A. *Time-Dependent Changes in Progressing Cell Lines*

Having elucidated some of the shape characteristics of epithelial cell populations, we wanted to conduct further studies on cell lines that could serve as *in vitro* models of transformation. The criteria for selecting these lines were stipulated in advance such that the loss of growth control could be substantiated by empirical tests, especially for tumorigenicity, and the alterations took place on time scale permitting the collection of populations at widely spaced intervals. Several of the F344 rat respiratory tract epithelial lines developed by Marchok and Steele (Marchok *et al.*, 1978; Steele *et al.*, 1979) fulfilled these criteria. In addition, cell lines originated at IARC (International Agency for Research on Cancer) from the liver of BD-IV and BD-VI rats (Montesano *et al.*, 1975) also appeared promising. The 1000 W and IAR 6-7 lines were used in our early attempts to detect shape changes correlated with the development of tumorigenicity. Data on the tumorigenicity of these lines are summarized in Table VI.

The preliminary methodological studies described in Section II were useful in indicating an approach to the problem of classifying progressing cell lines. While early- and late-passage populations from the 1000 W line could be resolved partially on bivariate plots, it seemed likely that the analysis could be optimized by taking advantage of the information content of numerous descriptors. Our early pairwise comparisons of cell lines suggested that no particular disadvantages were encountered in subjecting closely related populations to statistical evaluation. Whereas only a fraction of the descriptor values showed a similar variance when dissimilar cell lines were compared (Table V), we expected a larger proportion to have a similar variance when comparisons were conducted on sequential passages of a single cell line. By

TABLE VI

RESULTS OF TUMORIGENICITY TESTS ON CELL LINES SELECTED FOR STUDIES OF SHAPE CHANGE

	Time in culture (weeks)									
Cell line	20		30		40		50		60	
IAR 6–7 (1–5 × 10^6 cells sc)[a]		0/4 0/3		2/8			9/9 7/7			
1000 W (0.5–1 × 10^6 cells im)[b]						0/2		0/2		4/4

[a] Derived from untreated liver (BDIV strain of rats); Montesano *et al.* (1975).

[b] Derived from 7,12-dimethylbenz[*a*]anthracene-treated tracheal tissue (F344 strain of rat); Marchok *et al.* (1978).

determining the differences in mean and variance for several populations sampled from the 1000 W line, we found that about one-third of the dimensionless descriptors differed in variance. A large number of these descriptors (SHPF, PTOM, NONC, FRNC, SDFD, MAXP, ASHR, PSHR, and ACAV) differed in either variance or mean, or both, for all the populations under consideration. Of the variables that showed equivalent variance in pairwise comparisons, some differed in mean, but there was no single variable set that seemed to provide a basis for resolving more than two populations. For example, the differences between the earliest and latest passages examined could be recognized based on the mean values of CSQD, ALTI, FOCI, and FINE. When middle- and late-passage populations were studied, however, only the descriptors AFRN, DCNT, and CAVS had means which varied while their variance remained the same. In rederiving a sample separately from an adjacent passage of the middle population, however, we found that it differed from the late population in the values of SDNC, MINP, ASHR, and CAVS (Table VII).

Another area in which problems of interpretation could be anticipated was in determining the causal relationships among shape changes. Cell shape was no doubt influenced by such factors as nutritive conditions, the rate of cell attachment and spreading, and the phase of the cell cycle—factors which were difficult either to monitor or to control. That the subset of descriptors differing in variance was highly specific suggested striking changes in overall shape, whether caused by trivial or nontrivial events. It was important to determine whether a gradation of changes, such as those suggested by our electron microscopic observations, could be detected and quantified. The extent of the summed alterations in shape could be quantified by means of the linear least-squares algorithm, which fits variable values

TABLE VII

CHI-SQUARE AND t TESTS PERFORMED ON VARIOUS POPULATIONS DERIVED FROM THE 1000 W CELL LINE

	Descriptor	Class	χ^2					t test			
			W_{18}–W_{50}	W_{39}–W_{50}	W_{38}–W_{50}	W_{50}–WT_9	W_{50}–W_{50}	W_{18}–W_{50}	W_{39}–W_{50}	W_{38}–W_{50}	W_{50}–WT_9
1. OCNT	Number of contours	Dimensioned parameters	—	—	—	—	—	—	—	—	—
2. AREA	Area, A		—	0.005	0.001	0.005	—	0.01	0.01	0.001	0.001
3. PERI	Perimeter, P		0.001	0.001	0.001	0.001	—	0.001	0.001	0.001	0.001
4. MJAX	Major axis, M		—	0.01	0.001	0.001	—	0.01	0.001	0.001	0.001
5. SHPF	P^2/A	Shape factors	0.001	0.001	0.001	0.001	—	0.001	0.001	0.001	0.001
6. PTOM	P/2M		0.001	0.001	0.001	0.001	0.02	0.001	0.001	0.001	0.001
7. AXRT	M/m, ratio of major and minor axes		—	—	—	—	—	—	—	—	0.01
8. AFRN	A/AO, area of contour/area of largest contour	Spatial comparisons	—	—	—	—	—	—	0.001	—	—
9. DCNT	Distance from top of cell to centroid/M		—	—	—	—	—	—	0.01	—	—
10. ANGL	Angle beween M and line joining DCNT		—	—	—	—	—	—	—	—	—
11. CURV	Curvature, Σc	Measures related to curvature	—	—	0.01	0.005	—	—	—	—	—
12. CSQD	Curvature squared, or "bending energy," Σc^2		—	0.005	0.001	0.001	0.02	0.01	0.001	0.001	0.001
13. NONC	Number of regions with negative curvature	Measures related to negative curvature	0.01	0.001	0.001	0.001	—	0.002	—	—	0.01
14. FRNC	Fraction of perimeter in these regions		0.01	0.001	0.001	0.001	—	0.1	0.001	0.001	0.001
15. LNNC	Mean length of negative regions/M		—	—	—	—	—	—	—	—	—
16. SDNC	Coefficient of variation of length		0.02	—	—	—	—	—	—	0.02	0.01
17. BMPS	Number of "bumps"		—	0.001	0.001	—	—	—	—	0.02	0.001
18. MEDN	Mean median/M		—	—	—	—	—	—	—	—	—

19. SDMD	Coefficient of variation of median	Measures related to minor irregularities of the contour	—	0.01	—	—	—	—	—	—	—
20. ALTI	Mean altitude/M		—	—	—	0.01	—	0.01	—	—	—
21. SDAL	Coefficient of variation of altitude		—	—	—	—	—	—	—	—	—
22. WDTH	Mean width/M		—	0.001	0.001	—	—	—	0.001	0.001	0.001
23. SDWD	Coefficient of variation of width		—	—	—	—	—	—	—	—	0.01
24. MDAL	Median/altitude (skewness)		0.02	0.001	0.01	—	—	—	0.001	0.002	—
25. ARAT	Area of ellipse/area of contour		—	—	—	0.005	—	—	—	—	0.01
26. CENT	Mean distance from centroid/M	Measures relating to deviation from circularity and ellipticity	—	—	—	0.01	—	—	—	—	0.001
27. SDCD	Coefficient of variation of distances from centroid		—	—	—	0.001	—	—	—	—	0.001
28. FOCI	Mean distance from foci/M		—	—	—	—	—	0.02	—	—	—
29. SDFD	Coefficient of variation of distance		0.005	0.01	0.005	0.005	—	0.001	0.01	0.01	0.001
30. FINE	Fraction of contour area contained in ellipse		—	—	—	0.001	—	0.02	—	—	0.001
31. MAXP	Area of polygon formed by joining maxima/A	Model figure comparisons	0.001	0.001	0.001	0.001	0.02	0.001	0.001	0.001	0.001
32. MINP	Area of polygon formed by joining minima/A		—	—	—	—	—	—	—	0.01	—
33. ASHR	Area of shrink figure/A		0.005	0.001	0.001	0.001	—	0.001	0.001	0.001	0.001
34. PSHR	Perimeter of shrink figure/P		0.001	0.001	0.001	0.001	0.005	0.001	0.001	0.001	0.001
35. CAVS	Number of major concavities in contour/P	Measures related to major concavities	—	—	—	—	—	—	0.001	0.001	0.001
36. ACAV	Sum of concavity areas/A		0.02	0.02	0.02	0.001	—	—	—	0.02	0.001
37. SDCV	Coefficient of variation of concavity areas		—	—	—	—	—	—	—	—	—
38. LCAV	Area of largest concavity/A		—	0.02	0.02	—	—	—	—	—	0.001

to the vector of predicted properties. This algorithm, LEAST (Begovich and Larson, 1977), was applied to three of the populations examined in the previous analysis (Table VII) to see whether they were separable by linear decision boundaries. By assigning to the vector of predicted properties the value of the cumulative time the population had been cultured, a solution of weight vectors for each of the sample classes was obtained. The fit of this solution to the training set could be visualized by reclassifying each of the objects as though it were an unknown. The resulting classification showed an approximately correct estimation of the mean periods of time elapsed since culturing these populations (Fig. 21).

Since classifications of the training set usually provided superior resolution in comparison to that obtained with unknown sets, we also wanted to analyze some data sets from additional 1000 W populations. In classifying a set of 50 cells taken from passage 39, using the weight vectors generated with known sets, we found that the distribution of predicted time values was broader for this sample than for passage 38 cells, but the means were similar. We also attempted to classify several populations from passage 50, using the time elapsed since plating as the vector of predicted properties, but this trial failed to provide a linear separation even among the training sets (data not shown). Thus, these analyses confirmed that populations collected at widely interspersed times differed with respect to cell shape and implied that these distinctions could not be induced readily by trivial alterations in culture conditions, such as the time allowed for attachment and spreading. Once it was established that meaningful changes occurred in the populations as a whole, we returned to the question of whether they could be attributed to incremental, directional changes in the values of certain descriptors. The possible advantage of identifying such descriptors would be to gain insight about the kinds of cellular properties that were modified in the course of oncogenic transformation. By calculating the coefficients of correlation of descriptor values with time, we found that certain variables were correlated (Table VIII). The values of these variables, however, sometimes failed to rank in the exact order of the progressive time values. Since the values of many variables, especially for the first interference contour, seemed susceptible to short-term influences, it was not feasible to rely solely on a few highly correlated descriptors to measure long-term change.

The first question to be raised about the correlated descriptors was whether their values were correlated with time as a consequence of being normalized to a single dimensioned descriptor which itself changed coordinately with time. Although the dimensioned descriptors

TABLE VIII

CORRELATION COEFFICIENTS OF INDIVIDUAL SHAPE DESCRIPTORS WITH TIME FOR TWO PROGRESSING CELL LINES

	1000 W		IAR 6-7	
	Contour 1	Contour 2	Contour 1	Contour 2
OCNT				
SHPF	(−)xx[a]	xx		
PTOM	(−)xx			
AXRT				
ARAT[b]	(−)x			
AFRN	x		xx	
DCNT				
ANGL				
CURV				
CSQD	xx	(−)xxxx	x	
NONC[c]			x	(−)x
FRNC	xx	(−)x	(−)x	(−)x
LNNC[d]			xxx	x
SDNC				
BMPS[c]	x		xx	
MEDN[d]			x	
SDMD			(−)x	
ALTI[d]			x	
SDAL			(−)x	
WDTH[d]	x	x	xx	
SDWD	(−)x			
MDAL				
CENT[d]				
SDCD	(−)x			
FOCI[d]				
SDFD	(−)x			
FINE[b]	x			
MAXP[b]	(−)x			
MINP[b]				
ASHR[b]	(−)xx	(−)x		
PSHR[c]	xxx	xx		x
CAVS[c]	xxx		xx	(−)x
ACAV[b]	(−)x			
CVSD				
LCAV[b]	(−)x			

[a] x = 0.3–0.4; xx = 0.4–0.5; xxx = 0.5–0.6; xxxx = 0.6–0.7.
[b] Normalized to contour area.
[c] Normalized to length of perimeter.
[d] Normalized to length of major axis of ellipse.

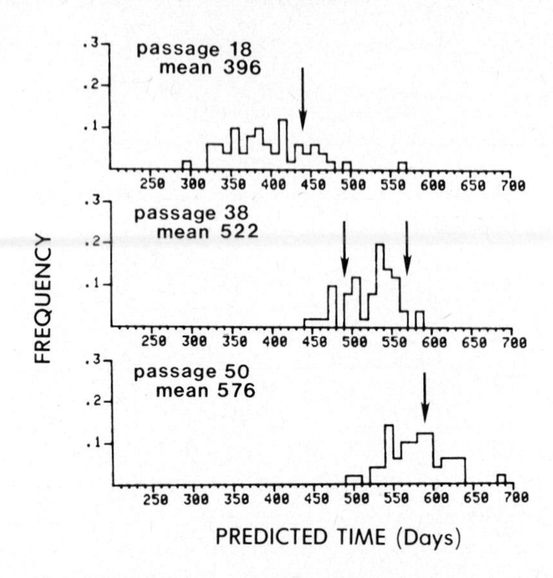

FIG. 21. Least-mean-square error analysis performed on early, middle, and late passages from the 1000 W cell line. The actual times elapsed since the populations were cultured are 369, 522, and 606 days, respectively. The figures shown represent the means recalculated for individual cells from each group. Although partially resolving the populations, the decision boundaries still resulted in substantial areas of overlap between the classifications (arrows).

were not used directly in the analyses, other variables were normalized to their values. It was clear from examining the highly correlated variables in the case of the 1000 W line that at least one descriptor was normalized to each dimensioned descriptor (Table VIII). For the IAR 6-7 cell line, only those descriptors normalized to the lengths of the perimeter or major axis appeared to be correlated with time. Although this suggested that the descriptor, AREA, may have been altered in a nonuniform way so that descriptors incorporating these values also became less well correlated, the values of AREA were actually found to decrease progressively with time. A very important question, aside from determining what factors affected the time-dependent values, was whether there was any similarity between the lists of correlated descriptors obtained for different cell lines. If so, this might indicate that consistent changes in cellular structure occurred in the course of oncogenic transformation. When the lists generated for the 1000 W and IAR 6-7 cell lines were compared, seven of the time-dependent variables found were the same, and the direction of change for six of them, AFRN, CSQD, BMPS, WDTH, PSHR, and CAVS, was identical for the respective contours of the two cell types. The seventh descriptor, FRNC, showed a difference in the direction of change for the comparable first contours of cells from the two lines. This result suggested that certain cellular features were affected in a uniform way during long periods of *in vitro* culture.

Many of the variables that were changed in common appeared to measure invaginations or evaginations of the contours. In the first contour, the proportion of the perimeter composed of straight lengths was reduced concomitantly with an increase in the relative number of major invaginations. The projections on the contour became wider at the base for cells of both lines and, for the IAR 6-7 line, appeared taller as well. Finally, the increasing values of AFRN indicated that cells from later passages rose more steeply from the substrate than those from early passages. Some time-dependent changes also occurred uniquely in one of the two cell lines. The shape factors and model figure comparisons, SHPF, PTOM, ARAT, MAXP, ASHR, and PSHR, for example, were important indicators for the 1000 W line but not for the IAR 6-7 line. More discrete structures, particularly reflected by the number and length of negative curvature regions (NONC, LNNC) and the values and statistics of projection measures (MEDN, SDMD, ALTI, SDAL, and WDTH), appeared more strongly correlated for the IAR 6-7 than for the 1000 W line. Whether the descriptors which changed in common were more generally and uniformly correlated with transformation than indicated here will be determined by analyzing additional cell lines. The comparison of correlation coefficients warranted the important conclusion that common alterations could be detected. A second point of some technical interest was that the majority of descriptors whose values varied as a function of time, especially for the IAR 6-7 line, could only be measured at the level of resolution provided by anodic oxide interferometry. Interestingly, the descriptors encompassing time-related changes in the 1000 W line included some for which this high level of resolution may not be required. The level of assurance anticipated in visualizing changes related to transformation, however, would be greater if the data were collected by interferometric methods.

In order to determine what factors affected the process of shape revision in epithelial cells, it was necessary to define quantitative criteria for change. Although it was expected that separate criteria for each cell line would be needed, it was also an intriguing possibility that a more comprehensive set of criteria, applicable to several lines, could be achieved. Although classification of populations by the LEAST algorithm showed how such criteria might be implemented, it became clear that this algorithm had some problematic aspects. As indicated by the reassignment of vector values to individual 1000 W cells in a preliminary classification trial (Fig. 21), the means of the predicted values did not always correspond to the actual elapsed time in culture. Additional trials conducted with this line showed that there was a trend for the means to be intermediate between the actual time and the midpoint of all the time values, presumably due to the fraction of early and late populations which are misclassified as cells of an intermediate phenotype.

To see if the solution to the quantification problem could be improved upon, we sought to compare the results obtained from the application of different classification methods to the same data set. Since it was obvious that 1000 W populations could be discriminated, the IAR 6-7 line, possibly a more exacting classification problem, was used for further analysis. Hierarchical clustering based on the values of the overall descriptor set was employed to see what type of separation would be achieved. When 50-cell samples from four widely spaced passage intervals were submitted to the clustering algorithm, the results varied depending on the distance criterion specified. Generally more than four major groupings were formed, and cells sampled from all the time points were represented in most groups (Fig. 22). The obvious similarity among these cells overall suggested that improved discrimination and ranking of subpopulations from the IAR 6-7 line would require the use of parametric analyses or probability density functions. A subset of descriptors might be employed in these more sophisticated methods of analysis, since the degree of overlap among populations would be minimized by the operators. To see whether the classification would be affected by choosing a subset of time-correlated values, the data set was again subjected to hierarchical cluster analysis, based in this case on a restricted set of 20 variables. While the cell sample was again assorted into more than four groupings, each was now composed mainly of cells from one or two time points (Fig. 22). This analysis yielded the additional information that the A and B populations were quite similar, although a small group of these cells more closely resembled the majority of C and D cells which were classified into a single large cluster. This kind of distribution would be anticipated if incremental shifts in the values of certain geometrical constructs occurred as a function of time.

These results suggested that, like the 1000 W populations, the IAR 6-7 populations could be classified as a function of time. While the LMSE procedure carried out for three passage intervals of the 1000 W cell line exemplified the kind of resolution that might be expected with a trainable pattern classifier, the first application to the 1000 W line

A

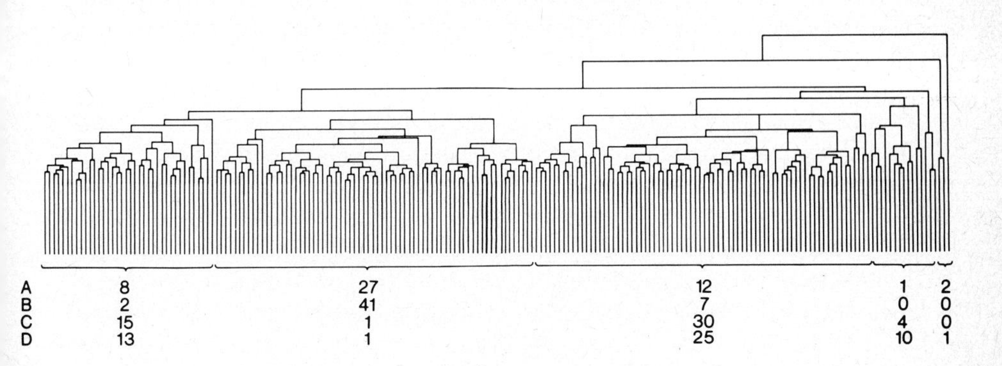

B

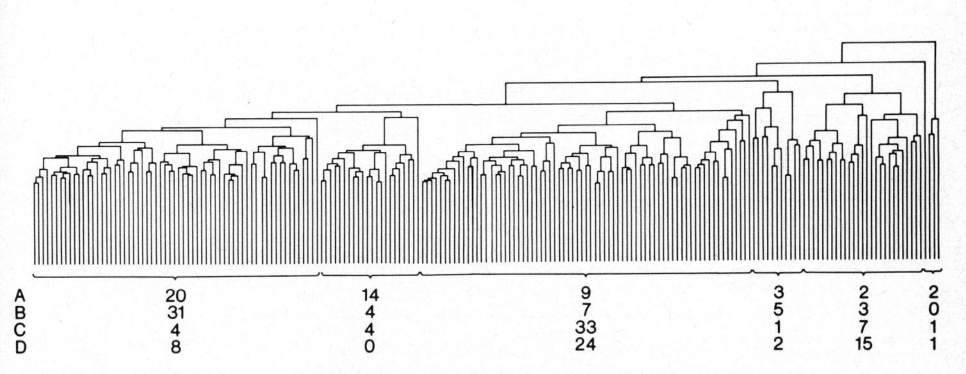

FIG. 22. Hierarchical cluster classification of four passages sampled from the IAR 6-7 cell line. The clustering criterion used was the group average. (A) Using the values of all descriptors, we found two large clusters, each containing two major groups. The first cluster has a group predominating in C and D cells and one with A and B cells. The second cluster has a group predominating in C and D cells and another with mainly D cells. (B) Using 20 descriptors whose values were highly correlated with time, we found three clusters. The first contained two major groups, one containing A and B cells and the other mainly A cells. The next cluster contained a group of C and D cells plus a small group of B cells. The third cluster consisted mainly of D cells.

had been based on the values of all the descriptors, some of which no doubt had little bearing on the analysis. Although there was no formal redundancy among the variables, it was probable that some of their values changed coordinately due to intrinsic properties of the biological structures being measured. For example, depending on the relationship among structural elements making up the projections at the cell periphery, the dimensions of these projections might vary in a coordinate way. Fortunately, this sort of biometric redundancy could be eliminated by the stepwise discriminant analysis, which is based on the covariance statistics of the descriptor values. This analysis employed the F statistic, a measure comparing the variance of the means for different groups with the variance of the combined samples from all groups to select a variable for inclusion in the list. Using the vector of means for all the other variables, excluding the one elected to the list, as a measure of the remaining differences among the samples, the program recalculated the F value for each variable. The second descriptor selected was therefore the one which accounted for the major variance not already corrected for by removal of the first descriptor. This procedure was repeated in a stepwise fashion until the F statistics calculated no longer exceeded the required level of significance, and a short list of nonredundant variables was attained.

When applied to cells sampled from the IAR 6-7 line, the stepwise discriminant analysis yielded a list of seven descriptors. Their identity and F values in the final iteration of the program are shown in Table IX. With the exception of FINE 1, the list consisted entirely of descriptors whose values were well correlated with time. This implied that the most dramatic changes seen in cell shape were a function of time in

TABLE IX

ROBUST DESCRIPTORS SELECTED BY STEPWISE DISCRIMINANT ANALYSIS, OPERATING ON IAR 6-7 CELLS

Descriptor	F value	Correlation coefficient
NONC 1	11.31	0.31
FRNC 1	8.46	−0.34
LNNC 1	15.90	0.55
FINE 1	11.02	0.00
AFRN 1	8.93	0.48
CAVS 2	6.35	−0.32
AFRN 2	4.90	0.50

culture. The exceptional descriptor, FINE 1, had similar mean values for samples derived from time points A, C, and D, and it appeared to discriminate between all of these samples and sample B. Its inclusion in the list was due to its efficacy in making this single discrimination, a feature of the analysis that we had also noticed during an earlier application for distinguishing unrelated populations (Olson *et al.*, 1980). Interestingly, three of the descriptors encompassing information about the negative curvature regions of the contour were selected as having complementary information (Table IX). This finding suggested the difficulty of making predictions about the types of descriptors which might exhibit biometric redundancy.

Although the most precise classifications could be expected from an analysis based on the values of many descriptors, it was clear that the populations of interest to us, including experimental unknowns, could be resolved by use of the list of robust descriptors. To evaluate the effectiveness of classifications based on a stringent descriptor set, we conducted a linear discriminant analysis. This analysis involved specifying a region representing the probability density distribution for each group of cells by solving for the coefficients of equations defining decision boundaries. Although the densities could be visualized easily in three dimensions as spheroidal distributions, the problem could only be approached analytically for n-dimensional patterns (Tou and Gonzalez, 1977). Objects from the training set or those from unknown sets could be classified by solving the equations defining each region in turn. The solution representing the smallest Mahalanobis distance between the unknown and the group mean has the greatest probability of being correct. The Mahalanobis distance (D) is

$$D = (x-m)^{\dagger} C^{-1} (x-m)$$

where x is the object vector, m is the group mean vector, $\dagger$ is the column vector, and C is the covariance matrix. Classification of 200 cells from the IAR 6-7 line by linear discriminant analysis based on the five most robust descriptor values is shown in Table X. The results confirm that the majority of cells from different passage levels can be reclassified correctly, indicating that 50-cell samples from unknown populations will probably be capable of being classified in the same way.

Since the linear discriminant analysis could be carried out based on information contained in a few robust descriptors, and most of these were correlated with time, it could be assumed that the major changes occurring in this line corresponded to some function of time. However, it was questionable whether a linear function was represented. Although the linear discriminant showed that the populations were de-

TABLE X

LINEAR DISCRIMINANT ANALYSIS FOR FOUR PASSAGE LEVELS FROM THE IAR 6-7 LINE

Group (weeks)	Class A	Class B	Class C	Class D
28	27	13	3	7
42	10	37	2	1
54	7	1	31	11
70	6	3	13	28

marcated by linear boundaries, it did not indicate that the boundaries themselves constituted a linear progression. The distinction among the groups may have arisen partly through a change in the population distributions, since it was not known whether these populations were normally distributed. If the error variance was not homogeneous, i.e., the populations differed in distribution, then further analysis would rely on an understanding of the multivariate distribution. Recently, computational resources have become available for addressing this question. If the data structure were represented by nonlinear functions, then the maximum likelihood estimation or other methods of nonlinear regression could be used to fit functions to the data. A few of these algorithms had no requirements as to the distribution of the populations sampled; others were applicable to distributions having any exponential form, such as normal, binomial, Poisson, or gamma distributions. In the likelihood analysis, the data vector extracted from the descriptor values and the parameter vector are used to define one of several time-dependent functions whose derivatives can be specified. In addition to specifying the derivatives of the functions, the source programs (Brown *et al.*, 1983) provide iterative algorithms for improving the estimates of the parameters. The parameter vector determined in the analysis can then be used to assign to each cell a value equivalent to its predicted time in culture. Thus, the IAR 6-7 cells could be classified by an approach similar to that employing vectors for prediction by the LEAST algorithm (Fig. 21). In the likelihood estimation, however, the analysis was halted after the first estimate was obtained so that we could assess the "native" structure of the data rather than the goodness of fit to various logarithmic functions.

When the likelihood estimation was performed for the IAR 6-7 cell samples, based on the parameter vectors obtained with the stringent set of seven descriptors (Table IX), the mean early and late time estimates deviated from the true means as they had in the LEAST analysis for 1000 W cells (Fig. 21). Furthermore, the intermediate predicted mean times showed remarkable deviations from the true means (Table XI). The results seemed to indicate a questionable fit of the data to a log-linear model. To speculate on the source of the discrepancies between actual and predicted times, it seemed likely that by excluding the vast majority of descriptors, we emphasized descriptors which were robust in the dual respects of making discriminations among relatively closely related populations and making time-dependent discriminations. An extreme example of the former capability was the selection of FINE 1 by the stepwise discriminant analysis (Table X), even though its correlation coefficient with time was zero. It was obvious from an examination of the mean values of descriptors on the stringent list that some of them changed in a nonprogressive way between certain time intervals. To see how the likelihood analysis would be affected by these considerations, a more inclusive descriptor list was made up by removing descriptors from the stringent list if their values showed a nonprogressive change for more than one time interval and by adding several descriptors which were also highly correlated with time.

A second likelihood estimation analysis, based on the expanded descriptor list, revealed less pronounced discrepancies between the predicted and true values (Table X). In order to visualize the time course of the changes, a frequency histogram showing the distribution of cells in various predicted time categories was plotted (Fig. 23). Although the greatest difference in means was again found between the intermediate time points, the range of predicted time values appeared to correspond to the ranking of actual times. A valuable aspect of the

TABLE XI

CLASSIFICATION OF IAR 6-7 SAMPLES BY LIKELIHOOD ESTIMATION BASED ON STRINGENT AND EXPANDED DESCRIPTOR SETS

	Time in culture (weeks)		
Sample	True value	Predicted value (stringent set)	Predicted value (expanded set)[a]
A	28	33	32
B	42	25	29
C	55	70	67
D	70	66	66

[a] Descriptors AFRN1, CSQD1, WDTH1, CAVS1, AFRN2, CSQD2, NONC2, MDAL2, CAVS2, and BMPS3.

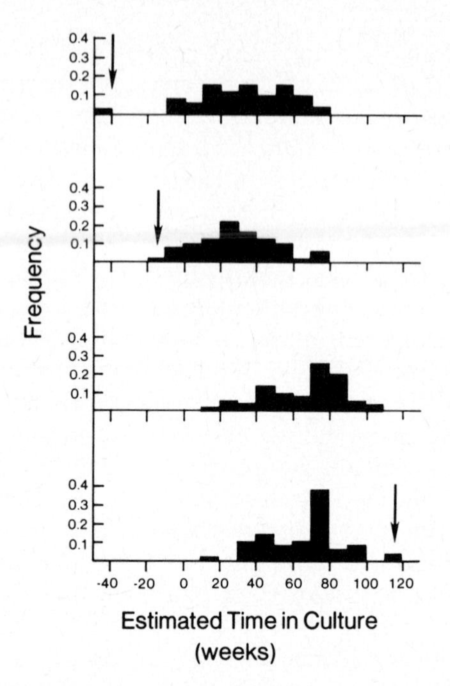

FIG. 23. Noniterative likelihood analysis performed on four passage intervals from the IAR 6-7 line. The progression of shape changes appeared nonlinear. Some of the variations might be attributable to technical factors, since the populations were maintained for differing lengths of time in the United States following origination in France (Montesano *et al.*, 1975). The predicted mean times in culture are shown in Table XI. Individual cells representing the earliest and latest phenotypes (arrows) could be recovered from the data sets for comparison.

reclassification exercise was that it allowed us to identify individual cells as representatives of early or late time points in a progressive series and to contrast the earliest phenotypes, assigned time values of −378 and −66 days, with the latest assigned values of 871 and 796 days (Fig. 23). When the profiles of these cells were reproduced (Fig. 24), they differed in several respects. The first contour extended a considerable distance beyond the second in the early cells but was curtailed in the late cells. The results were reminiscent of earlier studies elucidating the defects in lamellar spreading in transformed fibroblasts (Cherny *et al.*, 1975; Fox *et al.*, 1976). Since even the second and third contours were closer together in the late cells, it was evident that the cells became progressively more rounded at their margins. It seemed possible that this change in form was accompanied by a change in volume. Although reference to the original drawings for the cells shown in Fig. 24 indicated that the number of contours increased from four to five, the correlation coefficient of this descriptor (OCNT) with time was zero. Thus, the number of contours showed no consistent increase with time.

The changes exemplified by these extreme phenotypes could be related to the time-dependent changes in descriptor values. Although structures represented by the values of the major concavities (CAVS) and fine projections (BMPS) appeared more prominent in the contours from early cells, they were actually *less* prevalent when their numbers were expressed in proportion to the actual cellular dimensions. In addition, the shape of structures on the first contour changed to suggest a more continuous curvature, as indicated by increases in the summed curvature square values (CSQD) and in all the dimensions of the projections (WDTH, ALTI, MEDN). In view of the gradual changes in projected cellular area characteristic of the IAR 6-7 line, the descriptors normalized to area might be expected to show progressive changes. Considering that the area-normalized variables all related to the dimensions of major concavities or to model figure comparisons, the absence of changes in these values (Table VIII) suggested that the lamellar structure of the cells "shrank" in proportion to their projected

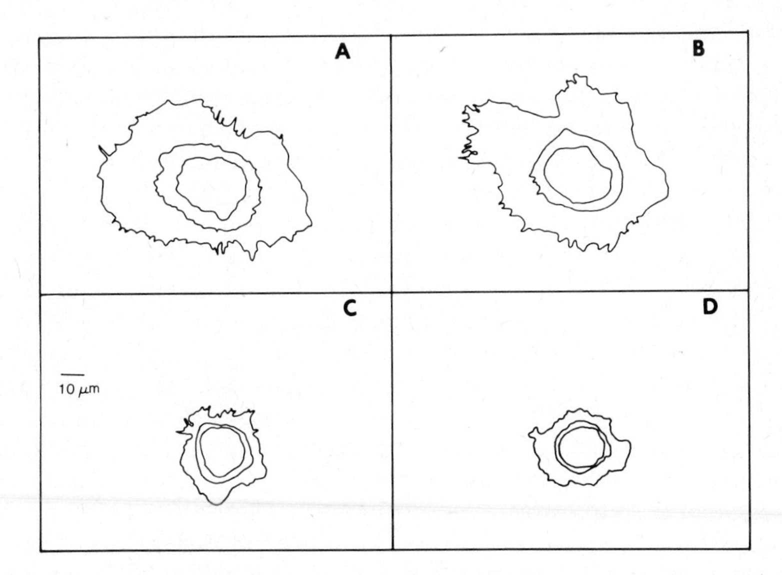

FIG. 24. Interference profiles obtained from cells representing the earliest (A,B) and latest (C,D) IAR 6-7 phenotypes. The position of the cells within the entire sample is indicated on Fig. 23.

area. An interesting implication of this finding with regard to the postulated mechanism of transformation was that the cellular centers dictating major cytoplasmic organization may not have been perturbed. The organizational controls in 1000 W cells, which showed little progressive change in area, may have been affected differently, however, since many of the area-normalized descriptors showed time-dependent changes in this line (Table VIII). To summarize the findings, the two progressing lines, though originating from widely disparate organ sources, showed common changes in shape. Smoothing of invaginations and projections at the cell periphery and rounding of the height dimensions were common responses to oncogenic transformation in the lines. The difference in trends for the variables reflecting more central cellular organization, namely, ASHR, PSHR, CAVS, ACAV, CVSD, and LCAV, indicated that global organization could be perturbed to a greater or lesser degree. Most of these variables changed in a time-dependent way only for the 1000 W line.

B. Global Differences between Normal and Transformed Epithelial Cells

The *in vitro* environment was an indispensable feature of progressing models of cellular transformation. While relieving some of the nutritional and spatial constraints placed on growth *in vivo,* the conditions of *in vitro* culture no doubt imposed a novel set of selective pressures; for example, cells with rapid plating times and short cycle times would tend to be self-selecting. Additional selective processes could be driven by deficiencies in hormones or nutrients, including those that were easily denatured or rapidly depleted by the cells. Due to the likelihood of specific processes occurring *in vitro* that had no *in vivo* counterpart, data on progressing lines were only considered interpretable in reference to *in vivo* studies. Even our initial experiments on the *in vitro* models included a consideration of their relationship to changes occurring *in vivo.* A subline derived from a tumor formed by the 1000 W line, for example, was examined by scanning electron microscopy, along with the progressing populations. Although the range of phenotypes represented among clonal populations of the tumor-derived line was extremely broad, the highly deformable cells that represented a "late" phenotype in the parent line were found among them. Thus, we could rule out the possibility that phenotypic changes were reversed by *in vivo* growth (Heckman and Olson, 1979).

Although we wanted to do similar experiments on phenotypic reversibility using the remaining three progressing lines available, the approach used with the 1000 W line had the drawbacks of being technically difficult and time-consuming without yielding detailed information about cell shape. Moreover, attempts to study the effects of *in vivo* environments on morphological phenotype were likely to encounter the well-known phenomenon of tumor heterogeneity, which implied a broad distribution of phenotypes in tumor-derived populations. We had already made observations consistent with this expectation while studying clonal populations from the tumor-derived 1000 WT line (see Section I,B). The validity of work on the progressing model systems, however, hinged on the assumption that the altered shape characteristics of cell lines was indissociable from their oncogenicity. Although studies on the degree of linkage between cell shape and oncogenicity were complicated by several experimental considerations, an inductive approach could be taken by postulating that the most highly transformed epithelial cells capable of being identified would show no consistent morphological features *vis-à-vis* nontransformed cells. This hypothesis had the virture of being easy to test, since we had already developed cell lines from transplantable tumors which had proved to be highly malignant (Heckman and Olson, 1979; Scott *et al.*, 1979; Manger and Heckman, 1982). More importantly, if the hypothesis was untenable, as was anticipated, then the experiment would provide a new model system that might be useful for elucidating the mechanisms of cytoskeletal or adhesive perturbation. Since this experiment employed a lengthy course of progression *in vivo* to reach a readily identifiable phenotype, it was equivalent to investigating the reversibility of morphological phenotypes under *in vivo* conditions. The results of the experiments clearly showed that pronounced morphological changes were induced in the course of *in vivo* as well as *in vitro* progression (Manger and Heckman, 1982).

Given the unexpected uniformity of lesions in keratin structure seen at the light microscopic level, the alterations characterizing the highly malignant lines appeared at least to involve mechanically interrelated structures. Studies at the ultrastructural level will be required for further clarification of the cellular basis of the anomalies. Recent data have indicated that the aberrant phenotypes showed distinct differences in motility and adhesion (see Section I,B). In conjunction with this behavior, there was extensive overlapping of the margins among tumorigenic, but not among nontumorigenic, cells (Manger and Heckman, 1984). We interpreted these data as an indication that the structural anomalies already recognized were related to the behavior of the cells. This relationship was no doubt implemented by mechanochemical factors mediating cell position and motility. The major challenges now facing us are first, to design experimental methods capable of elucidating the exact mechanistic components involved in this relationship, and second, to test the relationship of these components to the overall control of growth in epithelial cells.

Even further removed from these immediate goals is the elucidation of the genetic basis of transformation. While progress along these lines appeared to promise rapid solutions only a few years ago, it has recently become evident that the cellular basis of growth control has an unforeseen degree of complexity. Although great strides have been made in establishing the nucleotide sequence of the retroviral oncogenes, the mechanisms by which these viruses accomplish cell transformation are just beginning to be uncovered. For the most exhaustively studied tumors of human origin, transforming genes of this type have only been detected in 20% of them by the assays currently in use (Land *et al.,* 1983). Thus a multitude of causes may be invoked for transformation at the level of molecular genetics.

From our studies to date, it was possible to delineate some common perturbations of cell form that accompanied transformation in the epithelial model systems. Specific features affected in this process included a thickening of the lamellar cytoplasm and a corresponding relaxation in the shape of peripheral projections. Although the image analysis/pattern recognition methods offered the main approach to quantifying these changes, we noted similar structural modifications in examining cells by scanning electron microscopy (Manger and Heckman, 1982). The alterations appeared to extend to cell–cell interactions as well. In this case the changes found were suggestive of increased cell deformability (Heckman and Olson, 1979; Manger and Heckman, 1982), a property thought to relate to the reduction in adhesiveness and the increase in rates of motility. A single finding from the molecular biology of viral oncogene models (see Section I,A) appeared to have a direct bearing on our observations. The membrane localization and subsequent activation of the *src* gene product suggested that membrane-bound substrates could occupy a pivotal position in implementing an altered growth program. In fact, the entire spectrum of shape and behavioral changes in the epithelial model system was consistent with a structural lesion arising in the plasma membrane. Thus, the configuration of the membrane is considered a fruitful area for future experimental inquiry.

IV. Conclusions

Evidence from several laboratories suggests that the growth of normal epithelial cells is more stringently constrained than the growth of mesenchymal cells. Although a rich nutritional environment supported the net growth of epithelia from certain organs of the rat, this condition was not sufficient to support growth in all cases. Even those epithelia which grew in isolation from other tissue constituents often failed to grow when isolated from adjacent cells. The constraints that account for this effect may involve specific spatial or inductive requirements. Although these growth restrictions were relieved in populations exposed to chemical carcinogens or were occasionally spontaneously relieved, the resulting cell lines had an indeterminate (or immortal) life span. The vast majority of cell lines originating from rat respiratory tract and liver became tumorigenic over the course of *in vitro* growth and maintenance, suggesting a relationship between *in vitro* longevity of epithelial lines and their tumorigenicity.

The *in vitro* properties of cells from the rat liver and respiratory tract have been thoroughly characterized beginning with the studies of Montesano *et al.* (1975) and Marchok *et al.* (1978). Even when not frankly tumorigenic, these lines tended to progress under *in vitro* conditions of growth, i.e., they eventually became tumorigenic. This process bears an interesting resemblance to the period of *in vivo* latency which precedes the rapid development of a tumor mass. In studying the process, we found that progressive changes in tumorigenicity were accompanied by a subtle revision of cell form. This finding was of interest in light of the pronounced changes in cell shape that occurred in mesenchymal models for cell transformation. Compelling reasons for finding these changes of interest also arose from the implications of experiments on mesenchymal cell growth by Folkman and Moscona (1978), Westermark (1978), and Ponten and Stolt (1980). These studies described a relationship between the ability of cells to spread out on artificial surfaces and their growth rate. The inhibition of growth by steric restrictions was shown to be independent of actual cell–cell contact. The data of Ponten and Stolt, in particular, suggested an alternative explanation, that cells have a refined and self-contained mechanism for "sensing" the space allotted for spreading.

In defining phenotypic extremes for the progressing epithelial lines, we drew a parallel with the phenotypic expression of the temperature-sensitive *src* gene in models of viral transformation. The extreme phenotypic forms are referred to as "early" and "late," in reference to their course of appearance in cultures. The late progressing phenotype could also be identified after a prolonged course of *in vivo* selection. Populations of this type showed radical changes in cell configuration and behavior—in particular, adhesive behavior. These results were consistent with what one might predict based on the hypothesis that cytoarchitectural organization was linked to growth control. Early on, this idea was termed the gestalt hypothesis (G. Blumenthal, personal communication). The term admirably suggests the consensus-forming function of the cytoarchitectural configuration in relation to a growth control center. Considering the subcellular organization of anchorage-dependent cells, a change in the coordinated structure could be induced by changes in the adhesive plaques, the plasma membrane, or the interior cytoskeletal components. The quantitative assay we have developed to detect transformation-related changes in cell configuration may also be useful for elucidating the cellular basis of these changes. Although it must be recognized that the tie between oncogenic potential and morphological alterations in cells of epithelial origin may be quite indirect, it is equally possible that a mechanistic link exists such as has been demonstrated for mesenchymal cells.

The results summarized here indicate a continuous process of shape change in progressing epithelial cell lines. The two lines analyzed to date exhibit patterns of change that are similar although not identical. The cells of later passages tend to rise more steeply from the substrate, and the features of their perimeters become more relaxed. The changes most consistently correlated with transformation would be difficult to detect except by the high resolution conferred by use of interference principles. It is possible that these techniques, which have proved effective for discriminating among similar populations of cultured cells, will lead to a mathematical basis for categorizing cells within a gradient of progressive changes.

In experiments planned to shed light on the biophysical integration of cytoarchitectural structures, population statistics similar to those used for macroscopically identifiable individuals appear to be applicable to populations sampled from cultured cell lines. To date, experiments such as those of Ponten and co-workers have been difficult to perform with cells of epithelial origin due to their more stringent general requirements for growth. However, an improved understanding of growth requirements may permit the parallel experiments to be done. Thus it is difficult to determine at this time whether the cytoplasm of epithelial cells also includes an organizational network capable of receiving input about the nature of neighboring surfaces. The questions of how spatial information is interpreted in mesenchymal cells, whether a failure of the sensory mechanism can account for uncontrolled growth, and whether the concepts apply equally to the regulation of growth in the epithelia remain to be answered.

ACKNOWLEDGMENTS

I am grateful to my co-workers, Drs. Ann Campbell Olson, Ronald L. Manger, and Nancy M. Larson, for generously providing specific items of data and to Drs. A. Marchok,

R. Montesano, and V. Steele for providing cell lines. Early phases of this work were conducted at Oak Ridge National Laboratory. Stella Perdue, Aguan Wei (ORNL), Pia Jackson-Eshiett, Mark Stone, Christina McHenry, Jane Trumbull, Pat Geller (BGSU), and the Instructional Media Center provided excellent assistance in preparing materials for publication. This research was supported by Grants PCM 81-10597 and 82-18087 from the National Science Foundation and by BGSU Biomedical Research Grant 8317. Studies at the International Agency for Research on Cancer, Lyon, France, were supported by the International Cancer Research Data Bank of the National Cancer Institute, National Institutes of Health, U.S.A. (Contract No. NO1-CO-65341, International Cancer Research Technology Transfer with the Union Internationale Contre le Cancer). Work at Oak Ridge National Laboratory was supported by the U.S. Department of Energy under contract W-7405-eng-26 and by the Environmental Protection Agency under interagency agreement 40-484-74 with the Department of Energy.

References

Adkins, B., Hunter, T., and Sefton, B. M. (1982). *J. Virology* **43,** 448–455.

Allen, T. D., Iype, P. T., and Murphy, M. J. (1976). *In Vitro* **12,** 837–844.

Ambros, V. R., Chen, L. B., and Buchanan, J. M. (1975). *Proc. Natl. Acad. Sci. U.S.A.* **72,** 3144–3148.

Amini, A., and Kaji, A. (1983). *Proc. Natl. Acad. Sci. U.S.A.* **80,** 960–964.

Amos, H., Levanthal, M., Chu, L., and Karnovsky, M. J. (1976). *Cell* **7,** 97–103.

Anderson, S. M., and Hanafusa, H. (1982). *Virology* **121,** 32–50.

Asch, B. B., Medina, D., and Brinkley, B. R. (1979). *Cancer Res.* **39,** 893–907.

Bader, J. P. (1972). *J. Virology* **10,** 267–276.

Ball, E. H., and Singer, S. J. (1981). *Proc. Natl. Acad. Sci. U.S.A.* **78,** 6986–6990.

Bannikov, G. A., Guelstein, V. I., Montesano, R., Jint, I. S., Tomatis, L., Troyanovsky, S. M., and Vasiliev, J. M. (1982). *J. Cell Sci.* **54,** 47–67.

Barbacid, M., and Lauver, A. V. (1981). *J. Virology* **40,** 812–821.

Barbacid, M., Beemon, K., and Devane, S. G. (1980). *Proc. Natl. Acad. Sci. U.S.A.* **77,** 5158–5162.

Begovich, C. L., and Larson, N. M. (1977). A User's Manual for the Pattern Recognition Code RECOG-ORNL. ORNL/CSD/TM-21. Natl. Tech. Info. Serv., Springfield, Virginia.

Bister, K., Lee, W. H., and Duesberg, P. H. (1980). *J. Virology* **36,** 617–621.

Borek, C., and Fenoglio, C. M. (1976). *Cancer Res.* **36,** 1325–1334.

Branton, P. E., Lassam, N. J., Downey, J. F., Yee, S. P., Graham, F. L., Mak, S., and Bayley, S. T. (1981). *J. Virology* **37,** 601–608.

Breitman, J., Neil, C., Moscovici, C., and Vogt, P. K. (1981). *Virology* **108,** 1–12.

Brinkley, B. R., Beall, P. T., Wible, L. J., Mace, M. L., Turner, D. S., and Cailleau, R. M. (1980). *Cancer Res.* **40,** 3118–3129.

Brown, M. B., Engelman, L., Frane, J. W., Hill, M. A., Jennrich, R. I., and Toporek, J. D. (1983). *In* "BMDP Statistical Software" (W. J. Dixon, ed.), pp. 318–322. Univ. Calif. Press, Berkeley.

Brugge, J. S., Erikson, E., and Erikson, R. L. (1981). *Cell* **25,** 363–372.

Butel, J. S., Dudley, J. P., and Medina, D. (1977). *Cancer Res.* **37,** 1892–1900.

Cassman, M., and Vetterlein, D. (1974). *Biochemistry* **13,** 684–689.

Chacko, S., Conti, M. A., and Adelstein, R. S. (1977). *Proc. Natl. Acad. Sci. U.S.A.* **74,** 129–133.

Cherny, A. P., Vasiliev, J. M., and Gelfand, I. M. (1975). *Exp. Cell Res.* **90,** 317–327.

Cloyd, M. W., and Bigner, D. D. (1977). *Am. J. Pathol.* **88,** 29–44.
Collett, M. S., and Erikson, R. L. (1978). *Proc. Natl. Acad. Sci. U.S.A.* **75,** 2021–2024.
Collett, M. S., Purchio, A. F., and Erikson, R. L. (1980). *Nature (London)* **285,** 167–169.
Cooper, J., and Hunter, T. (1983). *J. Biol. Chem.* **258,** 1108–1115.
Cooper, J. A., Bowen-Pope, D. F., Raines, E., Ross, R., and Hunter, T. (1982). *Cell* **31,** 263–273.
Courtneidge, S. A., and Bishop, J. M. (1982). *Proc. Natl. Acad. Sci. U.S.A.* **79,** 7117–7121.
Daniel, J. C., and Adelstein, R. S. (1976). *Biochemistry* **15,** 2370–2377.
David-Pfeuty, T., and Singer, S. J. (1980). *Proc. Natl. Acad. Sci. U.S.A.* **77,** 6687–6691.
Doolittle, R. F., Hunkapiller, M. W., Hood, L. E., Devare, S. G., Robbins, K. C., Aaronson, S. A., and Antoniades, H. N. (1983). *Science* **221,** 275–277.
Downward, J., Yarden, Y., Mayes, E., Scrace, G., Totty, N., Stockwell, P., Ullrich, A., Schlessinger, M., and Waterfield, D. (1984). *Nature (London)* **307,** 521–523.
Eckhart, W., Hutchinson, M. A., and Hunter, T. (1979). *Cell* **18,** 925–933.
Edelman, G. M., and Yahara, I. (1976). *Proc. Natl. Acad. Sci. U.S.A.* **73,** 2047–2051.
Emerman, J. T., and Pitelka, D. R. (1977). *In Vitro* **13,** 316–328.
Emerman, J. T., Enami, J., Pitelka, D. R., and Nandi, S. (1977). *Proc. Natl. Acad. Sci. U.S.A.* **74,** 4466–4470.
Emerman, J. T., Burwen, S. J., and Pitelka, D. R. (1979). *Tissue and Cell* **11,** 109–119.
Feldman, R., Hanafusa, T., and Hanafusa, H. (1980). *Cell* **22,** 757–765.
Feldman, R. A., Wang, L. H., Hanafusa, H., and Balduzzi, P. C. (1982). *J. Virology* **42,** 228–236.
Folkman, J., and Greenspan, H. P. (1975). *Biochim. Biophys. Acta* **417,** 211–236.
Folkman, J., and Moscona, A. (1978). *Nature (London)* **273,** 345–349.
Fox, C. H., Dvorak, J. A., and Sanford, K. K. (1976). *Cancer Res.* **36,** 1556–1561.
Franke, W. W., Schmid, E., Brietkreutz, D., Luder, M., Boukamp, P., Fusenig, N.E., Osborn, M., and Weber, K. (1979). *Differentiation* **14,** 35–50.
Ghysdael, J., Neil, J. C., Wallbank, A. M., and Vogt, P. K. (1981). *Virology* **111,** 386–400.
Goldman, R. D., Chang, C., and Williams, J. F. (1974). *Cold Spring Harbor Symp.* **39,** 601–614.
Gospodarowicz, D. (1983). *Proc. 41st Ann. Meeting Elect. Micros. Soc. Am.,* Phoenix, Arizona, pp. 664–667.
Gospodarowicz, D., Greenburg, G., and Birdwell, C. R. (1978). *Cancer Res.* **38,** 4155–4171.
Hanafusa, H. (1977). Cell transformation by RNA viruses. *In* "Comprehensive Virology 10" (H. Fraenkel-Conrat and R. P. Wagner, eds.), pp. 401–483. Plenum, New York.
Handleman, S. L., Sanford, K. K., Tarone, R. E., and Parshad, R. (1977). *In Vitro* **13,** 526–536.
Heckman, C. A. (1983). *In Vitro* **19,** 31.
Heckman, C. A., and Olson, A. C. (1979). *Cancer Res.* **39,** 2390–2399.
Heckman, C. A., Vroman, L., and Pitlick, F. A. (1977). *Tissue and Cell* **9,** 317.
Heckman, C. A., Marchok, A. C., and Nettesheim, P. (1978). *J. Cell Sci.* **32,** 269–292.
Heckman, C. A., Manger, R. L., and Greenway, C. (1985). (Submitted).
Hennings, H., Michael, D., Cheng, C., Steinert, P., Holbrook, K., and Yuspa, S. H. (1980). *Cell* **19,** 245–254.
Hunter, T., and Sefton, M. B. (1980). *Proc. Natl. Acad. Sci. U.S.A.* **77,** 1311–1315.
Kalckar, H. M., Ullrey, D., Kijomoto, S., and Hakomori, S. (1973). *Proc. Natl. Acad. Sci. U.S.A.* **70,** 839–843.

Karasaki, S., Simard, A., and deLamirande, G. (1977). *Cancer Res.* **37,** 3516–3525.
Kawai, S., and Hanafusa, H. (1971). *Virology* **46,** 470–479.
Kawai, S., Yoshida, M., Segawa, K., Sugiyama, H., Ishizaki, R., and Toyoshima, K. (1980). *Proc. Natl. Acad. Sci. U.S.A.* **77,** 6199–6203.
Kraehenbuhl, J. P., Suard, Y., Racine, L., and Hauptle, M. T. (1982). *Prog. Clin. Biol. Res.* **91,** 389–401.
Krueger, J. G., Garber, E. A., Goldberg, A. R., and Hanafusa, H. (1982). *Cell* **28,** 889.
Land, H., Parada, L. F., and Weinberg, R. A. (1983). *Science* **222,** 771–778.
Levinson, A. D., Opperman, H., Levintow, L., Varmus, H. E., and Bishop, J. M. (1978). *Cell* **15,** 561–572.
Levinson, A. D., Courtneidge, S. A., and Bishop, J. M. (1980). *Proc. Natl. Acad. Sci. U.S.A.* **78,** 1624–1628.
Manger, R. L., and Heckman, C. A. (1982). *Cancer Res.* **42,** 4591–4599.
Manger, R. L., and Heckman, C. A. (1984). *Cancer Res.* **44,** 688–696.
Marceau, N., Noel, M., and Deschenes, J. (1982). *In Vitro* **18,** 1–11.
Marchok, A. C., Rhoton, J. C., and Nettesheim, P. (1978). *Cancer Res.* **38,** 2030–2037.
Martin, G. M., Ogburn, C. E., and Sprague, C. A. (1981). *In* "Aging: A Challenge to Science and Society" (D. Danon, N. W. Shoek, and M. Marois, eds.), Vol 1, pp. 124–135. Oxford University Press, New York.
Marx, J. (1984). *Science* **223,** 673–676.
Michalopoulos, G., and Pitot, H. C. (1975). *Exp. Cell Res.* **94,** 70–78.
Miettinen, A., and Virtanen, I. (1978). *In* "Protides of the Biological Fluids. Proceedings of the Twenty-Sixth Colloquium" (H. Peeters, ed.), pp. 511–515. Pergamon Press, New York.
Montesano, R., Saint Vincent, L., Drevon, D., and Tomatis, L. (1975). *Int. J.Cancer* **16,** 550–558.
Montesano, R., Drevon, C., Kuraki, T., Saint Vincent, L., and Handleman, S., Sanford, K. K., DeFeo, D., and Weinstein, I. B. (1977). *J. Natl. Cancer Inst.* **59,** 1651–1656.
Neil, J. C., Ghysdael, J., and Vogt, P. K. (1981). *Virology* **109,** 223–228.
Nigg, E. A., Cooper, J. A., and Hunter, T. (1983). *J. Cell Biol.* **96,** 1601–1609.
Olson, A. C., N. M. Larson, and Heckman, C. A. (1980). *Proc. Natl. Acad. Sci. U.S.A.* **77,** 1516–1520.
Oppermann, H., Levinson, A. D., and Varmus, H. E. (1981a). *Virology* **108,** 47–70.
Oppermann, H., Levinson, W., and Bishop, J. M. (1981b). *Proc. Natl. Acad. Sci. U.S.A.* **78,** 1067–1071.
Pickett, P. B., Pitelka, D. R., Hamamoto, S. T., and Misfeldt, D. S. (1975). *J. Cell Biol.* **66,** 316–332.
Ponten, J., and Stolt, L. (1980). *Exp. Cell Res.* **129,** 367.
Porter, K. R., Todaro, G., and Fonte, V. (1973). *J. Cell Biol.* **59,** 633–642.
Radke, K., Gilmore, T., and Martin, G. S. (1980). *Cell* **21,** 821–828.
Radke, K., Carter, V. C., Moss, P., Dehazya, P., Schliwa, M., and Martin, G. S. (1983). *J. Cell Biol.* **97,** 1601–1611.
Reynolds, F. H., Jr., Van de Ven, W. J. M., and Stephenson, J. R. (1980). *J. Biol. Chem.* **255,** 11040–11047.
Rheinwald, J. G., and Green, H. (1975). *Cell* **6,** 331–343.
Richman, R. A., Claus, T. H., Pilkis, S. J., and Freidman, D. L. (1976). *Proc. Natl. Acad. Sci. U.S.A.* **73,** 3589–3593.
Rohrschneider, L. R. (1980). *Proc. Natl. Acad. Sci. U.S.A.* **77,** 3514–3518.
Rubsamen, H., Saltenberger, K., Friis, R. R., and Eigenbrodt, E. (1982). *Proc. Natl. Acad. Sci. U.S.A.* **79,** 228–232.

Sanford, K. K., Barber, B. E., Woods, M. W., Parshad, R., and Law, Z. W. (1967). *J. Natl. Cancer Inst.* **39,** 705–733.
Schneider, E. L., and Mitsui, Y. (1976). *Proc. Natl. Acad. Sci. U.S.A.* **73,** 3548–3588.
Scordilis, S. P., Anderson, J. L., Pollack, R., and Adelstein, R. S. (1977). *J. Cell Biol.* **74,** 940–949.
Scott, C. C., Heckman, C. A., Nettesheim, P., and Snyder, F. (1979). *Cancer Res.* **39,** 2390–2399.
Sefton, B. M., Hunter, T., Ball, E. H., and Singer, S. J. (1981). *Cell* **24,** 165–174.
Sefton, B. M., Trowbridge, I. S., Cooper, J. A., and Scolnick, E. M. (1982). *Cell* **31,** 465.
Shannon, J. M., and Pitelka, D. R. (1981). *In Vitro* **17,** 1016–1028.
Shih, T. Y., Papageorge, A. G., Stokes, P. E., Weeks, M. O., and Scolnick, E. M. (1980). *Nature (London)* **287,** 686–691.
Smart, J. E., Oppermann, H., Czernilofsky, A. P., Purchio, A. F., Erikson, R. L., and Bishop, J. M. (1981). *Proc. Natl. Acad. Sci. U.S.A.* **78,** 6013–6017.
Steele, V. E., Marchok, A. C., and Nettesheim, P. (1978). *Cancer Res.* **38,** 3563–3565.
Steele, V. E., Marchok, A. C., and Nettesheim, P. (1979). *Cancer Res.* **39,** 3805–3811.
Stoner, G. D., Harris, C. C., Myers, G. A., Trump, B. F., and Conner, R. D. (1980). *In Vitro* **16,** 399–406.
Summerhayes, I. C., Cheng, Y- S. E., Sun, T-T., and Chen, L. B. (1981). *J. Cell Biol.* **90,** 63–69.
Temin, H. M., and Rubin, H. (1958). *Virology* **6,** 669–688.
Tou, J. T., and Gonzalez, R. C. (1977). "Pattern Recognition Principles." Addison-Wesley, Reading, Massachusetts, 377 pp.
Vogt, P. K. (1977). The genetics of RNA tumor viruses. *In* "Comprehensive Virology" (H. Fraenkel-Conrat and R. P. Wagner, eds.), Vol. 10, pp. 341–455. Plenum, New York.
Voyles, B. A., and McGrath, C. M. (1976). *Int. J. Cancer* **18,** 498–509.
Wang, L. -H., Feldman, R. A., Shibuya, M. Hanafusa, H. Notter, M. F. D., and Balduzzi, P. C. (1981). *J. Virology* **40,** 258–267.
Westermark, B. (1978). *Exp. Cell Res.* **111,** 295–299.
Wetzel, B., Sanford, K. K., Fox, C. H., Jones, G. M., Westbrook, E. W., and Taron, R. E. (1977). *Cancer Res.* **37,** 831–842.
Wigley, C. B., and Franks, L. M. (1976). *J. Cell Sci.* **20,**149–165.
Witte, O. N., Dasgupta, A., and Baltimore, D. (1980). *Nature (London)* **283,** 826–831.
Yuspa, S. H., Hawley-Nelson, P., Koehler, B., and Stanley, J. R. (1980). *Cancer Res.* **40,** 4694–4703.

HYBRIDOMA TECHNOLOGY

Paul J. Price

Hybridoma Sciences, Inc.
Atlanta, Georgia

I. Introduction

When a foreign element such as a bacterium, virus, parasite, or tumor cell enters our body, some of our body cells should respond by producing and secreting into the bloodstream proteins called immunoglobulins. The foreign element is called an antigen and the globulin which it elicits is called an antibody. The serum fraction of blood containing antibodies against specific antigens is called antisera. One antibody specifically recognizes and attaches to only one antigen. The antibody-producing cells are the B lymphocytes, which are found in abundance in the spleen. When the antigen enters our body some of the B lymphocytes replicate and become plasma cells which secrete the specific antibody. Because of this specificity, antibodies have been used to identify, quantify, purify, and classify antigens. The classical method for producing antibodies for clinical and diagnostic studies has been to repeatedly inoculate an animal such as a goat, rat, or mouse with a specific virus, bacterium, pharmacological agent, parasite, etc. The

ISBN 0-12-007904-6

animal will respond by producing antibodies, and small samples of blood taken at intervals can be examined for the presence of the specific antibody. When the amount of specific antibody is high (the quantitative measure of antibody is called titer), the animal is bled and the antibody molecules can be isolated and concentrated. Either directly or after tagging dyes or other agents to the antibody, the antibody can be used to rapidly identify the invading organism. This is the basis of the immunodiagnostic reagent marker. The disadvantages of this classical method of producing antibody are numerous. (1) The quality and quantity of antibody varies between animals and even between bleedings using the same animal. (2) Results obtained with one antisera may not be reproducible with another. (3) Extensive and repeated absorption of conventional antisera is cumbersome and results in a loss of titer. (4) Even purified antisera contain antibodies of different affinities with different subspecificities and cross-reactivities. (5) In addition the antisera may contain antibodies of different classes and subclasses that vary in their ability to carry out effector functions such as to agglutinate or precipitate antigens or to fix complement. It is this heterogeneity and lack of predictability inherent in the classical method that makes it virtually impossible to continually generate large amounts of antibody having constant properties (Kwan *et al.*, 1980).

A new technology which emerged in 1975 now allows for the production of pure antibody specifically targeted to a single antigen. This new technology is called hybridoma (Kohler and Milstein, 1975). In short, one first inoculates an animal (usually a mouse) with the antigen. The animal's blood is tested for specific antibody and titer and when high, the spleen or lymph nodes are removed from the animal and the lymphocytes separated. Each lymphocyte has the ability to produce a specific antibody but is unable to live for very long outside the animal's body. To keep these antibody-producing cells from dying, they are chemically fused to a mouse tumor cell of white cell lineage which, due to its tumor cell characteristics, can live forever outside the body. The cell used in the fusion is a mutant lacking an enzyme such as hypoxanthine phosphoribosyltransferase (HPRT) or thymidine kinase (TK). Because it is lacking one of these enzymes, it cannot detoxify a cell poison called aminopterin. Aminopterin is a folic acid antagonist and its presence blocks the main pathway for DNA synthesis by blocking *de novo* synthesis of purines and pyrimidines. If aminopterin is included in the growth medium, the mutant cells die (Littlefield, 1964). The spleen cell, however, has the HPRT enzyme and since during the fusion the genetic material (DNA) of the antibody-producing cell is mixed with the genetic material of the myeloma cell, the resulting hybrid cancer cell (hence the term hybridoma) can grow in the presence of aminopterin by utilizing the salvage pathway which allows the cells to incorporate exogenous hypoxanthine and to synthesize purines. The hybridoma can also produce a specific antibody and can live forever. The nonfused lymphocytes and the lymphocyte–lymphocyte fusions die out because they cannot grow for very long in culture. The unfused myeloma cells and myeloma–myeloma fusions die out because they cannot live in the presence of aminopterin.

The fusing agent most often used is polyethylene glycol (PEG) (Galfre *et al.*, 1977). PEG induces a reduction in free water and a closer opposition of cell membranes resulting in membrane fusion of adjacent cells. A second method of hybridization which may find applicability in the near future is electrofusion (Claude and Justin, 1983).

The hybrid cells are grown for several weeks in plastic dishes and the fluid medium that the cells are grown in is tested for antibody. Hybridomas producing antibody of desired specificity are grown and separated into single cells (clones). The clones are grown for a few more weeks in dishes and tested for antibody. Cultures producing desired antibodies are frozen and stored in a viable state for future reference (Goding, 1980; Yelton *et al.*, 1980; Zola and Brooks, 1982; Galfre and Milstein, 1981).

Since the hybrid cell (hybridoma) has the cancer cell characteristic inherited from its myeloma parent, it will also grow in the mouse and produce even more antibody per unit volume than if grown using standard cell culture techniques. Artificial capillary cultures (Calabresi *et al.*, 1981), dialysis cultures (Adamson *et al.*, 1983), microencapsulation of the hybridomas (Jarvis *et al.*, 1982), the use of ceramic chambers with pH and oxygen control (Lydersen *et al.*, 1983), and other high batch methods can yield at least as much antibody as is possible using mice over the same time period, without the problem of contaminating mouse proteins. The antibody-containing fluids (called antisera and ascitic fluids if collected from the mouse, and supernatant fluids if collected from cell cultures) are collected and the antibodies isolated.

The major advantages of monoclonal antibodies are the following:

1. You are now dealing with a chemically defined analytical reagent.
2. The titer of antibody obtainable from a hybridoma clone is 100 to 1000 times greater than is obtainable by the simple inoculation of an antigen into a mouse.
3. There is a continuous supply of antibody.
4. Cell supernatants contain no unwanted antibodies. (Since each hybridoma culture originated from a single myeloma cell and a single spleen cell, the antibody is called monoclonal.)

5. Pure antibody and multiple antibodies can be obtained from a nonpurified antigen.

6. Antibodies having desired effector functions, specificity, or avidity can be selected (Diamond *et al.*, 1981).

The major disadvantages of the system include the following:

1. Assays requiring cross-linking of molecules may not be possible with antibodies recognizing a single epitope.

2. Unexpected cross-reactions may be encountered because of polyfunctional binding sites.

3. Antibodies may not produce the desired effector function or avidity.

4. Heterologous antibody when used as a drug can elicit serum sickness, anaphylaxis, and renal failure.

These disadvantages are also true in varying degrees with polyclonal antibodies produced by the classical method.

II. Utilization

1. Disease Diagnosis and Treatment for Both Man and Animals

Using one of many immunodiagnostic assays, monoclonal antibodies can be used both qualitatively and quantitatively as clinical reagents for the rapid and specific identification of the causative organism or agent of a disease syndrome (Sevier *et al.*, 1981) and to separate an active from a transient infection. Since monoclonal antibodies are able to detect subtle antigenic differences, such as amino acid substitutions in a virus or bacterium, they can be used to (1) monitor antigenic drift, (2) decide the proper viral strains to be used in a vaccine (such as influenza), (3) classify organisms by species, (4) separate different serotypes of closely related organisms, and (5) probe the physical, biochemical, and antigenic characteristics of organisms. When a human hybridoma system is developed, monoclonal antibodies may have application as a means of passive immunity to augment the patient's own immune system.

2. Antibodies Reacting with Receptors for Hormones and Drugs

Such antibodies may be used to (1) localize and quantitate nanomolar blood levels of hormones and drugs, (2) treat hormone overproduction and drug overdoses, (3) differentiate between an active form of a drug and its metabolites, (4) study drug receptors and thereby enhance understanding of pharmacokinetics and drug metabolism (Butler *et al.*, 1977; Aeberhard *et al.*, 1980).

3. Cancer Diagnosis and Treatment

We can produce monoclonal antibodies against tumor-associated antigens, which can be used to (1) classify malignant diseases, (2) understand the basic mechanisms of cancer biology, and (3) detect cancer earlier. Tumor-specific antibodies can be tagged with a radioactive label and then be used to detect or image the location of the tumor cells (Ballou *et al.*, 1979), and to monitor the disease during and after treatment (Farrands *et al.*, 1982). If an antibody can be isolated that recognizes only the patient's tumor cells, it may be able to be used directly if cytotoxic (IgG_{2a} isotype has been shown to specifically inhibit the growth of human tumors) or it can be tagged with a specific cell toxin (chemotherapeutic agent or bacterial or plant toxin such as the Ricin A chain or diptheria toxin A chain) and used directly for treatment or following surgery, irradiation, or transfusion in a search-and-kill mission to destroy any tumor cells missed by the treatment (Gililand *et al.*, 1980; Strand *et al.*, 1984). By tagging the cell toxin or radioactive label to an antibody which recognizes only the tumor cell, or by binding drug-loaded liposomes (Barbet *et al.*, 1981) to a tumor-associated antibody, the tumor may be able to be destroyed while at the same time leaving the natural defense mechanisms of the patient intact. Monoclonal antibodies have already been used clinically with varying degrees of success in animals and humans with melanoma (Steplewski, 1980), neuroblastoma (Momoi *et al.*, 1980), colorectal carcinoma (Herlyn *et al.*, 1980), breast cancer (Schlom, 1980), several leukemias (Ritz *et al.*, 1980), lymphomas (Nadler *et al.*, 1980), and prostate carcinoma (Ware *et al.*, 1982). Bone marrow can be removed from patients with cancer and treated *in vitro* with a tumor-specific cytotoxic antibody. The patient can then be treated with toxic doses of irradiation or chemotherapy that would normally be lethal due to destruction of the bone marrow and rescued by injection of the autologous bone marrow now free of cancer cells (Bast, Jr., *et al.*, 1983). Clinically, however, there are problems of antigenic modulation, blocking antigens (Nadler *et al.*, 1980), and the production of antimouse antibody by the recipient (Miller *et al.*, 1981) The use of conjugated α-emitting isotopes may circumvent the problem of modulation if some of the cells in the region of the tumor bear the specific antigen. Multiple antibodies recognizing different antigens would also be helpful. Problems of anaphylaxis should be reduced or eliminated when adequate *in vivo* or *in vitro* human-human systems are developed (Olsson and Kaplan, 1983). If a tumor produces or concentrates specific products, labeled antibody to the product should be effective for imaging and treatment [i.e., CEA (Steplewski, 1980), AFP (Uotila *et al.*, 1980), and HCG (Gupta and Talwar, 1980)].

For clinical application many questions still need to be answered and most studies reported are preliminary. The importance of antibody class, immunoconjugates, immunogenicity, schedule, dose, and route of administration need to be better evaluated. For radioimaging and tumor localization radiolabeled $F(ab')_2$ bivalent antibody fragments appear to delineate tumors earlier, produce less background radioactivity than whole antibody at comparable time points, and are less immunogenic (Wahl *et al.*, 1983). Both for imaging and therapy, mixtures of monoclonal antibodies may prove superior to single monoclonal antibodies (Rosen *et al.*, 1983).

4. Organ Transplant or Blood Transfusion

Since these antibodies recognize subtle antigenic differences, they can be utilized in organ, tissue, or blood typing. By improving the antigenic profile, the chance of rejection of an organ by the recipient or the wrong blood being transfused would be greatly reduced. Specific antiimmunoglobulin antibodies (antibodies which destroy other antibodies) may be used to enhance the successful transplantation of a tissue or organ by inactivating invading helper/inducer T cells (Cosimi *et al.*, 1981; Goldstein, 1984). These antibodies will also be used in forensic medicine, for the screening of tissues for contamination with pathogenic organisms, and eventually will be used for the diagnosis, treatment, and understanding of autoimmune diseases (Cuello *et al.*, 1980; Littman *et al.*, 1983).

5. Isolation and Purification

Monoclonal antibodies can be absorbed to a solid matrix such as beads or immunoabsorbent columns for the isolation of specific organisms or antigens. The target organism or antigen can then be eluded from the matrix and collected in a purified form. Examples of this utilization would include the isolation of specific antigens or developmental stages of parasites and the isolation and purification of rare or costly biological materials (Van Heyningen *et al.*, 1983).

6. Cell Surface Antigens

Antibodies can be developed to detect, delete, or separate antigens in order to study development and differentiation (Billing *et al.*, 1981) and as substrates for the adherence of differentiated cells *in vitro* (MacLeish *et al.*, 1983).

7. *Genetics*

The antibodies can be used to map antigentic determinants on chromosomes and to isolate and study mutant proteins.

III. Immunization

Immunization is necessary to increase the number of antibody-producing clones and to stimulate the B and T cells to divide and differentiate. A typical regimen using intact cells would be to inject the mice intravenously with 0.1 m1 containing 1 or 2 $\times$ 10^7 cells on day 1, 3 $\times$ 10^7 cells 3 weeks later, followed by removal of the spleen for fusion 3 days later. A typical regimen using soluble foreign proteins, viruses, other microorganisms, or haptencarrier conjugates of polysaccharides would be to immunize with from 1 to 100 μg of antigen emulsified in 0.1 m1 of complete Freund's adjuvant. Viruses in infected cells can be sonicated and emulsified in the adjuvant. Primary immunization is subcutaneously and intraperitoneally with intraperitoneal boosts at about 1-month intervals (doubling the dose each time) in incomplete adjuvant or antigen in saline. Three days before fusion a relatively large dose is given in saline intravenously or intraperitoneally. The best time for fusion is always from 3 to 4 days after the last boost (Goding, 1980). When using soluble antigens we boost 3, 2, and 1 day prior to fusion. This increases the number of blast cells. Fusion at the peak of antibody production (usually from 7 to 8 days after boosting) results in having less cells in proliferation and hence less hybridization. Two tricks which have been published to enhance the frequency of antigen-specific spleen cells are (1) to culture the spleen cells for 3 or 4 days *in vitro* on a rocking platform at 37°C in a mixture of 83% N_2, 7% O_2, and 10% CO_2 and in the presence of the immunizing antigen and then to do the fusion, and (2) to remove the spleen from the immunized animal, isolate the cells, and then inoculate them iv into X-irradiated syngeneic recipients, followed immediately by an ip inoculation of antigen plus adjuvant followed by removal of spleen 3 to 4 days later for fusion. One supposedly can expect up to a 50-fold increase in specific antibody-producing clones (Siragarian *et al.*, 1983).

It is advantageous to purify the antigen prior to immunization if the antigen can be purified without reduction of immunogenicity. Immunogenic contaminants can be reduced by mixing the antigen with specific antibody against the contaminant and then using this mixture for immunization.

Do not waste your time fusing a spleen from a mouse with no or low titer. Since most of the available myelomas are of BALB/c origin, BALB/c mice or heterozygous BALB/c strains are the spleen-donor animal of choice. This is especially important if one wishes to grow the hybridoma *in vivo*. An alternative method, if rats or other mouse strains are used, would be to grow the hybridoma in athymic nude mice. Cross-species hybridization does not result in reduced fusion, but there are often problems with chromosome loss.

Each mouse spleen contains about 10^8 nucleated cells. Four- to twelve-week-old mice are usually used for immunization. Using a good immunogen, up to 30 specific hybrids can be expected. Using a poor immunogen, one may need to pool the spleens of 3–5 mice and screen several thousand wells to find one cell producing the desired antibody. Since mice respond differently to immunization, it is always a good idea to immunize 5–10 mice and select those having the best antibody response.

IV. Fusion[1]

1. *Myeloma Cells*

The most popular myeloma lines lack the enzyme HPRT, are of BALB/c origin, are resistant to killing by 20 μg/m 8-azaguanine (8-AZG), and are killed in a medium containing hypoxanthine, aminopterin, and thymidine (HAT). Early myelomas used in fusion secreted Ig. Later variants were chosen which contained only light chains. The lines now most often used for mouse cell fusions do not produce either their own light or heavy chain and are designated SP2–Ag14 (Shulman *et al.*, 1978), P3-NS1-Ag4-1 (Kohler *et al.*, 1976), and P3 X63 Ag-8-6.5.3. (Kearney *et al.*, 1979). Prior to fusion the myeloma cells are grown in a standard medium (RPMI 1640 FBS 10 or DME FBS 15, supplemented with $10^{-5}M$ 2-mercaptoethanol (2-ME), 2 m*M* L-glutamine, 1.0m*M* sodium pyruvate, and antibiotics). We use a medium called LoSM which is formulated and sold by Hybridoma Sciences and only needs to be supplemented with 2–4% FBS plus 2-ME, L-glutamine, and antibiotics. 2-ME oxidizes easily and concentrates should be stored refrigerated, tightly sealed, and protected from light. The

[1]This procedure was adapted from one developed by J. Kearney.

normal transfer ratio is 1:10 weekly. The medium should be further supplemented with 15 μg/ml 8-AZG 1 week prior to fusion to kill any $HPRT^+$ revertants which can grow in HAT. Two days later the cells are transferred 1:4 in the presence of 20 μg/ml 8-AZG. The next day or 4 days prior to fusion the cells are refed in the standard medium (not containing 8-AZG). The next day (day 2) the cultures are divided 1:4 and refed again the following day (day 3). This regimen assures exponentially growing cells on the day of fusion (very important—especially with human myelomas). A culture grown to 80% confluency in a 150-cm^2 flask should yield about $1–2.5 \times 10^7$ cells and, since about 5×10^7 to 1×10^8 cells are needed, a minimum of 4 flasks should be planned. On the day of fusion the myeloma cells are washed twice with a serum-free growth medium, counted using Trypan Blue, washed again (room temperature centrifugation at 800 rpm), and resuspended to 5×10^7 to 1×10^8 cells in 1 m1 of serum-free media.

If the myeloma cells lose their capacity to form hybrids consider contamination with mycoplasma (check for mycoplasma weekly). Low fusion frequencies may also result from overcrowding of cells. The cells must be kept in log-phase growth and cultures replaced as needed from frozen stocks.

An interesting mouse myeloma cell is designated FOX-NY and is deficient in adenine phosphoribosyltransferase (APRT). The fusion partner for this cell is a spleen cell from a mouse having a Robertsonian 8.12 translocation chromosome. In this mouse the heavy-chain Ig locus on chromosome 12 is genetically linked to the selection locus (APRT) on chromosome 8. Selection for the $APRT^+$ cells eliminates $APRT^-$ myelomas and $APRT^-$ hybridomas. This results in approximately two-thirds of the hybridomas being antibody producers. In contrast, in the $HPRT^+$ system the hybridomas must retain 2 chromosomes to survive in HAT and produce antibody (the X chromosome for HPRT and the heavy-chain Ig loci). Using the HAT selection method less than one-third of the hybridomas are antibody producers (Taggart and Samloff, 1983).

2. Spleen Cell Suspension

The mice are killed by cervical dislocation or CO_2 intoxication. The mouse skin is disinfected with 70% ethanol. With the mouse placed on its back and using sterile instruments and gloves, the skin on the left side is grasped and a 0.5-in. cut made just below the ribs. The skin on each side of the cut is held with the fingers and the skin pulled firmly toward head and tail. This skins the mouse, revealing the spleen below a membranous tissue. Using fresh sterile instruments the spleen is removed and transferred to a petri dish containing a few milliliters of serum-free media. Adhering tissue is teased away from the spleen. Using 2 tuberculin syringes with 22-gauge, 1.5-in. needles bent at 90° about 0.5 in. from their tip, the end of the spleen is cut. While holding the spleen with one needle, the spleen is massaged with the flat side of the other needle in the direction of the open end. This pushes the spleen cells from the spleen, leaving only the membranous sack. The cells are transferred to a 15-ml conical with 10 ml of media, aspirated to break up clumps, and the clumps allowed to settle out of solution (about 5 minutes). The supernatant is transferred to a 50-ml conical, topped to 50 ml with serum-free media, centrifuged at 800 rpm for 5 minutes, and washed 2 more times with serum-free media. All procedures and centrifugations are at room temperature. Between the first and second wash the cells are diluted to 10 ml and counted using a Crystal Violet–citric acid stain. One mouse should yield about 1×10^8 nucleated cells. It has been our experience that extremely large spleens fuse poorly. The spleen and myeloma cells are mixed in a 50-ml conical in a ratio of 10:1 to 1.5:1 and centrifuged at 600–800 rpm for 5 minutes (Kennett *et al.*, 1980). The supernatant is poured off and, using a Pasteur or 1-ml pipette, all residual fluid removed. One milliliter of PEG (we use a MW of 1500 or 4000) at 41°C is added slowly down the side of the tube (Gefter *et al.*, 1977). PEG ranging in MW from 500 to 6000 has been used. After disrupting the pellet by tapping or by a quick Vortex touch or by aspirating once, the suspension is allowed to sit for a total of 45 seconds (timed from the addition of PEG). Then 10 ml of serum-free media are added drop wise, slowly, and with gentle mixing so that the entire procedure of adding the media takes about 3 minutes and 15 seconds. Many groups dilute even slower—10 ml over 15 minutes. The tube is then topped to 50 ml with the serum containing hybridoma growth media. We prepare our PEG by adding to 20 g of autoclaved PEG cooled to 50°C 28 ml of Dulbecco's PBS (warmed to 41°C and containing 15% DMSO) for PEG 4000 or 37 ml for PEG 1500. DMSO stabilizes the phospholipids in the cell membrane, reducing PEG toxicity (Tilcock and Fisher, 1982). PEG should be pretested for fusion efficiency as lots differ in toxicity. PEG should be stored below 20°C in the absence of light and should be odorless (Kadish and Wenc, 1983). The pH of the PEG solution should be adjusted to 8.0 prior to being put in vials (De St. Groth and Scheidegger, 1980; Sharon *et al.*, 1979). After fusion, the suspension is spun at 600 rpm for 5 minutes, the supernatant decanted, and the cells resuspended in the HAT selection media at a concentration of between 2 and 5×10^5 myeloma cells/ml and 1.0 ml/well/24-well dish. An alternative procedure is to place the suspension in a petri dish and allow the cells to incubate at 37°C for 1–3 hours in the CO_2 incubator, wash once, and plant in HAT. This gets rid of contaminating fibroblasts, removes more PEG and DMSO, and may increase the number of fused cells. Pretreating the spleen suspension with phytohemagglutinin and other mitogens has been reported to increase the fusion efficiency at least twofold (Woloschak and Senitzer, 1983). You can also clone directly by planting 0.5×10^5 cells/ml and 0.2 ml/well in 96-well dishes. If you clone directly, you should add to each well 1×10^4 peritoneal exudate cells prior to planting the suspension. Some laboratories add spleen cells or conditioned media (25–50%) from myeloma, spleen, or peritoneal exudate cells instead (see Section V). A dish of myeloma control cells must be planted in the same concentration of HAT as an HAT control. We add 1×10^4 peritoneal exudate cells to a 96-well dish 1 day prior to fusion at 100 μl/well in LoSM FBS 4. The next day the cells are added in 100 μl/well in LoSM FBS 4 with 2X HAT.

The 1X HAT selection media is made by supplementing the growth medium with 1×10^{-4} *M* hypoxanthine, 4×10^{-7} *M* aminopterin (or 10^{-4} *M* amethopterin), and 1.6×10^{-5} *M* thymidine. Hypoxanthine and thymidine are dissolved in DI H_2O with heating to 45°C for about 1 hour. A 100× stock solution is prepared by dissolving 136 mg of hypoxanthine and 38.7 mg of thymidine in 50 ml of 0.1 *N* NaOH. Adjust the pH to 9.0 with 1 *N* HCl and dilute to 100 ml with deionized water. The aminopterin is dissolved in 0.1 *N* NaOH and then neutralized with 1 *N* HCl. A 1000 stock solution is prepared by dissolving 17.6 mg in 50 ml of 0.1 *N* NaOH. Adjust the pH to 7.2 with 1 *N* HCl and dilute to 100 ml with deionized water. Amethopterin and aminopterin are light sensitive and concentrates should be stored frozen and protected from light. This is also true for the 100X HAT medium.

Between 7 and 10 days incubation, colonies of hybrid cells should be visible and the medium should be turning yellow. If the myeloma control cells have not all been killed by the HAT medium on day 7, remove half the volume of medium and add an equal volume of hybridoma growth media containing fresh HAT to the fusion plates. If the myeloma cells in the myeloma control plate are not dead on day 10 refeed with HAT and check the cultures again on day 12. Many groups refeed 2 or 3 times with HAT prior to dilution without HAT. As soon as colonies are visible macroscopically (usually about the tenth or eleventh day), screening of supernatants should begin. When good growth is observed and it is obvious that all the parental myeloma cells were killed, one-half the HAT medium is withdrawn and replaced with an equal volume of HT medium. The cells can now be fed and transferred

1X per week in hybridoma growth media. At this stage most groups continue to supplement the medium with HT and some continually add HT. The logic for keeping HT in the medium is that until all the aminopterin is sufficiently diluted the cell must continue to use the salvage pathway for nucleic acid production. LoSM is formulated with H and T as two of its ingredients so that no further supplementation is necessary. To reduce contamination wrap the cell culture plates singly or in small stacks in plastic film. The CO_2 incubators are maintained at 36.5°C, 90% humidity, and 6% CO_2. Some laboratories set their incubators as high as 10% CO_2. All reagents must be pretested for toxicity. This can be done by a comparative plating efficiency assay using myeloma or hybridoma cells. If the cultures are growing but not producing antibody consider contamination with mycoplasma. Mycoplasma can cause the myeloma cells to bypass the HAT block, thus allowing them to survive and overgrow the hybridoma colonies. Mycoplasma spreads rapidly, so immediately discard positive lines and sanitize the incubator. When transferring cells from a 24-well dish to 25-cm^2 flasks, it is advantageous to add media to the flasks 24 hours prior to inoculation and then to discard the media and add fresh media with the cell suspension or, even better, to use peritoneal macrophage feeder cells (De St. Groth and Scheidegger, 1980).

V. Cloning

After detecting a desired hybridoma, one-half of the cells should be frozen and the other half immediately cloned. It is important to clone early and often to minimize overgrowth by unwanted hybrids. In mouse–mouse hybrids chromosome loss is somewhat random and gradually stops (Croce *et al.*, 1979). Two methods are used to isolate single cells. In the limiting-dilution method (Goding, 1980), cells are diluted so that each well of a 96-well plate contains theoretically 0.5 cells/well. Because of counting variations, plant dishes at 5.0, 1.0, and 0.5 cells/well and screen for and mark wells containing single colonies from 3 to 4 days later. An alternative procedure would be to examine single droplets of media (1 droplet per well) microscopically and mark wells containing just 1 cell. Media is then added to each selected droplet (Sijens *et al.*, 1983). The cloning efficiency can be raised by adding either primary spleen cells (1×10^7/ml), peritoneal exudate cells (1.0×10^5/ml) (De St. Groth and Scheidegger, 1980), or complete culture media to the wells 24 hours prior to cloning (incubated during this time) or by adding to the medium 25% peritoneal exudate supernatant fluid or supernatant fluids from primary human endothelial cells (Astaldi *et al.*, 1980) or an endothelial cell growth supplement (Pintus *et al.*, 1983) or human umbilical cord serum (Westerwoudt *et al.*, 1983). The second method of cloning is to use soft agar or methylcellulose (Goding, 1980; Kennett, 1980). At day 10 the clones should be large enough to pick with a micropipette and to transfer to a well of a 24-well dish containing 0.5 ml of media. Antimouse immunoglobulin can be added to the soft agar so that a precipitate forms around those clones secreting immunoglobulin.

Due to random chromosome loss stable clones can gradually lose antibody production because of overgrowth of nonsecretors. Recloning usually solves this problem.

VI. Freezing and Reconstitution

It is important to freeze early to lessen the chance of complete loss of the experiment due to contamination, chromosome loss, overgrowth by nonsecretors, or cell death. Exponentially growing cells are collected by centrifugation and concentrated to about 10^7 cell/ml. One milliliter of cells in cold freeze media [fetal bovine serum (FBS) supplemented with 10% DMSO] is added to a sterile plastic vial (2-ml cryotubes). Many laboratories freeze their cells in RPMI 1640 or DME supplemented with 7.5% DMSO and 20% FBS. The freeze media and cell suspensions are kept at wet-ice temperature throughout the procedure and immediately transferred to a −40°C freezer overnight. They are then transferred directly to storage in a liquid nitrogen (LN_2) vapor phase freezer. If more than about 2 minutes elapses between removal from the −40°C and storage in liguid nitrogen vapor, the vialed suspensions should be placed directly in liquid nitrogen and then transferred to the LN_2 freezer. To thaw, the vials are removed from the freezer and placed directly and quickly into a 37°C water bath. When thawed, and before the temperature of the suspension can rise to ambient, the suspension is added slowly and dropwise to 25 cm^2 cell culture flasks containing 5 ml of prewarmed growth media. Many groups add HT to the reconstitution media and wash at least once prior to incubation. This is not necessary if LoSM is used. The cultures are incubated and refed the next day (Galfre and Milstein, 1981).

You can also freeze the clones directly in the 96-well or 24-well dish. Draw off the medium, add cold freeze media, wrap in plastic, and place at $-40°C$ overnight and then into the LN_2 vapor freezer. We have had 100% recovery viability and no loss of antibody production (Wells and Price, 1983).

VII. Production of Antibody

Once the hybridoma has been cloned and characterized, it can then be used to produce large quantities of monoclonal antibody. The easiest method is to grow the cells in bulk using roller bottles, spinner flasks, or shaker cultures at an initial planting concentration of 3×10^5 cells/ml. The medium for spinner flasks and shaker cultures should be Ca^{2+} free and buffered with HEPES. Typical yields would be between 10 and 100 μg of antibody/ml of fluid. Far higher yields are obtainable using newer, more exotic *in vitro* procedures (see Section I). The easiest way to produce large amounts of antibody is to inoculate adult syngeneic mice, pretreated with 0.5 ml of pristane intraperitoneally (ip) at least 2 weeks earlier, with 1×10^6 cells/0.1 ml ip. Tumor growth should be evident in 2–4 weeks. Ascites can be harvested or transferred to other mice at a ratio of 1:5. IgG levels in the serum can be as high as 25 mg/ml. The initial tumor tends to produce only about 5 ml of ascitic fluid. The hybridoma may grow as a solid tumor and the ascitic fluid collected is usually quite bloody. With passage the hybridoma cells grow faster, are cleaner, and produce gradually more ascitic fluid (up to 10 ml per mouse). One animal can be tapped several times. Once collected, antibody can be concentrated by salt fractionation (50% saturated ammonium sulfate followed by dialysis against PBS) or by ultrafiltration followed by dialysis (Heide and Schwics, 1978). This can be followed by affinity chromatography. Affinity columns such as cyanogen bromide-activated Sepharose are made with purified goat or rabbit antibodies to mouse immunoglobulins. These are then used to absorb immunoglobulin from the supernatants. The antibody is eluted with mild acid or alkali or with chaotropic agents such as $MgCl_2$ (3.2 *M*). Ion exchange chromatography can also be used on DEAE columns for IgG immunoglobulins (Hudson and Hay, 1976). Support matrices such as acrylamide, agarose (DEAE-Sepharose), or beaded cellulose (DEAE-Sephacel) can be used to absorb antibody followed by salt elution (Seeher and Burke, 1980; McMaster and Williams, 1979). This method is not good for IgM. IgM antibodies can be partially purified by 45% ammonium sulfate precipitation followed by concanavalin A affinity chromatography. Most mouse immunoglobulins of the IgG class bind to protein A so that protein A-Sepharose affinity columns (pH 8.2) can be used to isolate mouse IgG followed by acid elution. IgG_1 binds poorly (Pearson and Anderson, 1980). Because of potential bacterial contamination all columns should be run in the cold (Yelton, *et al.*, 1980).

VIII. Characterization of Immunoglobulin Classes

Knowing the class or subclass of a monoclonal antibody is important for choosing the method of subsequent purification and for picking an antibody to give a specific response, i.e., fix complement, induce complement-dependent cytotoxicity, specific binding, etc. One commonly used procedure is the Ouchterlony analysis whereby concentrated antibody is in a central well and subclass specific goat or rabbit antimouse antibodies are placed in the surrounding wells (μ, λ, 2α, etc.). Precipitin lines will develop between the antibody and the specific isotype. A second procedure would be immunofluorescence using the hybridoma cells and fluorescein-labeled goat or rabbit antimouse isotype antisera. The use of nitrocellulose membrane as a carrier of antigen or monoclonal antibody offers a fast, convenient test system. In the immunodot procedure rabbit antiisotype and light-chain specific antibodies are immobilized in specific sequence as dots on a strip of nitrocellulose filter paper. The monoclonal antibodies bind to the appropriate antiisotype dot and are detected with an enzyme-conjugated goat antimouse immunoglobulin reagent. A substrate is added which catalyzes the enzyme to give a specific color reaction. Class and subclass are determined using supernatant fluids since the ascitic fluids may be contaminated with mouse immunoglobulins (McDougal *et al.*, 1983; Horejsi and Hilgert, 1983).

IX. Monoclonality, Epitope Screening, and Affinity

Isoelectric focusing and polyacrylamide gel electrophoresis (Renart *et al.*, 1979) are often used as a proof of antibody purity. However, if an antibody shows the required specificity it should be satisfactory proof of true cloning (Yelton *et al.*, 1980).

Methods to test whether different monoclonal antibodies recognize the same or different epitopes can be determined by separation of the antigen or by immune precipitation (Bordenave *et al.*, 1979), by a competitive radioimmune assay (Tsu and Herzenberg, 1980), or by an ELISA double antibody binding system (Friguet *et al.*, 1983). The latter method is the easiest in that two monoclonal antibodies are added to the antigen coated on the microtiter plate and the amount of bound antibody measured quantitatively by ELISA.

High-affinity antibodies are needed for immunoassays, while low- or variable-affinity antibodies are preferred for immunopurification systems. Comparison of dilution curves for antibodies directed to the same antigen at different pHs allows for the ranking of the order of affinity (Van Heynigen *et al.,* 1983).

X. Screening Assays

The screening assays occupy the greatest amount of time and must be developed prior to hybridization. The work load increases almost exponentially as positive wells are found and cloned so that unwanted cultures need to be eliminated as quickly as possible. Serological tests for antibody which depend on antigen cross-linking such as precipitation do not work well with single monoclonals. Unless there is more than one reacting epitope present on the same molecule, monoclonal antibodies cannot form a precipitin lattice. In this case it would be desirable to use a mixture of monoclonal antibodies recognizing different epitopes. The screening assays that do work well with single monoclonal antibodies all utilize the ability of proteins to stick tightly to certain plastics such as polyvinyl chloride. Binding affinity depends on pH and temperature so that physiochemical conditions must be controlled prior to adding the hybridoma supernatants (Galfre and Milstein, 1981).

A. ELISA (Enzyme-Linked Immunosorbent Assay)

In this method (Suddith *et al.,* 1980; Douillard and Hoffman, 1983) the antigen is coated onto the wells of a plastic plate (polyvinyl) or grown (if whole cells) in a plastic plate and then fixed (0.25% glutaraldehyde in PBS) or centrifuged. To detect intracellular antigens the cells are pretreated with acetone. Unreactive sites are blocked with bovine serum albumin (BSA) prior to adding the hybridoma supernatants. The hybridoma supernatant fluids are added and the plates incubated for 1 hour at 37°C or at 4°C overnight. The wells are then thoroughly washed to remove unbound antibody. Conjugated anti-mouse class-specific or heterologous antibody (i.e., with horseradish peroxidase or alkaline phosphatase) is then added. After washing to remove unbound conjugates, the enzyme substrate is added and the substrate is hydrolyzed yielding a colored product that can be quantitated spectrophotometrically or screened by eye for a yes or no type assay.

B. *Radioimmunoassay*

This method is the same as the ELISA but the antimouse immunoglobulin is labeled with a radioactive marker (Ivanyi, 1980; Tsu and Herzenberg, 1980). After washing, the individual wells are sliced with a hot wire and counted in a gamma counter.

C. *Immunofluorescence*

In this test (White *et al.*, 1978; Johnson *et al.*, 1978) the antimouse immunoglobulin is conjugated to a fluorescent dye (i.e., fluorescein or rhodamine) and the reaction monitored in a fluorescent microscope or by using a fluorescence-activated cell sorter. Biotin can also be coupled to the antibody and detected using a fluorochrome-labeled egg white protein called avidin.

Each assay requires a negative control (medium only and antibody to an unrelated antigen) and positive control (conventional mouse antiserum). If all clones are negative from a previously positive clone or mother well, then either the clones have not grown up enough and need to be retested, the cells were overgrown by a nonpositive clone (need to reclone), or Ig production has stopped due to chromosome loss. Another problem can be the prozone effect. This is especially true with ascitic fluid. To solve this you need to test serial dilutions of the antibody.

Immunoperoxidase staining of frozen tissue sections or fixed paraffin imbedded tissues can also be used to test specificity of an antibody to a particular tissue or cell type (i.e., tumor specificity) (Curran and Gregory, 1977).

XI. Storage of Antibodies

Antibodies in general have a wide range of stability or fragility. Most culture supernatants can be kept for weeks at 4°C or months at −20° with the addition of 0.01% sodium azide. The highest titers are obtained from cells which have been allowed to overgrow. In general, antibodies store better at higher concentration (mg/ml) (Pearson, 1982). Repeated freezing and thawing reduces activity. Sera and ascitic fluid should be aliquoted into single use quantities and stored frozen, ammonium sulfate precipitated and stored at 4°C or frozen and stored at −20°C. For optimal ascitic fluid titers, bleed out the mouse and add the blood to the ascitic fluid. This will cause the ascitic fluid to clot and reduce the chance of clotting during storage (Kohler, 1981). The tumor cells can be spun out and reinoculated into other mice. Complement can be inactivated (as well as proteolytic enzymes) by heating at 56°C for 30 minutes in glass containers.

XII. Human Monoclonal Antibody Production

Animals are primed with the selected antigen and their antibody response monitored. With humans this is for the most part impossible. Peripheral blood lymphocytes and in some cases lymph nodes can be obtained from individuals naturally immunized to a limited group of antigens (tumor, Rh, vaccine, specific diseases, etc.). The percentage of specific lymphocytes in these cases is small and the percentage of specific blast cells even smaller. Attempts to prime peripheral blood lymphocytes (PBL) *in vitro* have been disappointing (Cavagraro and Osband, 1983). Stimulation by mitogen appears to be necessary with pokeweed mitogen at 2.5–25 μg/ml optimal (Olsson *et al.*, 1983). Even with stimulation one does not get a secondary immune response after *in vitro* stimulation. The problem of suppressor cells must also be considered. Mouse–human hybridization leads to genetically unstable hybrids that selectively lose their human chromosomes (Croce *et al.*, 1979). Epstein–Barr virus transformation of PBL results in immortality but poor immunoglobulin production (Steinitz *et al.*, 1977). At this point in time there is no human fusion partner available with all of the attributes of the mouse lines used as fusion partners (i.e., high fusion efficiency, HAT sensitivity, high antibody production after fusion, active mitotic state, and stability). Optimal conditions have obviously not been defined. The following procedure has been somewhat successful (Olsson and Kaplan, 1983). After fusion to an HAT-sensitive human myeloma or lymphoma, the cells are planted in 96-well microtiter plates and HAT treatment using 2×10^{-8} *M* aminopterin begun 2 days later. Human peripheral blood monocytes (buffycoat) at about 2×10^4 cells/well or human or mouse thymocytes at 5×10^5 cells/well can be used as feeder cells. Fresh feeder cells should be added every 2 weeks.

Another problem reported with some human myeloma–spleen fusions is that in some cases what appears to be a stable antibody producer permanently stops secreting antibody after recovery from the freezer. Human lymphomas and myelomas also seem to be more susceptible to fetal bovine serum toxicity. One needs to screen each lot prior to purchase (Cote *et al.,* 1983).

XIII. Laboratory Setup

The major equipment needed for the technology includes a laminar flow work station, a humidified CO_2 incubator, a centrifuge, an inverted microscope. and a liquid nitrogen freezer. The proper CO_2 and humidity levels are essential. Do not trust digital readout gauges. CO_2 levels need to be tested weekly with a fye-rite and the digital gauge readjusted. Humidity must be tested with a humidistat and temperature with a thermometer.

References

Adamson, S. R., Fitzpatrick, S. L., Behie, L. A., Gaucher, G. M., and Lesser, B. H. (1983). *Biotech. Lett.* **5,** 573.

Aeberhard, P., Butler, V. P., Smith, T. W., Haber, E., Tse Eng, D., Brau, J., Chalom, A., Glatt, B, Thebaut, J. F., Delangenhagen, B., and Morin, B. (1980). *Arch. Mal. Coeur* **73,** 1471.

Astaldi, G. C. B., Janssen, M. C., Landsdorp, P., Williems, C., Zeijlemaker, W. P., and Oasterhof, F. (1980). *J. Immunol.* **125,** 1411.

Ballou, B, Levine, G., Hakala, T. R., and Solter, D. (1979). *Science* **206,** 844.

Barbet, J., Machy, P., and Leserman, L. D. (1981). *J. Supramolec. Struct. Cell. Biochem.* **16,** 243.

Bast, R. C., Jr., Ritz, J., Lipton, J. M., Feeney, M., Sallan, S. E., Nathan, D. G., and Schlossman, S. F. (1983). *Cancer Res.* **43,** 1389.

Billing, R., Terasaki, P. I., Sugich, L., and Foon, K. (1981). *J. Immunol. Meth.* **47,** 289.

Bordenave, G. R., Pages, J., Stoltz, F., and Bussard, A. E. (1979). *Immunol. Lett.* **1,** 165.

Butler, V. P., Schmidt, D. H., Smith, T. W., Haber, E., Ragnov, B. D., and Demartini, P. (1977). *J. Clin. Invest.* **59,** 345.

Calabresi, P., McCarthy, K. L., Dexter, D. L., Cummings, F. J., and Rotman, B. (1981). *Proc. AACR ASCO* **1200,** 302 (abs.).

Cavagraro, J., and Osband, M. E. (1983). *Biotechniques* **1,** 30.

Claude, B., and Justin, T. (1983). *Biochem. Biophys. Res. Comm.* **114,** 663.

Cosimi, A. B., Colvin, R. B., Burton, R. C., *et al.* (1981). *N. Engl. J. Med.* **305,** 308.

Cote, R. J., Morrissey, D. M., Houghton, A. N., Beattie, E. J., Jr., Oettgen, H. F., and Old, L. J. (1983). *Proc. Natl. Acad. Sci. U.S.A.* **80,** 2026.

Croce, C. M., Shander, M., Martinis, J., Cicurel, L., D'Ancona, G. C., Dorby, T. W., and Koprowski, H. (1979). *Proc. Natl. Acad. Sci. U.S.A.* **76,** 3416.

Cuello, A. C., Milstein, C., and Priestley, J. V. (1980). *Brain Res. Bull.* **5,** 575.

Curran, R. C., and Gregory, J. (1977). *Experientia* **33,** 1400.

De St. Groth, S. F., and Scheidegger, D. (1980). *J. Immunol. Meth.* **35,** 1.

Diamond, B. A., Yelton, D. E., and Scharff, M. D. (1981). *New Engl. J. Med.* **304,** 1344.

Douillard, J. Y., and Hoffman, T. (1983). *Meth. Enzymol.* **92,** 168.

Farrands, P. A., Perkins, A., Pimm, M. V., Embleton, M. J., Hardy, J. D., Baldwin, R. W., and Hardcastle, J. D. (1982). *Lancet* **11,** 397.

Friguet, B., Djavadi- Ohaniance, L., Pages, J., Bussard, A., and Goldberg, M. (1983). *J. Immunol. Meth.* **60,** 351.

Galfre, G., and Milstein, C. (1981). *Meth. Enzymol.* **73,** 1.

Galfre, G. S. C., Howe, C., Milstein, C., Butcher, G. W., and Howard, J. C. (1977). *Nature (London)* **266,** 550.

Gefter, M. L., Marqulies, D. H., and Scharff, M. D. (1977). *Somatic Cell Genet.* **3,** 231.

Gililand, D. G., Steplewski, Z., Collier, R. J., Mitchel, K. F., Chang, T. H., and Koprowski, H. (1980). *Proc. Natl. Acad. Sci. U.S.A.* **77,** 4539.

Goding, J. W. (1980). *J. Immunol. Meth.* **39,** 285.

Goldstein, G. (1984). *In* "Advances in Inflammation Research" (I. Otterness, R. Capetola, and S. Wong, eds.), Vol. 7, p. 115. Raven, New York.

Gupta, S. K., and Talwar, G. P. (1980). *Ind. J. Exp. Biol.* **13,** 1361.

Heide, K., and Schwics, H. G. (1978). *In* "Handbook of Experimental Immunology" (D. M. Weiss, ed.), p. 6. Blackwell, Oxford.

Herlyn, D. M., Steplewski, Z., Herlyn, M. F., and Koprowski, H. (1980). *Cancer Res.* **40,** 717.

Horejsi, V., and Hilgert, I. (1983). *J. Immunol. Meth.* **62,** 325.

Hudson, L., and Hay, F. C. (1976). *In* "Practical Immunology," p. 152. Blackwell, Oxford.

Ivanyi, J. (1980). *In* "Protides of the Biological Fluids" (H. Peeters, ed.), p. 471. Pergamon Press, New York.

Jarvis, A. P., Spriggs, T. A., Chipura, W. R., Sullivan, M., and Koch, G. A. (1982). *In Vitro* **18,** 276.

Johnson, G. D., Holbrow, E. J., and Dorling, J. (1978). *In* "Handbook of Experimental Immunology" (D. M. Weiss, ed.), 3rd ed. Ch. 15, 16. Blackwell, Oxford.

Kadish, J. L., and Wenc, K. M. (1983). *Hybridoma* **2,** 87.

Kearney, J. F., Redbruch, A., Liesegang, B., and Rajewsky, K. (1979). *J. Immunol.* **123,** 1548.

Kennett, R. H. (1980). *In* "Monoclonal Antibodies (R. H. Kennett, T. J. McKearn, and K. B. Bechtol, eds.), p. 372. Plenum, New York.

Kennett, R. H., McKearn, T. J., and Bechtol, K. B. (1980). *In* "Methods for Production and Characterization of Monoclonal Antibodies." Plenum, New York.

Kohler, G. (1981). *Immunol. Meth.* **44,** 286.

Kohler, G., and Milstein, C. (1975). *Nature (London)* **256,** 495.

Kohler, G., Howe, C. S., and Milstein, C. (1976). *Eur. J. Immunol.* **6,** 292.

Kwan, S. P., Yelton, D. E., and Scharff, M. D. (1980). *Monoclonal Antib.* **1,** 31.

Littlefield, J. W. (1964). *Science* **145,** 709.

Littman, B. H., Muchmore, A. V., Steinberg, A. D., and Greene, W. C. (1983). *J. Clin. Invest.* **72,** 1987.

Lydersen, B. K., Pugh, G. G., Duncan, E. C., Overman, K. T., Johnson, D. M., and Sharma, B. P. (1983). *In Vitro* **19,** 260.

MacLeish, P. R., Barnstable, C. J., and Townes-Anderson, E. (1983). *Proc. Natl. Acad. Sci. U.S.A.* **80,** 7014.

McDougal, S. J., Browning, S. W., Kennedy, S., and Moore, D. D. (1983). *J. Immunol. Meth.* **63,** 281.

McMaster, R., and Williams, A. (1979). *Immunol. Rev.* **47,** 117.

Miller, R. A., Maloney, D. G., McKillop J., and Levy, R. (1981). *Blood* **58,** 78.

Momoi, M., Kennett, R. H., and Glick, M. C. (1980). *J. Biol. Chem.* **255,** 11914.

Nadler, L. M., Stashenko, P., Hardy, R., Kaplan, W. D., Button, L. N., Kufe, D. W., Antman, K. H., and Schlossman, S. F. (1980). *Cancer Res.* **40,** 3147.
Olsson, L., and Kaplan, H. S. (1983). *Meth. Enzymol.* **92,** 3.
Olsson, L., Kronstrom, H., Cambon de Mouzon, A., Honsik, C., Brodin, T., and Jakobsen, B. (1983). *J. Immunol. Meth.* **61,** 17.
Pearson, T. W. (1982). UNDP, World Bank, World Health Organization, 47.
Pearson, T. W., and Anderson, N. L. (1980). *Anal. Biochem.* **101,** 377.
Pintus, C., Ransom, J. H., and Evans, C. H. (1983). *J. Immunol. Meth.* **61,** 195.
Renart, J., Reiser, J., and Stark, G. R. (1979). *Proc. Natl. Acad. Sci. U.S.A.* **76,** 3116.
Ritz, J., Pesardo, J. M., Notis-McConarty, J., and Schlossman, S. F. (1980). *J. Immunol.* **125,** 1506.
Rosen, S. T., Winter, J. N., and Epstein, A. L. (1983). *Ann. Clin. Lab. Sci.* **13,** 173.
Schlom, J., Wunderlich, D., and Teramoto, Y. A. (1980). *Proc. Natl. Acad. Sci. U.S.A.* **77,** 6841.
Seeher, D. S., and Burke, D. C. (1980). *Nature (London)* **285,** 446.
Sevier, E. D., David, G. S., Martinis, J., Desmond, W. J., Bartholomew, R. M., and Wang, R. (1981). *Clin. Chem.* **27,** 1797.
Sharon, J., Morrison, S. L., and Kabat, E. A. (1979). *Proc. Natl. Acad. Sci. U.S.A.* **76,** 1420.
Shulman, M., Wilde, C. D., and Kohler, G. (1978). *Nature (London)* **276,** 269.
Sijens, R. J., Thomas, A. A. M., Jackers, A., and Boeye, A. (1983). *Hybridoma* **2,** 231.
Siragarian, R. P., Fox, P. C., and Berenstein, E. H. (1983). *Meth. Enzymol.* **92,** 17.
Steinitz, M., Klein, G., Koskimies, S., and Makel, O. (1977). *Nature (London)* **269,** 420.
Steplewski, Z. (1980). *Transplant. Proc.* **12,** 384.
Strand, M., Scheinberg, D. A., and Gansow, D. A. (1984). *In* "Cell Fusion: Gene Transfer and Transformation" (R. F. Beers, Jr. and E. G. Bassett, eds.), p. 385. Raven, New York.
Suddith, R. L., Townsend, C. M., and Thompson, J. C. (1980). *Surg. Forum* **31,** 185.
Taggart, R. T., and Samloff, I. M. (1983). *Science* **219,** 1228.
Tilcock, C. P. S., and Fisher, D. (1982). *Biochem. Biophys. Acta* **685,** 340.
Tsu, T. T., and Herzenberg, L. A. (1980). *In* "Selected Methods in Cellular Immunology" (B. B. Mishell and S. M. Shiigi, eds.), p. 373. W. H. Freeman, San Francisco.
Uotila, M., Engvall, E., and Rusolahti, E. (1980). *Mol. Immunol.* **17,** 741.
Van Heyningen, V., Brock, D. J. H., and Van Heyningen, S. (1983). *J. Immunol. Meth.* **62,** 147.
Wahl, R. L., Parker, C. W., and Philpott, G. W. (1983). *J. Nucl. Med.* **24,** 317.
Ware, J. L., Paulson, D. F., Parks, S. F., and Webb, K. S. (1982). *Cancer Res.* **42,** 1215.
Wells, D. E., and Price, P. J. (1983). *J. Immunol. Meth.* **59,** 49.
Westerwoudt, R. J., Blom, J., Naipal, A. M., and Van Rood, J. J. (1983). *J. Immunol. Meth.* **62,** 59.
White, R. A. H., Mason, D. W., Williams, A. F., Galfre, G., and Milstein, C. (1978). *J. Exp. Med.* **48,** 664.
Woloschak, G. E., and Senitzer, D. (1983). *Hybridoma* **2,** 341.
Yelton, D. E., Margulies, D. H., Diamond, B., and Scharff, M. D. (1980). *In* "Monoclonal Antibodies" (R. H. Kennett, T. J. McKearn, K. B. Bechtol, eds.), p. 3. Plenum, New York.
Zola, H., and Brooks, D. (1982). *In* "Monoclonal Hybridoma Antibodies" (J. G. R. Hurrell, ed.), p. 1. CRC Press, Florida.

ADVANCES IN CELL CULTURE, VOL. 4

MORPHOGENESIS OF MITOCHONDRIA DURING SPERMIOGENESIS IN DROSOPHILA ORGAN CULTURE

Winfrid Liebrich, Karl Heinz Glätzer, and Norbert Kociok

Institut für Genetik
Universität Düsseldorf
Düsseldorf, Federal Republic of Germany

I. Introduction

It is a widespread phenomenon that during animal spermiogenesis mitochondria of male germ cells undergo a metamorphosis. During this process the typical mitochondrial structure is modified or completely lost (reviewed by Bacetti, 1970; Beatty and Glueckson-Waelsch, 1972; Bacetti and Afzelius, 1976; Fawcett and Bedford, 1979). The most drastic alterations occur in insects (for literature see Perotti, 1973).

Probably the most detailed information on spermiogenesis in general is available for *Drosophila* (Meyer, 1968; Tates, 1971; Stanley *et al.*, 1972; Rosen-Runge, 1977; Lindsley and Tokuyasu, 1980). During this phase the mitochondria of the male germ cells change their shape and

ISBN 0-12-007904-6

form the so-called nebenkern. Little is known about the complete morphogenic process because only small aspects of single stages have been described so far (e.g., the nebenkern; see Meyer, 1968; Tates, 1971; Tokuyasu, 1975). Also the dynamics of aggregation or mechanisms leading to the nebenkern derivatives are poorly understood.

For about two decades it has been possible to isolate germ cell tissue of *Drosophila* and cultivate it *in vitro* (Kuroda, 1972, 1974; Fowler and Uhlmann, 1974; Fowler and Johannisson, 1976). This presents the opportunity to directly observe the morphogenesis of the mitochondria and select single, well-defined stages for further investigations by, e.g., transmission electron microscopic, autoradiographic, or immunologic techniques.

Surprisingly, the last complete survey on the fate of the mitochondria during male germ cell differentiation was published by Meves in 1900. Much new information has been accumulated in the past eight decades. In the present review an attempt is made to combine the data. Special attention is paid to results obtained from organ culture *in vitro*. For explanation of special terms concerning spermiogenesis and the morphogenesis of the mitochondria the reader is referred to Section III, A.

II. Historical Outline

A. Discovery of Mitochondria

Recently, an extensive article about the history of the discovery of the mitochondria appeared with special reference to their function (Ernster and Schatz, 1981). Therefore, only a summary of the detection of the mitochondria as a cell organelle is given.

First reports about the mitochondria were published more than 100 years ago. This was during the exciting time of the origin of cytology (see Hughes, 1959). Kölliker (1857, 1888) mentioned in his work muscle cell "Körnchen" (= granules; mitochondria) without attaching much importance to them. In the following years mitochondria have been described as elements of various cells. They were described as "Bioblasten," "Mikrosomen," "Cytosomen," etc. (reviewed in Meves, 1901, 1911; Benda, 1902). Benda was the first to discover that the differently named structures represent the same cell organelle. In 1894, he found that the Körnchen (= granules) are a specific component of each animal cell. Later, in 1898, he introduced the term "mitochondria" because the "granules" (χόνδριον = chondrion) joined together in two "threads" (μίτος = mitos) during spermiogenesis of insects (cited after Benda, 1902).

B. Mitochondria as "Nebenkern" in Insect Spermiogenesis

During his study of spermiogenesis in meal beetles A. v. la Valette St. George discovered in 1867 that "einzelne Hodenkugeln eine auffallende Art von Zellen enthalten, welche neben einem blassen Kerne einen eigenthümlichen mehr oder weniger glänzenden Körper enthalten."[1] He found that this body had some kind of relationship to the sperm tail during the formation of spermatozoa. In later papers he termed this organelle "Nebenkörper." During the following years it was detected in many insect species. However, each scientist created his own word to name the same structure (e.g., Benda, in 1902, used the word "Fadenkörper" meaning "thread body"). Bütschli (1871a,b) established the expression "Nebenkern" (= accessory nucleus) which became adopted by other scientists during the following years. Even v.la Valette St. George accepted it. In 1886 he threw light on the relation between the "Cytomikrosomen" (= mitochondria) and the Nebenkern and explained in the introduction to his work that he actually had detected the Nebenkörper. However, he would use the term Nebenkern to avoid confusion. In his paper he also demonstrated for the first time that mitochondria change their appearance in the course of spermiogenesis, a concept we now know to be true. In 1900 Meves published his admirable precise cytological description of the complete differentiation of the mitochondria during spermiogenesis of Lepidoptera including all the relevant literature on the nebenkern.

Probably the first papers referring to morphogenesis of mitochondria during spermiogenesis in *Drosophila* appeared in the early 1930s (Metz, 1927; Huettner, 1930). Surprisingly, in the following years little attention was paid to the formation of the nebenkern. There were only a few papers dealing with the alterations of the mitochondria during spermiogenesis (e.g., Ito, 1960). With the development of electron microscopy, however, came new discoveries about the origin of the nebenkern and its architecture in *Drosophila* (Bairati, 1967; Tates, 1971; Tokuyasu, 1974a,b, 1975).

All investigations described so far had been carried out with dissected testes. In particular the fate of the mitochondria during meiosis and spermiogenesis had been reconstructed using squashed prepara-

[1]Single testicular spheres possess conspicuous kinds of cells (= cysts of spermatids), which contain a peculiar, more or less gleaming body (translated by the authors).

tions and serial sections. Using these techniques a general picture emerged about mitochondrial morphogenesis. However, some findings proved to be incorrect. Observations on differentiating living material has clarified some of these uncertainties (Liebrich, 1981a).

C. *In Vitro Culture of Insect Tissue*

Goldschmidt (1915, 1916, 1917) succeeded in cultivating insect tissue for the first time *in vitro*. He used isolated cysts of the pupal testes of *Samia cecropia* L. (Lepidoptera) and mounted them in a hanging drop of hemolymph in a special culture chamber. It was possible to obtain differentiation of a spermatocyte cyst "up to the full grown spermatozoa under the eye of the observer" (1915). This was the first time that spermiogenesis could be followed directly *in vitro*. However, Goldschmidt never repeated his experiments nor paid much attention to the fate of mitochondria of male germ cells.

"*In vitro* work on isolated cysts of pupal testes of *Drosophila hydei* was performed in the early 1970s by Fowler and his colleagues (Fowler and Uhlmann, 1974; Fowler and Johannisson, 1976) and by Kuroda (1972, 1974) and Cross and Shellenbarger (1979) on *Drosophila melanogaster*. The technique of Fowler was modified by Liebrich (1981a). It is now possible to cultivate single cysts from testes isolated from third instar larvae, pupae, and imagines of any *Drosophila* species.

III. Mitochondrial Morphogenesis during Spermiogenesis of *Drosophila In Vitro*

A. *Spermatogenesis in Drosophila*

The whole cycle of male germ cells resulting in free spermatozoa during the life span of an organism is called spermatogenesis. Spermiogenesis in contrast describes the postmeiotic differentiation of the spermatids.

1. *Cell Lineage*

The cell lineage of male germ cells has so far been investigated exclusively in *D. melanogaster* (Cooper, 1950; Fullilove *et al.*, 1978; Warn, 1979). According to these studies there are so-called stem cells of the germ cells; they grow in the testes to spermatogonia. Each spermatogonium is enveloped by two cyst cells of somatic origin (Aboim,

1945; Cooper, 1950; Rosen-Runge, 1977; Lindsley and Tokuyasu, 1980), thus forming a cyst. The encysted germ cell divides partly asynchronously. The number of these premeiotic divisions varies from species to species (Hanna *et al.*, 1982; Liebrich *et al.*, 1982). Thereafter, the germ cells, now called spermatocytes, pass through the meiotic divisions. In the ensuing process, spermiogenesis, they are called spermatids. All these events take place within the cyst.

2. Spermiogenesis

The development of the spermatids can be divided into several stages (see also Figs. 1–4). (a) Phase of nebenkern formation: The mitochondria of the germ cells aggregate after meiosis to form the spherical body, the nebenkern. (b) Phase of nuclei movement: Spermatids become oriented such that the nuclei are situated at one end of the now ellipsoid cyst. (c) Phase of cyst elongation: The ellipsoid cyst with the spermatids (spermatid bundle) becomes narrow and extended. The elongating spermatids are still connected via cytoplasmic bridges. (d) Phase of individualization: The spermatid bundle is completely elongated. A mechanism starts to remove the cytoplasmic bridges and, as a consequence, each spermatid is invested with its own cytoplasmic membrane. (e) Phase of coiling: The whole bundle adhers to the terminal epithelium of the testis and is wound into a disk-like body. After that phase the cyst cells degenerate and release free motile spermatozoa into the *vasa efferentia.*

Detailed descriptions of the single phase as well as their genetics may be found in the reviews of Meyer (1968), Tates (1971), Kiefer (1973), Hess (1980), or Lindsley and Tokuyasu (1980).

B. Technique of Cultivating Isolated Cysts

To obtain isolated cysts Kuroda (1972, 1974) dissected testes of 48-hour old pupae of *D. melanogaster* and cut them into several pieces. He cultivated the fragments in tissue culture flasks using his special insect medium K-17 (Kuroda, 1974). Mostly those cysts that had been in the process of elongation proceeded in their differentiation within 24 hours of culture. Cysts with spermatogonia, spermatocytes, or early spermatids increased only slightly in size. Since Kuroda used culture flasks he was not able to follow the differentiation of one single cyst. About the same time in our Institute Fowler (1973) started to cultivate whole testes of *D. hydei.* He observed that the sheet of some testes accidentally opened during culture and cysts were released into the medium of the culture chamber (chamber after Cruickshank *et al.*,

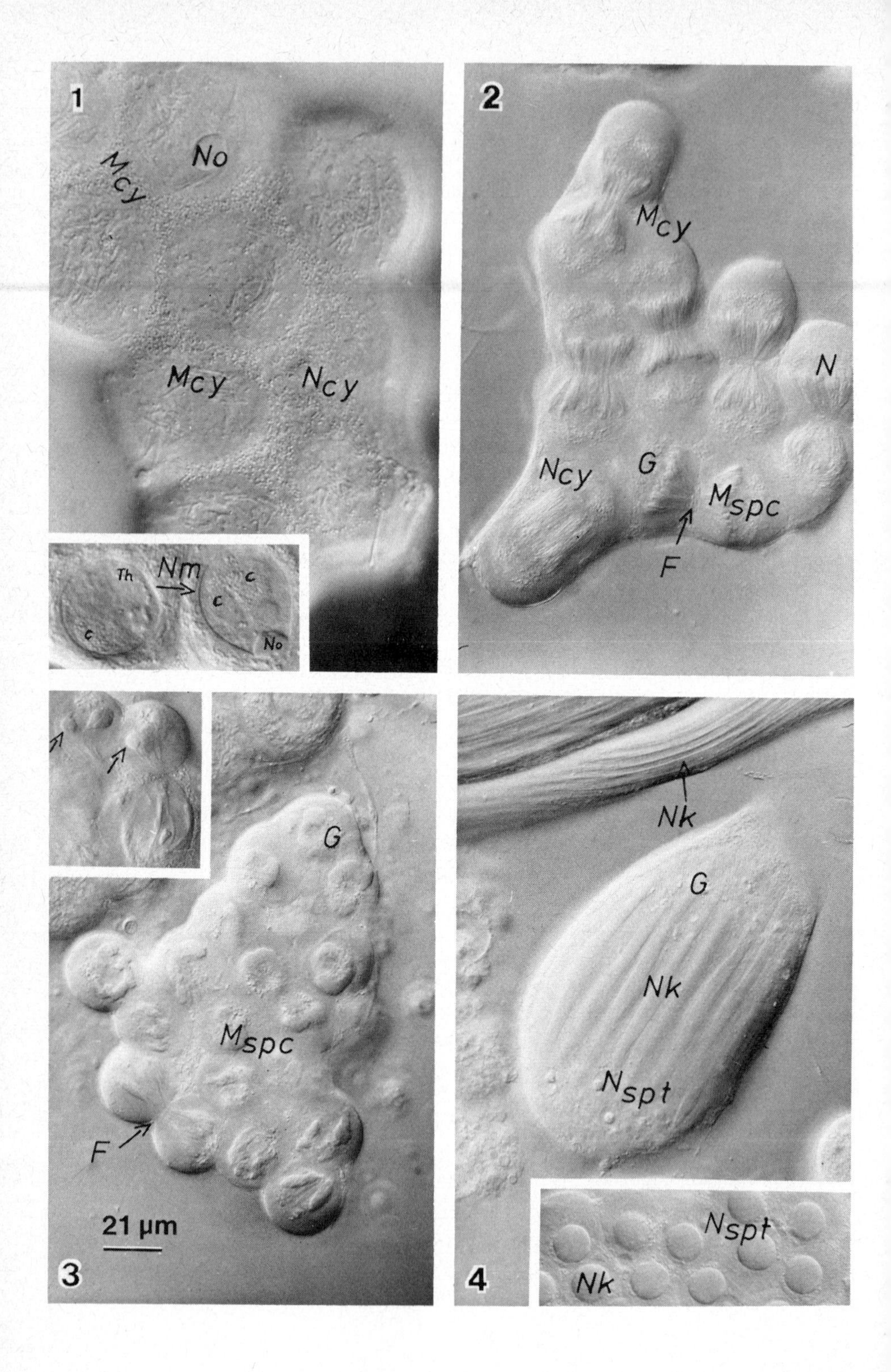
1
No
Mcy
Mcy
Ncy
Th
Nm
c
c
c
No
2
Mcy
N
Ncy
G
Mspc
F
3
G
Mspc
F
21 µm
4
Nk
G
Nk
Nspt
Nspt
Nk

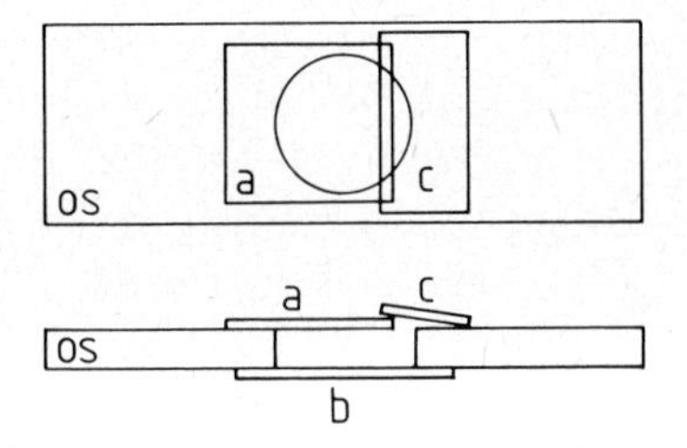

FIG. 5. Culture chamber. os, Object slide; b, a, c, sequence according to which the coverslips are mounted.

1959). Thus, it was possible to follow the differentiation of a single cyst during a short period of time. In later experiments Fowler isolated testes and disrupted their wall to obtain free single cysts (Fowler and Uhlmann, 1974; Fowler and Johannisson, 1976).

It is worth mentioning that neither Kuroda nor Fowler has utilized culture chambers that are used to perform living observations of single cells. Therefore, both authors could only apply low magnification objectives to study spermiogenesis *in vitro*. The culture chamber described below takes advantage of the high resolution of oil immersion objectives (Liebrich, 1981a). Its basic part is a glass object slide either 2.2 mm (for long-term culture) or 1.0 mm thick (for short-term culture) with a hole of 15.0 mm in diameter (Fig. 5). A coverslip is mounted from one side (Fig. 5b) and the resulting chamber filled partly with medium. An isolated testis is transferred into the medium and punctured by means of thin tungsten needles prepared according to Fjeld (1972). In general between 40 and 80 cysts were released undamaged through the opening. Thereafter the chamber is closed with another

FIGS. 1–4. Normal differentiation of male germ cell mitochondria. (Interference-contrast optics, ×360.) Fig. 1. Prophase I. Mitochondria of the germ cells (M_{spc}) begin metamorphosis. Inset: Same cyst, level of the nuclei (No, nucleolus of a primary spermatocyte; Nm, nuclear membrane) of the primary spermatocytes. Some of the Y chromosomal loops can be seen. Th, Threads; C, clubs; for nomenclature, see Hess (1980); M_{cy}, mitochondria of a cyst cell; N_{cy}, nucleus of a cyst cell. Fig. 2. Postmetaphase stages of meiosis I. Mitochondria of the spermatocytes (M_{spc}) pass through the fusom (F) [for nomenclature, see Meyer (1961)] from one daughter cell to the other. At that time, a granular material (G) appears over the nucleus (N) of the secondary spermatocyte. It could be the RNP material of the disintegrated Y chromosomal loops (Liebrich 1981). Fig. 3. Phase of nebenkern formation. Inset shows the thick ends of the aggregating mitochondria (arrows), The thin middle parts of the mitochondria (M_{spc}) pass through the fusom (F) of the two daughter cells. Fig. 4. Phase of cyst and nebenkern (Nk) elongation. In the upper part of the photograph are two sectioned elongating spermatid bundles. Inset shows the early "onion" nebenkern (Nk) stage (N_{spt}, spermatid nucleus). From Liebrich (1982).

coverslip (Fig. 5a), leaving a small slit through which the chamber is now finally filled with medium. The slit is sealed with a piece of coverslip (Fig. 5c). Vaseline served in all cases as glue.

A suitable medium to cultivate isolated cysts is that developed by Shields and Sang (1977) supplemented with 20% fetal calf serum (Liebrich, 1981a).

C. *Observations in Organ Culture*

Because there are nearly no data for other *Drosophila* species available, only results obtained with *D. hydei* are referred to here (Liebrich, 1981a, 1982).

1. *Meiosis*

The mitochondria of germ cells and of cyst cells are approximately the same size in cysts with spermatogonia and young spermatocytes (Fig. 6). The mitochondria of the germ cells are randomly distributed in spermatogonia and spermatocytes in *D. hydei* (Fig. 1). This is in contrast to *D. melanogaster* where only in spermatogonia a random distribution is observed (Tates, 1971). The first visible steps of differentiation can be observed during the early prophase I; the mitochondria of the spermatocytes thicken and are now easily distinguishable from those of the cyst cells (Fig. 1). They are equally distributed throughout the cytoplasma of the germ cells until the spindle apparatus begins to form. Then the mitochondria become ordered in the centriolar region. As soon as the two poles separate the remaining mitochondria become arranged in the spindle apparatus as well. During the first meiotic division mitochondrial "threads" cover the nuclear region more or less completely and form a Chinese lantern-like structure. In the course of cytokinesis they extend from one daughter cell to the other. This configuration changes rapidly during the second meiotic

---

FIGS. 6–11. *Drosophila hydei,* light microscopy. Fig. 6. Larval testis, cross section; to the left end (apical) are spermatogonia, to the right end (terminal) spermatocytes. Inset, left: enlarged part of the spermatogonia region. Mitochondria of the cyst cells and the spermatogonia are of the same size (dark dots represent mitochondria). Inset, right: enlarged part of the spermatocyte region. Mitochondria of spermatocytes are clearly larger and thicker than those of the cyst cells. Figs. 7–11. *In vitro* differentiating cysts (detail). Fig. 7. Metaphase II. Mitochondria are arranged from pole to pole. Fig. 8. Telophase II. Figs. 9–11. Postmeiotic nebenken formation. Fig. 9. Mitochondria form a funnel-like structure. Fig. 10. The aggregates of mitochondria resemble orange slices. Fig. 11. Mitochondria are aggregated into two bodies (upper right corner), which later on fuse to form only one body (e.g., lower left corner).

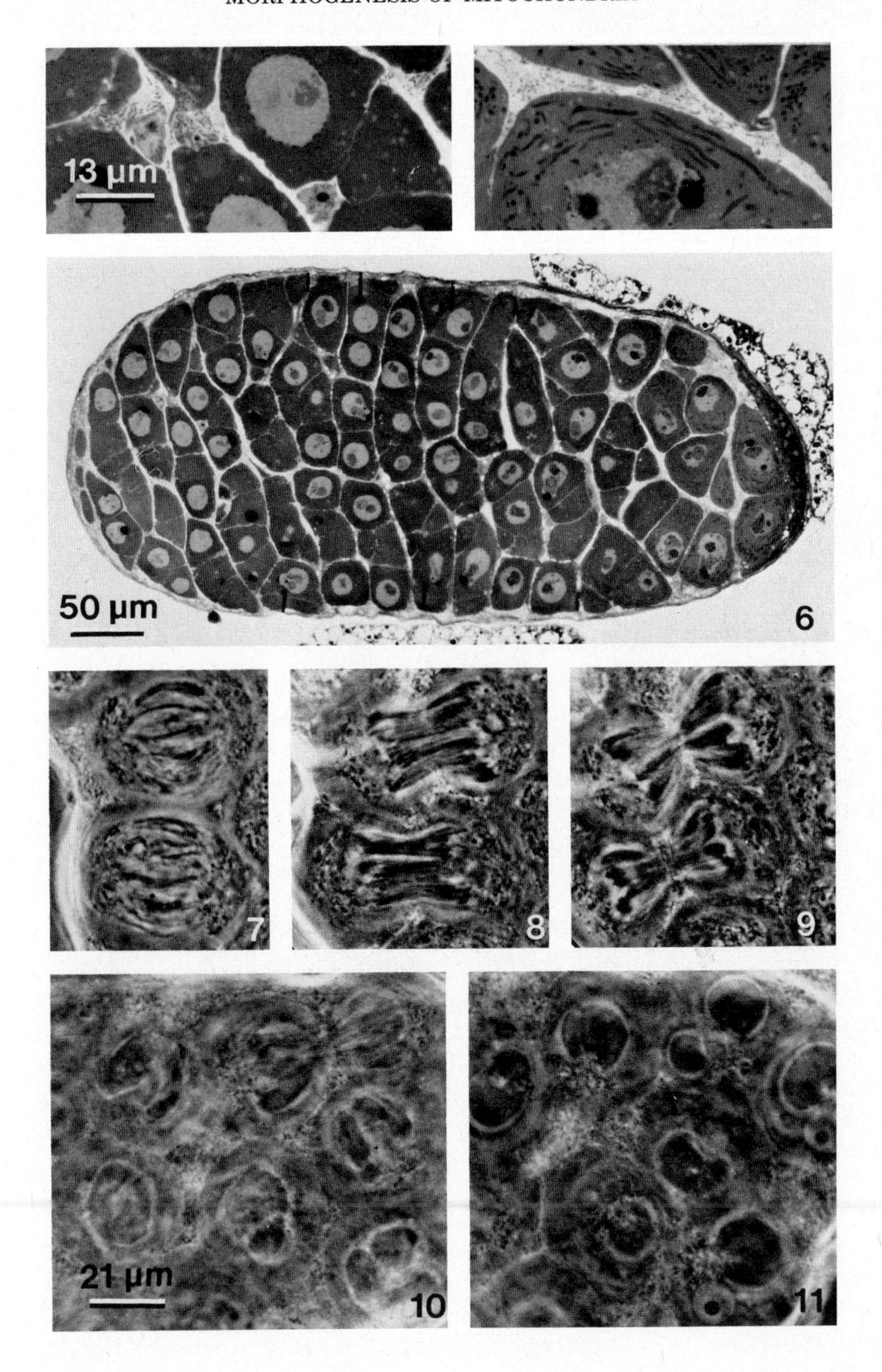
13 µm
50 µm
6
7
8
9
21 µm
10
11

division when the mitochondria again become oriented beside the new spindle apparatus (Fig. 3).

2. Nebenkern Formation

The most dramatic event represents the ensuing phase of nebenkern formation (Figs. 7–12). The linearly arranged mitochondria now aggregate into several groups and appear to fuse with each other. The result is a funnel-shaped body with the spermatid nucleus in its center (Fig. 3 inset). Subsequently the mitochondrial aggregates change their shape and resemble that of a group of orange slices (Fig. 10) which finally aggregate into a "peeled orange" whose slices are slightly separated at one end. The spermatid nucleus is now situated near this opening, outside the former funnel. Often two or more bodies can be seen during this phase (Fig. 11). Later on they fuse to form a single spherical nebenken (Fig. 4 inset). This stage is called the "phase of onion nebenkern" (Tokuyasu, 1975). There is no visible change for about 8 hours in culture. At the end of this period two nebenkern derivatives can be detected.

3. Nebenkern Elongation

After the onion phase the nuclei of the spermatids move to one end of the cyst. At the same time the nebenkern derivatives start to elongate as well. They become ellipsoid and extended. In well-elongated spermatid bundles the nebenkern derivatives appear as thin threads extending from the nucleus to nearly the end of the spermatid (Fig. 4). They parallel the axoneme.

D. Ultrastructure

Many papers deal with the mitochondria of male germ cells of *Drosophila*. However, only limited aspects of mitochondrial morphogenesis are covered in these reports. Most investigated is the phase of onion nebenkern or the sperm ultrastructure. In other insects [e.g., Lepidoptera (André, 1959) and locusts (Beams *et al.*, 19540] the morphogenesis of the mitochondria in male germ cells has been better investigated than in *Drosophila*.

1. Nebenkern Formation

Only a few ultrastructural data are known about the early stages of mitochondrial morphogenesis, aggregation, and nebenkern formation. These dramatic events take place within a few hours (Fowler and Johannisson, 1976; Cross and Shellenbarger, 1979; Liebrich, 1981a).

Investigations so far have used dissected material with its disadvantage that suitable stages are rare and found only by chance. Electron microscopic studies on cysts from *in vitro* cultures have not yet been performed.

In general, findings with the electron microscope confirm light-microscopic observations (see Section III,C). In addition, some new details about the ultrastructure were discovered: mitochondria of male germ cells are less electron dense than mitochondria of cyst cells (Fowler and Johannisson, 1976; W. Liebrich, unpublished observations). Furthermore, electron-dense granules are conspicuous in the mitochondrial matrix of growing germ cells (Fig. 12). During prophase I the number of mitochondria multiplies. According to Tates (1971) there is a more than fivefold increase for *D. melanogaster*. At the same time the mitochondria enlarge and become more than three times thicker than those of the cyst cells. Immediately after meiosis the mitochondria appear as two or more densely packed aggregates. Later on only one single body is seen. This stage is named "clew phase" by Tates (1971). In the subsequent onion phase (Tokuyasu, 1975) the mitochondria fuse and form a lamellar system that resembles a cross section of an onion. The end of this phase is characterized by two interlocked large "giant" mitochondria (Fig. 12). This onion phase is probably the most often described stage of spermiogenesis (Bacetti, 1970; for *Drosophila* see Stanley *et al.*, 1972; Kiefer, 1973; Lifschytz and Hareven, 1977; Brick *et al.*, 1979). In addition, several authors reconstructed the onion nebenkern using serial sections (Pratt, 1968; André, 1959; for *Drosophila* see Tates, 1971; Tokuyasu, 1974b).

2. Elongation and Crystallization of the Nebenkern

The mitochondrial fusion products separate into two differently sized derivatives (Meyer, 1961; Tates, 1971; Tokuyasu, 1974b). After separation the two derivatives elongate paralleling the axonem. In the course of further elongation the two derivatives become arranged in a distinct angle to the axonem (Meyer, 1964; Tokuyasu, 1974b). Thereafter the cristae disappear and simultaneously crystallization starts at that part of the mitochondrial membrane which is adjacent to the axonem (Fig. 13). While the paracrystallization proceeds the remaining content of the nebenkern derivatives becomes more electron dense. In *D. hydei* the two derivatives differentiate equally, whereas in *D. melanogaster* both differ markedly in size (cf. Meyer, 1961; Tokuyasu, 1974b). Microtubles are arranged between and around the nebenkern derivatives and exhibit a characteristic architecture (Meyer, 1964; Stanley *et al.*, 1972; Tokuyasu, 1974a,b, 1975). At about the period of

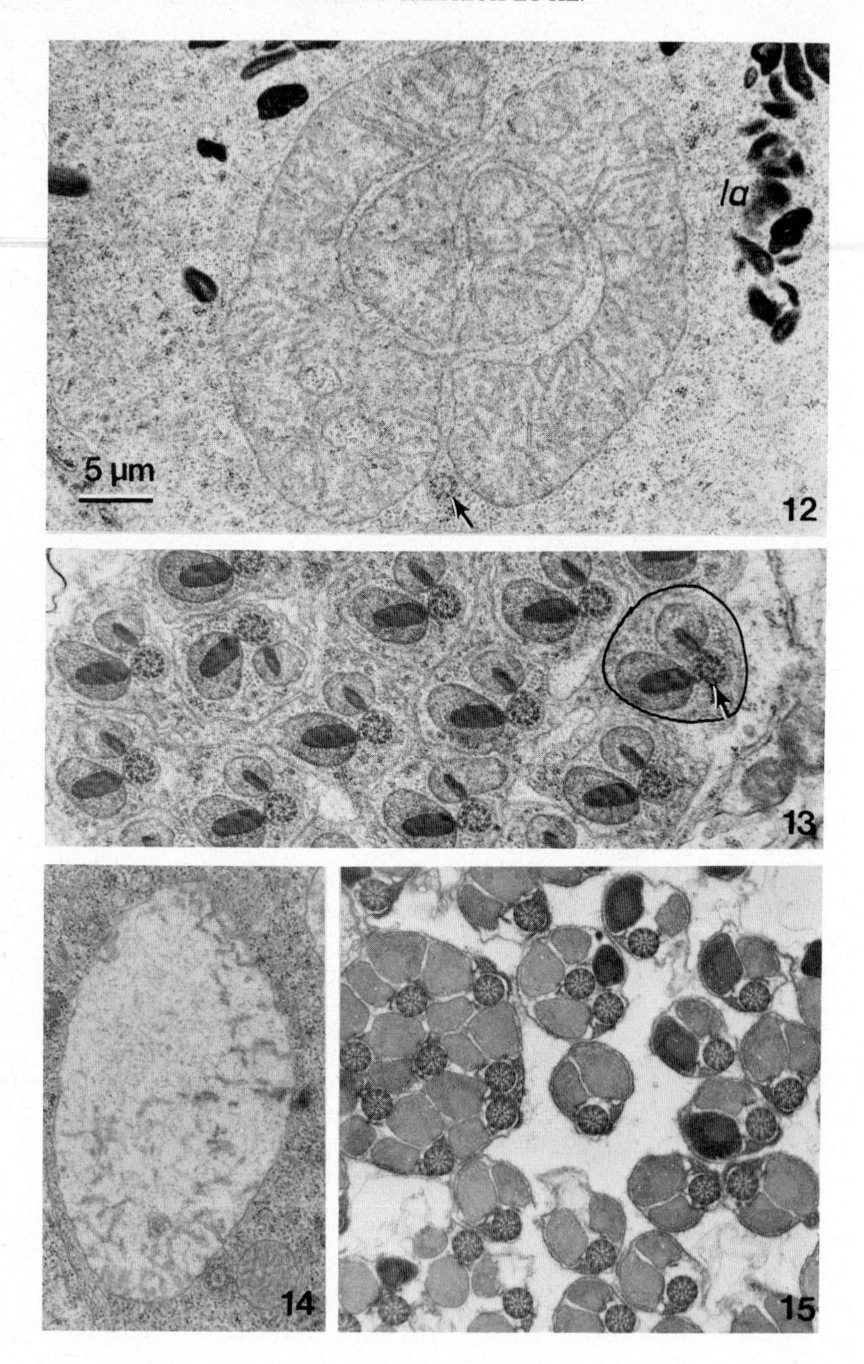
la
5 µm
12
13
14
15

individualization the transformation of the nebenkern derivatives is completed. They are now occupied by a highly ordered substance showing a characteristic crystal-like pattern (Meyer, 1964; Stanley *et al.*, 1972; Perotti, 1973; Tokuyasu, 1974b). A similar arrangement can also be found in spermatozoa from other insects (Bacetti, 1970; Itaya, 1980).

There are differences in the nebenkern transformation between *D. melanogaster* and *D. hydei*. In *D. hydei* the two derivatives differentiate equally, whereas in *D. melanogaster* both differ markedly in size (cf. Meyer, 1961; Tokuyasu, 1974b).

E. Molecular Features

1. Mitochondrial DNA

Much is known about the structure and function of the mitochondrial genome in eukaryotic cells (for literature see Levin, 1980). No detailed study of quantity, localization, and fate of the mitochondrial DNA during spermiogenesis is reported so far.

With specific fluorescent probes it is possible to detect DNA in cytological preparations. A good tool is the fluorescent dye 4',6-diamidino-2-phenylindole (DAPI) which is specific for adenine-thymidine pairs. It was introduced to detect and separate yeast mitochondrial DNA (Williamson and Fennel, 1975). It was asserted by the authors that the technique should permit the detection of one single mitochondrial DNA molecule provided it is packaged in a compact form.

About 25% of the mitochondrial DNA of *Drosophila* consists of an AT-rich region (Berninger *et al.*, 1979). We therefore used DAPI to follow the fate of the mitochondrial DNA during metamorphosis of the mitochondria. Some unpublished results obtained in *D. hydei* are shown in Fig. 16. During the meiotic divisions the mitochondria became concentrated and could be visualized by their fluorescence. Remarkably, during spermatid elongation and at later stages the fluorescence was equally distributed along the nebenkern despite the fact that the staining represented only the amount of AT compartments in the nebenkern (Fig. 17).

FIGS. 12–15. *Drosophila hydei,* transmission electron microscopy (all at the same magnification). Fig. 12. Two interlocked "giant" mitochondria at the end of the "onion phase" (la, lamellar body, arrow indicates the axonem). Fig. 13. Section through an elongating spermatid bundle (detail); encircled: a single spermatid showing the two nebenkern derivatives with starting crystallization (arrow: axonem). Figs. 14–15. Carnitine defects (for explanation, see text).

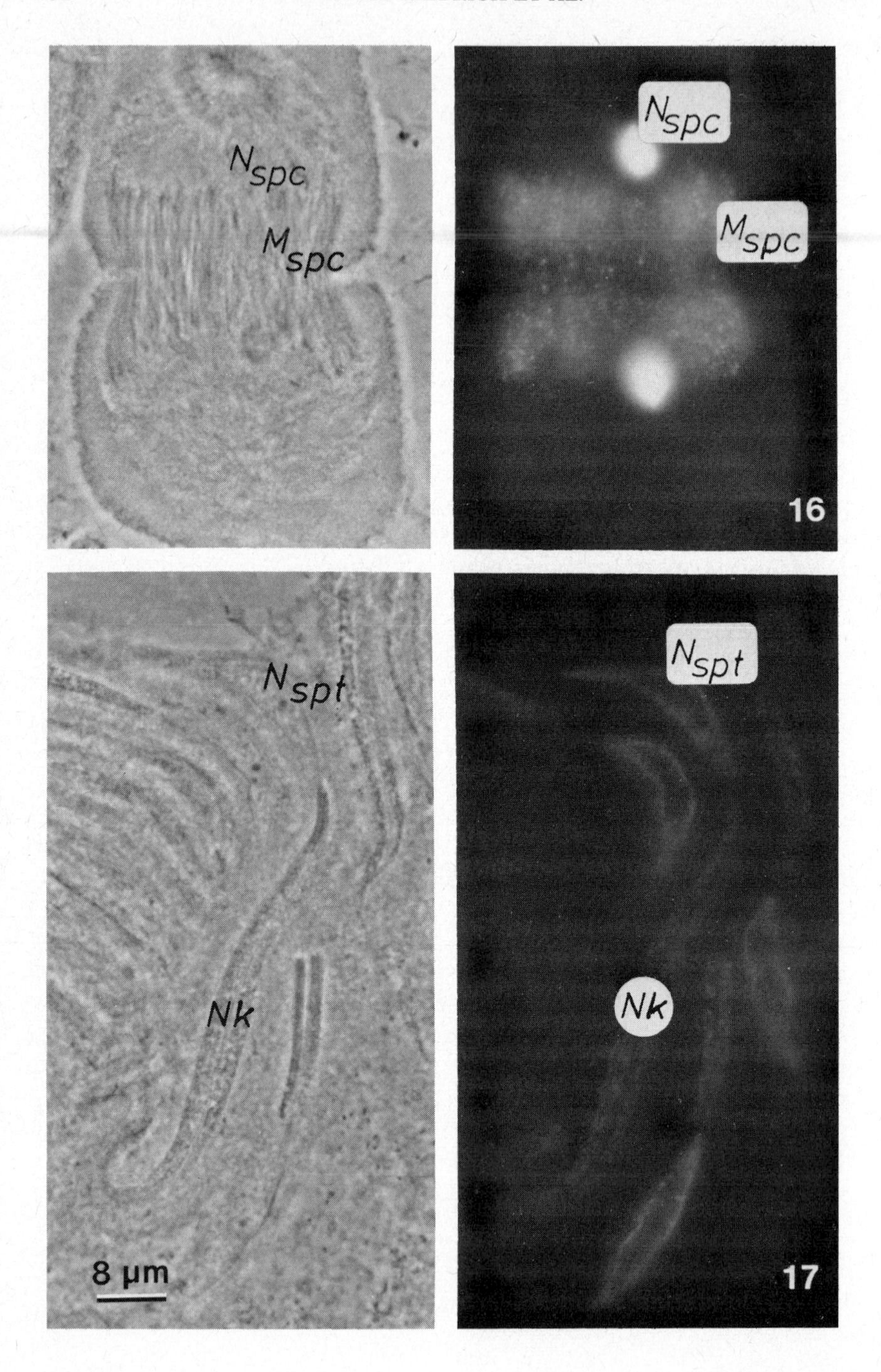

N spc
M spc
N spc
M spc
16
N spt
Nk
8 µm
N spt
Nk
17

These DAPI experiments proved high amounts of DNA in the nebenkern derivatives. The role of this DNA is not yet known. Two possible functions may be hypothesized. First, the DNA may be involved in the process of mitochondrial morphogenesis. Second, the DNA may play a role in paternal cytoplasmic inheritance. In fact, in the literature only the effect of the transmission of the genetic material has been investigated. The great excess of mitochondria in eggs suggests maternal inheritance. Because most sperms so far studied completely enter the egg at fertilization (cf. Perotti, 1973), the paternal mitochondrial genetic system also may contribute to the progeny mitochondria. Maternal inheritance was demonstrated in vertebrates by molecular hybridizations of the respective nucleic acids (Dawid and Blackler, 1972), and in *Drosophila* by the segregation of mitochondrial restriction sites in suitable crosses (Reilly and Thomas, 1980). Thus, it appears that the paternal mitochondrial genome does not contribute to cytoplasmic inheritance. This result and our experiments with DAPI suggest that the DNA in the nebenkern derivatives may function during transformation.

2. *Environmental Factors Affecting Nebenkern Differentiation*

The development of the nebenkern during spermiogenesis is a very sensitive process. It can be affected by genetic and environmental factors.

The effect of genetic factors located on the Y chromosome has been extensively studied and summarized in recent reviews (Lindsley and Tokuyasu, 1980; Hess, 1980). For example, *Drosophila* males lacking the complete Y chromosome already show defects at the very beginning of the mitochondrial transformation (Kiefer, 1973; Meyer, 1968).

Similar defects can be phenocopied by raising *Drosophila* on an artificial diet (Geer, 1967, 1972; Geer and Newburgh, 1970). For example, when carnitine is substituted for choline in the diet, only *Drosophila* males become sterile; the fertility of females is unaffected. Male sterility has a structural equivalent in abnormal nebenkern morphogenesis due to genetic defects. Often one of the nebenkern deriva-

---

FIGS. 16–17. *Drosophila hydei*. Mitochondria stained with DAPI (left: phase contrast optics: right: DAPI fluorescence). The mitochondrial DNA is marked by fluorescent dots. Fig. 16. Telophase I. N_{spc}, Nucleus of a spermatocyte; M_{spc}, mitochondria of the spermatocyte. Fig. 17. Elongating spermatids. N_{spt}, Spermatid nucleus; NK, nebenkern derivatives.

tives is enlarged (Fig. 14),[2] or more than two elements are found in one spermatid. The key role of choline is demonstrated by the reversibility of sterility when carnitine-raised males are fed choline (Geer, 1967). A hypothesis is put forward by the author that choline is required for the synthesis of phospholopid needed as an energy source for motility. Interestingly, crystallization of the nebenkern occurs despite severe morphogenetic disturbances, at least in *D. hydei* (Fig. 15).[2]

IV. Regulation of Morphogenesis and Function of Nebenkern Derivatives

A. *Cytological Aspects*

At least in *Drosophila* all mitochondria of male germ cells are involved in morphogenesis (Meyer, 1968; Tates, 1971; Stanley *et al.*, 1972; Kiefer, 1973; Brick *et al.*, 1979; Lindsley and Tokuyasu, 1980). This is in contrast to observations in Lepidoptera (André, 1959).

In the phase of nebenkern formation many cristae can be seen in the aggregating mitochondria. They gradually disappear during the elongation phase. This implies their possible function as a membrane reservoir at least for the inner membrane (Lindsley and Tokuyasu, 1980). At the outer mitochondrial membrane are "spheroid bodies" (membrane vesicles) which might serve as a supply source for this membrane (Tokuyasu, 1975). These structures obviously are a pool for the membrane necessary for elongation of nebenkern derivatives.

A few papers deal with structural components that are essential for the correct aggregation of the mitochondria. For example, only ring-like structures of mitochondria are found during nebenkern formation if the microtubule system is disrupted using colchicine (Wilkinson *et al.*, 1974, 1975; Liebrich, 1982). On the other hand, microfilaments sensitive to cytochalasin B are not involved in the aggregation process, at least *in vitro* (Liebrich, 1982).

The nebenkern itself could be a structure that promotes the elongation of the spermatids. However, in *D. hydei* cysts—in the nuclear movement phase—which are partly detached, it is possible to observe that (a) the early spermatids are already elongating while the nebenkern derivatives are still small ellipsoid bodies, and (b) in later elongation stages the nebenkern never extend up to the tip of the spermatid (Liebrich, 1981b). Furthermore, already elongated sper-

[2]Data and figures kindly provided by H. Küppers (unpublished).

matid bundles collapse *in vitro* if colchicine is added to the culture medium (Liebrich, 1982). The nebenkern seems to be an elastic structure (Tokuyasu, 1974b).

In this regard it is of interest that the crystallized derivatives show no typical mitochondrial enzyme activity. For example, cytochrome activity is missing (for extensive discussion of that problem see Perotti, 1973).

The above-mentioned data imply that the nebenkern might function, e.g., as a stabilizing element, as Meyer (1966) pointed out.

B. Genetic Aspects

Lifschytz and co-workers studied spermiogenesis in male sterile mutants of *D. melanogaster* whose mitochondria showed defects during the aggregation phase. They arrived at the conclusion that there might be two different types of mitochondria which later form the two nebenkern derivatives (Brick *et al.*, 1979). They also assumed the existence of "mitochondrial aggregation genes" (Lifschytz and Hareven, 1977). These should produce the specific information for the differentiation of the two mitochondrial types.

Autoradiographic studies using the light microscope indicate that most RNA required for spermatid development in *Drosophila* is synthesized in the nuclei before the meiotic divisions (Olivieri and Olivieri, 1965; Hennig, 1967; Gould-Somero and Holland, 1974). In contrast, studies using electron-microscopic autoradiography claim to detect transcription in the spermatid nucleus in Locusta and *D. melanogaster* (Curgy, 1972; Curgy and Anderson, 1972). Interestingly, the same authors also found activity in germ cell mitochondria as well as in the nebenkern derivatives even until the condensation of the crystalline material completed. These data are doubted (Gould-Somero and Holland, 1974; Lindsley and Tokuyasu, 1980). However, our own preliminary experiments with single isolated cysts of *D. hydei* indicate uridine incorporation into the nebenkern.

Still unsolved is the problem that morphogenesis might be a specific innate function of male germ cell mitochondria. In this respect the transcription of mitochondrial DNA in nebenkern derivatives (Curgy and Anderson, 1972) might play an essential role in morphogenesis.

All the available data are still insufficient to construct a general model for the regulation of morphogenesis or the actual function of the nebenkern derivatives. It is hoped that the use of *in vitro* culture of single isolated cysts may be of great value in learning more about the enigmatic role of the nebenkern.

Acknowledgment

The work carried out by the authors and described in this article was supported in part by the Deutsche Forschungsgemeinschaft (Grant 317/2-2 to W.L.; Grant 104/2-2 to K.H.G.).

References

Aboim, A. N. (1945). *Rev. Suisse Zool.* **52,** 53–154.

André, J. (1959). *Ann. Sci. Natl. Zool.* **12e Sér,** 283–307.

Baccetti, B. (ed.) (1970). "Spermatologia Comparata." Academia Nazionale dei Lincei, Quaderno No. 137, Rome. (Also published: "Comparative Spermatology." Academic Press, New York, 1970.)

Baccetti, B., and Afzelius, P. (1976). "The Biology of the Sperm Cell." Karger, Basel.

Bairati, A. (1967). *Z. Zellforsch.* **76,** 56–99.

Beams, H. W., Tahmisian, T. N., Devine, R. L., and Roth, L. (1954). *Biol. Bull.* **107,** 47–56.

Beatty, R. A., and Glueckson-Waelsch (eds.) (1972). "The Genetics of the Spermatozoon." Edinburgh, New York.

Benda, C. (1902). *Erg. Anat. Entwicklungsgesch.* **12,** 743–781.

Berninger, M., Cech, T., Fostel, J., Potter, D., Scott, M., and Pardue, M. L. (1979). *In* "Specific Eukaryotic Genes" (J. Engberg, H. Klenow, and V. Leick, eds.), Vol. 13, pp. 229–243. Munksgaard, Copenhagen.

Brick, D., Lifschytz, E., and Friedländer, M. (1979). *J. Ultrastruct. Res.* **66,** 151–163.

Bütschli, O. (1871a). *Z. Wiss. Zool.* **21,** 402–415.

Bütschli, O. (1871b). *Z. Wiss. Zool.* **21,** 526–534.

Cooper, K. W. (1950). *In* "The Biology of *Drosophila*" (M. Demerec, ed.), pp. 1–61. Wiley, New York.

Cross, D. P., and Shellenbarger, D. L. (1979). *J. Embryol. Exp. Morphol.* **53,** 345–351.

Cruickshank, C. N. D., Cooper, J. R., and Coran, M. B. (1959). *Exp. Cell Res.* **16,** 695–698.

Curgy, J.-J. (1972). *Exp. Cell Res.* **75,** 127–137.

Curgy, J.-J., and Anderson, W. A. (1972). *Z. Zellforsch.* **125,** 31–44.

Dawid, J. B., and Blackler, A. W. (1972). *Develop. Biol.* **29,** 152–161.

Ernster, L., and Schatz, G. (1981). *J. Cell Biol.* **91,** 227s–255s.

Fawcett, D. W., and Bedford, J. M. (1979). "The Spermatozoon." Urban and Schwarzacher, Baltimore, München.

Fjeld, A. (1972). *Develop. Biol.* **28,** 326–343.

Fowler, G. L. (1973). *Cell Different.* **2,** 33–42.

Fowler, G. L., and Uhlmann, J. (1974). *Drosophila Inform. Service* **51,** 81–83.

Fowler, G. L., and Johannisson, R. (1976). *In* "Invertebrate Tissue Culture" (E. Kurstak and K. Maramorosch, eds.), pp. 161–172. Academic Press, New York.

Fullilove, S. L., Jacobson, A. G., and Turner, F. R. (1978). *In* "The Genetics and Biology of *Drosophila*" (M. Ashburner and T. R. F. Wright, eds.), Vol. 2c, pp. 105–227. Academic Press, New York.

Geer, B. W. (1967). *Biol. Bull.* **133,** 548–566.

Geer, B. W. (1972). *In* "Insect and Mite Nutrition" (J. R. Rodriguez, ed.), pp. 471–491. North Holland, Amsterdam.

Geer, B. W., and Newburgh, R. W. (1970). *J. Biol. Chem.* **245,** 71–79.

Goldschmidt, R. (1915). *Proc. Natl. Acad. Sci. U.S.A.* **1,** 220–222.

Goldschmidt, R. (1916). *Biol. Zentralbl.* **36,** 160–167.

Goldschmidt, R. (1917). *Arch. Zellforsch.* **14,** 421–450.

Gould-Somero, M., and Holland, L. (1974). *Wilh. Roux Arch.* **174,** 133–148.

Hanna, P. J., Liebrich, W., and Hess, O. (1982). *Gamete Res.* **6,** 365–370.

Heidger, P. M. Jr. (1980). *Int. J. Insect. Morphol. Embryol.* **9,** 135–145.

Hennig, W. (1967). *Chromosoma (Berlin)* **22,** 294–357.

Hess, O. (1980). *In* "The Genetics and Biology of *Drosophila*" (M. Ashburner and T. R. F. Wright, eds.), Vol. 2d, pp. 1–32. Academic Press, New York.

Huettner, A. F. (1930). *Z. Zellforsch.* **11,** 615–637.

Hughes, A. (1959). "A History of Cytology." Abelard-Schuman, London, New York.

Itaya, P. W., Thompson, S. A., and Heidger, P. M., Jr. (1980). *Int. J. Morphol. Embryol.* **9,** 135–145.

Ito, S. (1960). *J. Biophys. Biochem. Cytol.* **7,** 433–440.

Kiefer, B. J. (1973). *In* "Genetic Mechanisms of Development" (F. H. Ruddle, ed.), pp. 47–102. Academic Press, New York.

Kölliker, A. (1857). *Z. Wiss. Zool.* **8,** 311–321.

Kölliker, A. (1888). *Z. Wiss. Zool.* **47,** 489–711.

Kuroda, Y. (1972). *Drosophila Inform. Service* **48,** 33.

Kuroda, Y. (1974). *J. Insect. Physiol.* **20,** 637–640.

Levin, B. (1980). "Gene Expression 2. Eucaryotic Chromosomes" (2nd ed.). Wiley, New York.

Liebrich, W. (1981a). *Cell Tissue Res.* **220,** 251–262.

Liebrich, W. (1981b). *Drosophila Inform. Service* **56,** 82–84.

Liebrich, W. (1982). *Cell Tissue Res.* **224,** 161–168.

Liebrich, W., Hanna, P. J., and Hess, O. (1982). *Int. J. Invert. Reprod.* **5,** 305–310.

Lifschytz, E., and Hareven, D. (1977). *Develop. Biol.* **58,** 276–294.

Lindsley, D. L., and Tokuyasu, K. T. (1980). *In* "The Genetics and Biology of *Drosophila*" (M. Ashburner and T. R. F. Wright, eds.), Vol. 2d, pp. 225–294. Academic Press, New York.

Metz, C. W. (1927). *Z. Zellforsch. Mikroskop. Anat.* **4,** 1–28.

Meves, F. (1900). *Arch. Mikroskop. Anat.* **56,** 553–606.

Meves, F. (1901). *Erg. Anat. Entwicklungsgesch.* **11,** 437–516.

Meves, F. (1911). *Arch. Mikroskop. Anat. Entwicklungsgesch.* **76,** 683–713.

Meyer, G. F. (1961). *Z. Zellforsch.* **54,** 238–251.

Meyer, G. F. (1964). *Z. Zellforsch.* **62,** 762–784.

Meyer, G. F. (1966). *Proc. 6th Int. Congr. Electr. Microsc., Kyoto,* Vol. II, pp. 629–630.

Meyer, G. F. (1968). *Z. Zellforsch.* **84,** 141–175.

Olivieri, G., and Olivieri, A. (1965). *Mutation Res.* **2,** 366–380.

Perotti, M. E. (1973). *J. Ultrastruct. Res.* **44,** 181–198.

Pratt, S. A. (1968). *J. Morphol.* **126,** 31–65.

Reilly, J. G., and Thomas, C. A. Jr. (1980). *Plasmid* **3,** 109–115.

Rosen-Runge, E. C. (1977). "The Process of Spermatogenesis in Animals." University Press, Cambridge.

Shields, G., and Sang, J. H. (1977). *Drosophila Inform. Service* **52,** 161.

Stanley, H. P., Bowman, J. T., Romrell, L. J., Reed, S. C., and Wilkinson, R. F. (1972). *J. Ultrastruct. Res.* **41,** 433–466.

Tates, A. D. (1971). Cytodifferentiation during spermatogenesis in Drosophila melanogaster. (Proefschrift). Pasmans, S.-Gravanhage.

Tokuyasu, K. T. (1974a). *J. Cell Biol.* **63,** 334–337.
Tokuyasu, K. T. (1974b). *Exp. Cell Res.* **84,** 239–250.
Tokuyasu, K. T. (1975). *J. Ultrastruct. Res.* **53,** 93–112.
Valette St. George, A.v.la (1867). *Arch. Mikroskop. Anat.* (*Berlin*) **3,** 263–273.
Valette St. George, A.V.la (1896). *Arch. Mikroskop. Anat.* **27,** 1–12.
Warn, R. M. (1979). *In* "Maternal Effects in Development" (D. R. Newth and M. Balls, eds.), pp. 199–219.University Press, Combridge
Wilkinson, R. F., Stanley, H. P., and Bowman, J. T. (1974). *J. Ultrastruct. Res.* **48,** 242–258.
Wilkinson, R. F., Stanley, J. P., and Bowman, J. R. (1975). *J. Ultrastruct. Res.* **53,** 354–365.
Williamson, D. H., and Fennel, D. J. (1975). *Meth. Cell Biol.* **12,** 335–351.

EFFECTS OF MYCOTOXINS ON INVERTEBRATE CELLS *IN VITRO*

J.-M. Quiot, A. Vey, and C. Vago

Centre de Recherches de Pathologie Comparée, INRA, CNRS, USTL. Saint-Christol, France

I. Introduction

Fungal toxins are attracting increasing interest. Thus, with respect to entomopathogenic fungi, work has recently developed in France, particularly at the Center for Research on Comparative Pathology of Saint-Christol, following the discovery by Kodaira and associates (Kodaira, 1962; Tamura *et al.*, 1964; Suzuki *et al.*, 1966) in Japan of cyclodepsipeptidic metabolites produced by the hyphomycete *Metarhizium anisopliae.*

The importance of cryptogamic toxins is due not only to their ability to cause lesions at the cellular level and to kill host animals but also to their antimicrobial (Boutibonnes, 1979; Boutibonnes and Lemarinier, 1981) and antiviral effects (Quiot *et al.*, 1980).

In France a close association between investigators of the National Center for Scientific Research (CNRS) and the National Institute for Agronomy Research (INRA) has led to the isolation of purified cryptogamic toxins and to detailed studies of their modes of action at the tissue and cellular level.

Cell culture of invertebrates provides an experimental system par-

ISBN 0-12-007904-6

ticularly favorable for the study of the action of these mycotoxins (Quiot, 1975; Vey and Quiot, 1975, 1976; Vey *et al.*, 1975; Belloncik and Gharbi-Said, 1977; Vey *et al.*, 1984). The work that we are doing in this field is motivated, on the one hand, by permanent research on antimicrobial substances and the analysis of immunosupressing factors in invertebrates and, on the other hand, by the determination of the mode of action of entomopathogenic fungi and of the role of substances produced by fungi for practical reasons. These motivations directed the choice of fungi toward those that have marked toxic action and show promise in biological control. This is the case with the hyphomycetes *Beauveria bassiana* and *Metarhizium anisopliae* and the zygomycete *Mucor hiemalis*.

Mycotoxins act by altering cells, blocking multicellular defense reactions, disturbing macromolecular syntheses, and producing antiviral effects all at once. Development of research in this field has passed through stages corresponding to the levels of purification, more and more advanced, of the mycotoxins. At first the source of toxins was a crude filtrate of cultures of fungi, then chloroform extracts of these filtrates, and finally toxins purified by Pais of the Institute of Chemistry of Natural Substances, CNRS, at Gif sur Yvette (France) (Pais *et al.*, 1981).

At present, to put them in more natural conditions, the actions of mycotoxins produced directly in the hemolymph of invertebrate hosts are tested and characterized.

II. Toxic Actions of Fungus Culture Filtrates

Filtrates come from cultures of *M. anisopliae* or *M. hiemalis*. Culture media are those used for cell culture such as D. 73 of Quiot and Paradis (1984) or Grace (1962). After 24 or 36 hours of culture the supernatant is filtered through a Millipore 0.45 μ membrane and the pH is returned to its normal value for cell culture. The filtrates are introduced into cultures at final dilutions of 1/4, 1/5, 1/10, 1/50, and 1/100.

A. Cell Lesions

Cell lesions were detected and characterized by phase microscopy, histology, and electron microscopy. For *Mucor* tests were conducted on the cells from the dictyopteran *Leucophaea maderae* (Vey and Quiot, 1976). *Metarhizium* filtrates were tested on the cell line of *Gromphadorhina laevigata* and that of *Bombyx mori*. Observed modifications of cell structure were mainly nuclear and consisted of significant chromatolysis expressing itself by an aggregation of chromatin at the periphery of the nucleus. The threshold of toxicity was about 1/50 dilution.

B. *In Vitro Effects of Fungus Culture Filtrates on the Multicellular Defense Reaction Process*

Among the defense reactions against microbial aggression or foreign agents the formation of "granulomas" may be considered one of the most important. The establishment of cultures of hemocytes of insects enabled us to follow *in vitro* the dynamics of the process of encapsulation continuously, even by microcinematography (Quiot, 1975; Vey *et al.*, 1975). For the initiation of the process, hemocytes of the coleopteran *Oryctes rhinoceros* were put in the culture medium with conidiospores of the fungus *M. anisopliae* inactivated by 30 hours of exposure to 45°C. The production *in vitro* of the multicellular defense reaction and the demonstration of a toxic action of filtrates of *Metarhizium* on cultured cells prompted us to test the action of these products diluted 1/4 and 1/5 on multicellular reactions. After 3 days there was no sign of cellular aggregation around the spore clusters. The blood cells (hemocytes) degenerated and showed no reaction against the spores. Thus, under the influence of filtrates of *Metarhizium,* the cells lost their capacity to congregate around spores to form granulomas.

III. Action of Substances Extracted with Chloroform from Culture Filtrates of Fungi or Hemolymph of Infected Insects

Most of the toxicological research on entomopathogenic fungi is done from cultures *in vitro* in artificial media, and the production of toxins or enzymes by these agents is strongly influenced by the composition of the medium. The toxinogenic action of a fungus growing in the host organism may be different from that observed in artificial culture.

These studies were conducted with strain Ma 51 of *M. anisopliae.* In the case of culture filtrates the chloroform extract was dissolved in a mixture of methanol and acetone, then diluted to 50, 25, 12.5, and 2.5 parts per thousand (ppm) and tested on cells of the dictyopteran *Gromphadorhina laevigata* and of the lepidopteran *Lymantria dispar.* (Quiot, 1976). After 4 days' exposure to 25 ppm the cytoplasmic projections consisted only of very fine long pseudopods and the cytoplasm became very granular.

The works of Kodaira and associates (Suzuki *et al.*, 1971) and Roberts (1966a,b) have each shown that *Metarhizium* produces identical substances or the same type of action *in vitro* and *in vivo*. In our laboratory we likewise approached this problem using different methods. Fungal infections were established by injecting conidia of strain 51 into the hemocoel of larvae of *O. rhinoceros*. Hemolymph of moribund insects was used for preparing a chloroform extract, using the technique applied to culture filtrates. The action of 50, 25, 12.5, and 2.5 ppm of this extract was studied on cells of *Lymantria* and of *Gromphadorhina*. At 25 ppm morphological changes are already apparent after only 24 hours of contact. The cytoplasm appears practically empty, except for granulations, and is partially ejected to the exterior in the form of vesicles. The *Gromphadorhina* line is affected even at the weakest dosage, 2.5 ppm. However, under these conditions only a part of the cells are modified, while many of them remain fixed, spread out, and of normal appearance. On the other hand, the *Lymantria* line shows no lesions with the same concentration of 2.5 ppm of extract. In conclusion, the pathotype 51 of *M. anisopliae* produces *in vivo* substances which may be considered very toxic given the activity manifested by the chloroform extract of infected hemolymph.

IV. Destruxin Action

Destruxins have been isolated from strain 120 of *M. anisopliae*. Destruxins are cyclodepsipeptides characterized by the presence in their molecules of 5 amino acids and a hydroxylated acid. Only the actions of destruxins A, B, and E majors and A1, A2, B1, B2, D, and desmethyl destruxin have been tested on cells of the lepidopterans *B. mori* and *L. dispar*, on the one hand, and of the dictyopteran *G. laevigata*, on the other.

A. Cell Lesions

Cell lesions were analyzed by phase microscopy, histology, and electron microscopy (Vey *et al.*, 1984).

As observed with phase contrast the action of compounds A, B, and E, appears early at a concentration of 5 ppm. In fact, after 24-hour exposure the cells are retracted, forming thin pseudopods, dense and branched or with large vesicular outgrowths with dense content. Even at 1 ppm the lesions express themselves. Under the effect of A and B the cells are contracted and of granular appearance; with destruxin E they agglomerate and float in the medium. Below this dose the destruxins have no effect on cells of the *Gromphadorhina* line while on the *Bombyx* line the action of the destruxin E is still evident down to a dose of 0.05 ppm. Destruxin A still has a weak activity at 0.25 ppm and compound B shows none below 0.5 ppm.

A comparative study of the action of destruxins E, A1, A2, B1, B2, D, and desmethyl on cells of the *Bombyx* line was made and it showed that at doses of 0.5 and 0.1 ppm destruxin E completely destroyed the cells and that the other toxins had no effect. Only a diminution of multiplication of cells in contact with destruxin A1 was noted. At the ultrastructural level in cells of the *Gromphadorhina* line the lesions were characterized by a considerable vacuolization of the cytoplasm arising from the dilatation of the endoplasmic reticulum, sometimes especially pronounced at the level of the perinuclear cisternae. The mitochondria were dilated and their cristae changed. In the nucleus the chromatin was greatly concentrated in dense masses, the nucleoli were changed, and the nucleoplasm became electron transparent.

The *Bombyx* cell line is more favorable for comparative ultrastructural studies of the three destruxins A, B, and E because it has a greater sensitivity to these compounds and the cell populations are more homogeneous. After 24 hours of contact at 1 ppm a strong and generalized toxicity of destruxin E was shown, characterized by the total disappearance of the typical pseudopods and the appearance at the surface of many cells of big round outgrowths. The cytoplasm was invaded by small vacuolar formations. The endoplasmic reticulum contained parallel or concentric structures with alternations of electron-clear and electron-dense zones. The nuclei were deformed and pycnotic. The chromatin formed masses often localized at the nuclear periphery.

After exposure to destruxins A and B, the lesions were discrete but slightly sharper in the case of B. They were characterized by slight vacuolization and retraction of the pseudopodia. At the nuclear level the chromatin had a tendency to group itself in masses.

At 0.25 ppm, only the lesions provoked by destruxin E remained visible and there was no action of compounds A and B. The results with weak doses make it appear that it is less a matter of a different mode of action than of a different intensity of action according to the destruxin considered.

Finally we have shown that, even at weak doses, destruxins have an inhibitory action on cell multiplication with only moderate cytotoxic effect. Thus, the cell line *Gromphadorhina,* after 4 days in contact with

1 ppm of destruxin E, had ⅓ to ⅕ as many cells as the controls. With the *Bombyx* cell line the same results were obtained at a dose of 0.05 ppm.

B. *Effect of Destruxins on Macromolecular Cell Synthesis*

In order to understand the mode of action of destruxins, especially at very weak doses, further studies were conducted on their effect on cell syntheses. In order to evaluate the modifications of these syntheses, the incorporation of specific radioactive precursors by the cells was followed. The radioactivity of the acid-insoluble precipitate obtained after lysis of the cells by sodium dodecyl sulfate was measured with a liquid scintillation spectrometer. A micromethod which requires only a very small quantity of radioactive product was used.

The first experiments were done with cultures of cells of *G. laevigata* and a high concentration (25 ppm) of each of the destruxins (A, B, and E). At this dose the inhibition of DNA by the 3 destruxins was early and strong. Nevertheless, no differences could be found between them. There was a clear inhibition of RNA synthesis with the 3 destruxins, with the strongest action by destruxin E and the weakest by A.

Comparative studies on the action of destruxins was facilitated by the use of weak doses. For this, the cell line of *Bombyx* was preferred to that of *Gromphadorhina* because it is more sensitive to toxins, the cell population is more homogeneous, and cell multiplication is very rapid, favoring the incorporation of radioactive precursors. The results from the liquid scintillation spectrometer were expressed as percentages of those from controls, which were taken as 100%.

1. *Action on DNA Synthesis (Fig. 1)*

The precursor used was [*methyl*-^{14}C] thymidine at a final concentration of 0.5 μCi/ml.

At a dose of 12.5 ppm the diminution in DNA synthesis was marked but the variation in incorporation among the 3 destruxins was negligible even after long periods of time (12 hours and 24 hours). At doses of 5 and 1 ppm, synthesis was weak in comparison to the controls, and differences in incorporation began to increase slightly but were not very significant.

At a dose of 0.1 ppm or less for destruxins A and B synthesis was normal up to 12 hours but it lowered slightly after 12 and 24 hours of contact. For destruxin E it remained at a very low level for the longest periods (12 and 24 hours). Differences in incorporation between destruxins A and B and destruxin E became significant. Below 0.1 ppm

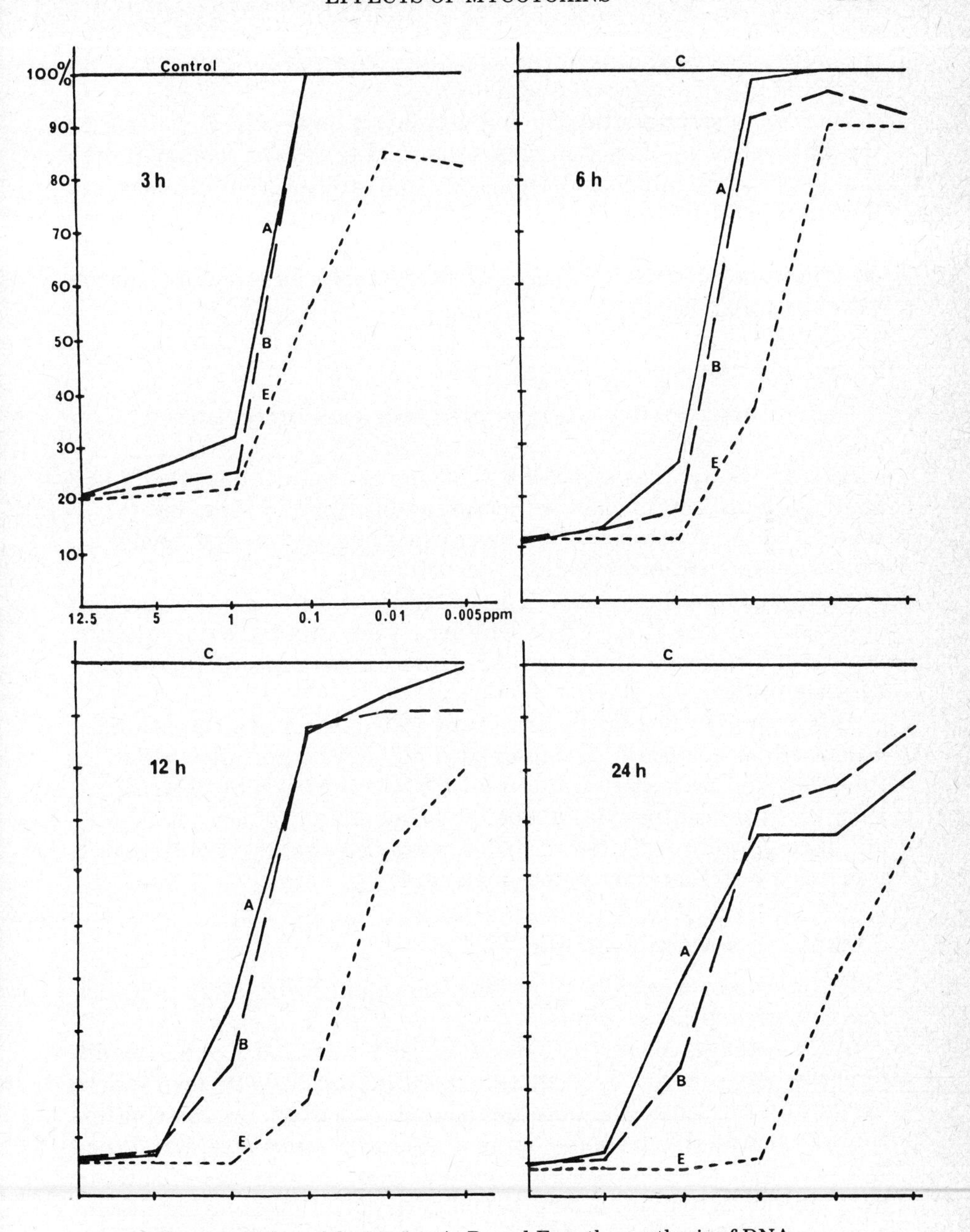

FIG. 1. Action of destruxins A, B, and E on the synthesis of DNA.

the only noteworthy observation was the progressive recovery in DNA synthesis of cells in contact with destruxin E.

In order to estimate the level of activity of destruxin E, considered the most effective, its action on synthesis of DNA was compared with that of cytosine arabinoside using the same radioactive precursors and the liquid scintillation spectrometer. After 24 hours of contact with the cells, destruxin E presented at the weakest concentration (0.0025 ppm) an inhibition of DNA synthesis about 50 times as strong as that of cytosine arabinoside.

2. *Action on RNA Synthesis (Fig. 2)*

The precursor used was [^{3}H]uridine at a final concentration of 2.5 μCi/ml.

As with the synthesis of DNA at strong concentrations (12.5 and 5 ppm), the synthesis of RNA was considerably affected, was very low in relation to the controls for whatever destruxin tested, and no significant difference in incorporation was observed.

Contrary to the case with DNA synthesis, below 1 ppm the effects of destruxins A and B on RNA synthesis were greatly reduced while destruxin E retained all its activity. Differences in incorporation were very significant.

At 0.1 ppm and below, the activity of destruxins A and B was almost absent while destruxin E remained inhibitory, especially for longer times (12 and 24 hours). Inhibition of RNA synthesis by destruxin E after 24 hours contact was about the same as that of actinomycin D, but at the lowest doses tested (0.005 ppm) the efficacy of destruxin E was much less than that of actinomycin D.

3. *Action on Protein Synthesis (Fig. 3)*

The precursor used was a ^{14}C labeled protein hydrolysate at a final concentration of 50 μ Ci/ml.

At the strongest doses, i.e., those between 12.5 and 1 ppm, the differences between the 3 destruxins were not great. The levels were essentially related to the duration of time of contact. At the shortest times (3 hours) the decreases in synthesis were almost zero and they were reduced to 40–50% at 24 hours. Below 1 ppm, the variations between the destruxins A and B and destruxin E became significant and remained significant at 0.005 ppm for the longest time tested (24 hours). Destruxin E continued to inhibit about 50% of the protein synthesis while A and B inhibited only about 10%.

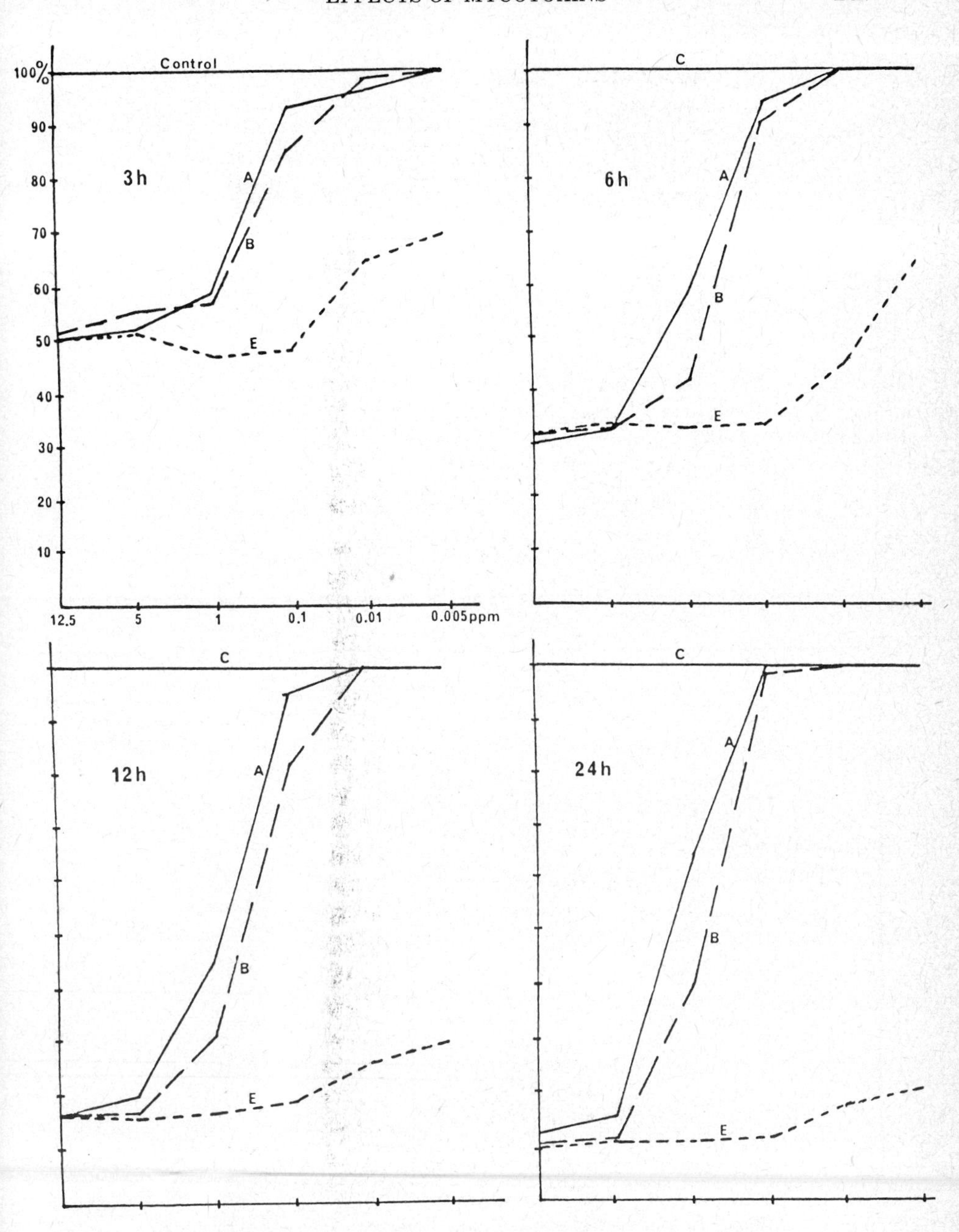

FIG. 2. Action of destruxins A, B, and E on the synthesis of RNA.

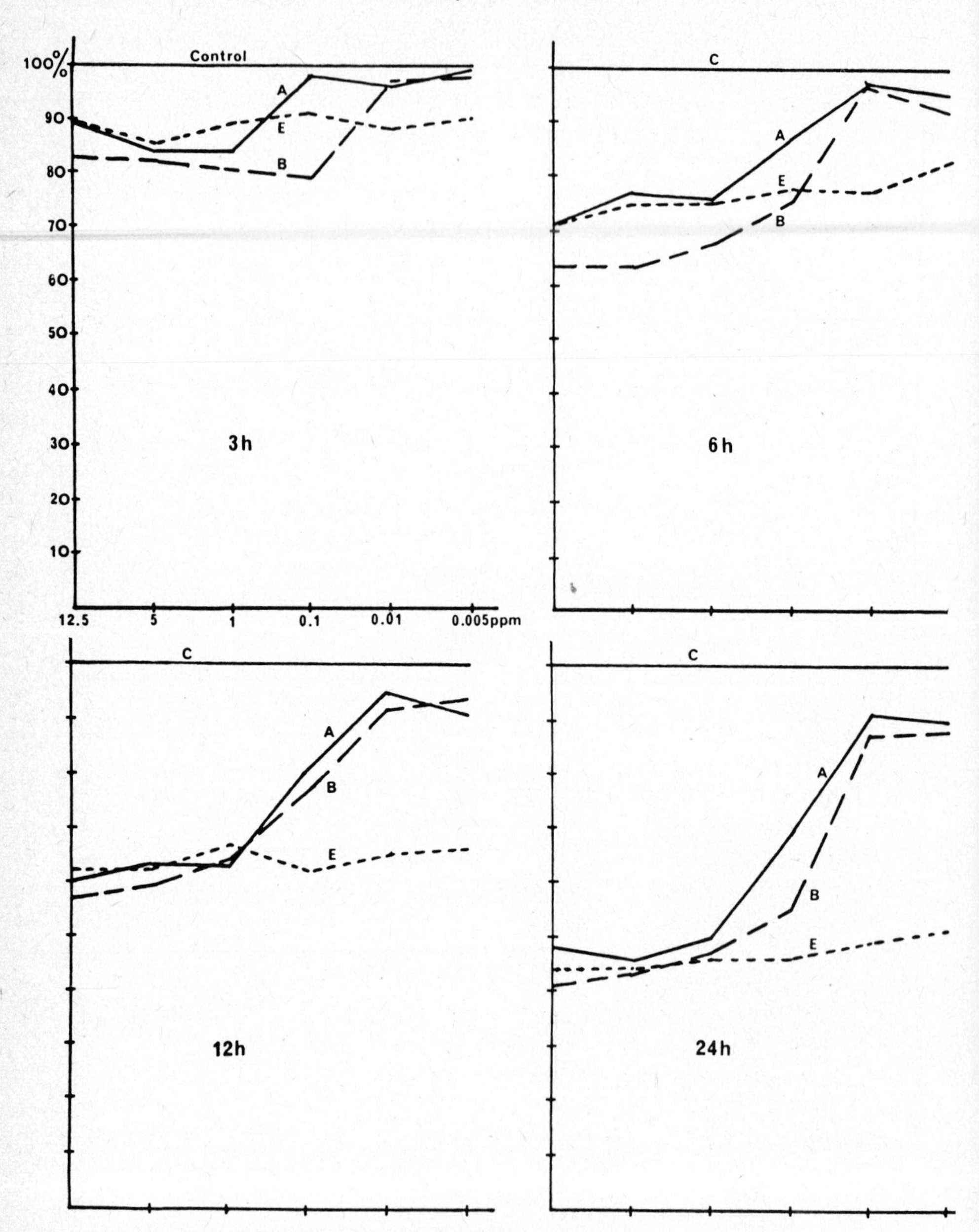

FIG. 3. Action of destruxins A, B, and E on the synthesis of proteins.

C. Antiviral Action of Mycotoxins

In medicine and in animal pathology in the past few years many studies have been undertaken to find antiviral substances without toxic effects on the cells. Since studies on destruxins have shown evidence of disturbance of nucleic acid synthesis in cultivated cells, the behavior of these compounds at the level of virus synthesis was examined. The DNA baculovirus of *Galleria mellonella* and the cell line from *B. mori* were used.

Viral inoculation took place 12 hours after the introduction of destruxin to the cultures. At doses of 0.01 or 0.005 ppm, destruxin E blocked viral multiplication but caused a slight alteration of the cells. At the weakest concentration of 0.0025 ppm, no viral inclusion bodies were observed in the nuclei, although the cells presented no apparent alterations.

At doses still weaker, 0.001 ppm, without effect on the cells, infection was not totally blocked but after 12 hours of contact only rare polymorphic protein bodies were observed.

Later a comparative study of the antiviral effects of the major destruxins A, B, and E was undertaken using a method of counting infected cells *in situ*. Culture dishes 35 mm in diameter were placed under the inverted microscope on a disk of cardboard the exact size of the bottom of the dishes and pierced by 12 holes placed at random, the holes corresponding exactly to the visual field of the microscope. Each hole was photographed with a Polaroid camera and the infected cells were counted.

With all 3 destruxins at the concentration of 1 ppm, infection was entirely blocked, no infected cell having been seen. At 0.5 ppm or less, a difference in activity was noted between the 3 substances, since no infected cell had been seen with destruxin E, 7 had been seen with A, and 69 with B. At 0.1 ppm, while destruxin E continued to block the infection, the destruxins A and B lost much of their efficacy. There were 248 infected cells in cultures containing destruxin A and 346 with B. Control cultures contained 380 infected cells per plate (dish). These results show that the degree of antiviral activity was tied to the degree of toxic activity, even if at antiviral doses the toxin did not seem to have an effect on the cells.

In using the methods described above with [*methyl*-^{3}H]thymidine and the liquid scintillation spectrometer, it has been possible to determine the action of destruxin E on synthesis of viral and cellular DNA during the course of infection of cells in culture with the baculovirus agent of a nuclear polyhedrosis. The action of the toxin manifested itself by a loss of DNA synthesis of about 50% as compared to a control infection.

The first results showed that a fungal toxin could, at very weak concentrations, inhibit the development and the pathogenesis of a DNA baculovirus. This inhibition took place even if no alteration due to the toxin was visible in the cells. Culture infection tests with the baculovirus which had been in contact with strong doses (1 ppm) of destruxins showed that inhibition of infection cannot be attributed to a degradation of virus but must be the consequence of disturbance at the level of macromolecular synthesis.

An experiment similar to that cited above with the baculovirus was done with reovirus of cytoplasmic polyhedrosis of *Euxoa scandens* on the *Bombyx* cell line, but up to the present time no clear antiviral action of destruxin E on this type of virus has been observed.

Given the interest of the phenomenon to general virology, similar experiments are under way with other RNA and DNA viruses of invertebrates and with viruses of vertebrates, particularly Herpes virus, on cell cultures of vertebrates. The effects of plant viruses are being studied in cell cultures of their arthropod vectors.

Moreover, since it would seem that such antiviral actions may be discovered in other fungal or microbial substances, natural or synthetic, it is necessary to test numerous compounds. For this, a standard method for a rapid and simple diagnosis of cytotoxicity and antiviral activity of a given substance has been devised. The experimental model chosen was the cell line of *B. mori,* destruxin E, and the baculovirus of nuclear polyhedrosis of *G. mellonella.* Viral infection was introduced on a cell lawn which after 1 hour of contact was covered with a layer of methylcellulose, 1.5%. On the surface of this layer was placed a disk of blotting paper impregnated with a known dilution of destruxin. After 6 or 7 days of infection, the cultures were separated into 2 lots. In one of the lots a 1% solution of Neutral Red was introduced. Subsequently, at the periphery of the disk of paper, a colorless aureola constituted of dead cells the size of which was related to the concentration of destruxin employed was found.

In the second lot, after partial elimination of the methylcellulose layer, the cells were stained with Methylene Blue. With the optical microscope one could observe that at the periphery of the zone stained with Neutral Red, corresponding to living cells (i.e., not altered by toxin), there was a heavy viral infection which diminished progressively toward the center and became zero in the most internal zone. These observations confirm the fact that destruxin E can block viral multiplication even in unaltered cells. The methods should be applicable to any virus and to any cell type that forms monolayers. For viruses with cytopathic effect the reading may be done directly after staining with Neutral Red.

V. Conclusions

These studies *in vitro* on the action of destruxins from the entomopathogenic fungus *Metarhizium* show that these substances in contact with cells cause in them profound disturbances, not only of their fine structure but also of their metabolism. The production of the toxic substances are closely related to the pathogenicity of the fungus.

One of the most interesting aspects is the demonstration of an action on macromolecular syntheses. We have shown that toxins, especially the purified destruxins, have a powerful inhibitory effect on the synthesis of DNA, of RNA, and of proteins even in weak dosages. We must also note the antiviral role of mycotoxins, particularly the destruxins that inhibit the development and pathogenesis of baculoviruses, which occur at very low concentrations without structural alteration of the host cells.

These studies show that research on microbial toxins seems particularly important in its relation to cell biology, yielding indispensable information for the comprehension of the mechanisms of action of pathogenic fungi and for the perspectives which the antiviral effect introduces at the comparative virology level. They also attract attention to the possibilities of utilizing such metabolites for technologies used in the microbiological struggle against insects that ravage food production in agronomy, or in medicine for treatment of viral diseases.

References

Belloncik, S., and Gharbi-Said, R. (1977). Effet toxique du champignon ascomycéte *Cordyceps militaris* sur les cellules du diptére *Aedes albopictus* cultivées *in vitro*. *Entomophaga* **22,** 243–246.

Boutibonnes, P. (1979). Mise en évidence de l'activité antibactérienne de quelques mycotoxines par utilization de *Bacillus thuringiensis* (Berliner). *Mycopathologia* **67,** 45–50.

Boutibonnes, P., and Lemarinier, S. (1981). Altérations cytologiques induites par la zéaralénone chez *Bacillus thuringiensis* (Berliner). *Mycopathologia* **74,** 107–111.

Grace, T. D. C. (1962). Establishment of four strains of cells from insect tissue grown *in vitro*. *Nature (London)* **195,** 788–789.

Kodaira, Y. (1962). Studies of the new toxic substances to insects, destruxins A and B, produced by *Oospora destructor*. I. Isolation and purification of destruxins A and B. *Agric. Biol. Chem.* **26,** 36–42.

Pais, M., Das, B. C., and Ferron, P. (1981). New depsipeptides from *Metarhizium anisopliae*. *Phytochem.* **20,** 715–723.

Quiot, J. M. (1975). Recherches sur la culture *in vitro* de cellules d'insectes et l'action de germes entomopathogénes en culture cellulaire. These doctorat d'Etat Montpellier.

Quiot, J. M. (1976). Etablissement d'une lignée cellulaire (SCLd135) á partir d'ovaires du Lépidoptére *Lymantria dispar l. C.R. Acad Sci. (Paris), Ser. D.* **282,** 465–467.

Quiot, J. M. (1982). Etablissement d'une lignée cellulaire (SPC Bm 36) á partir d'ovaires de *Bombyx mori* (Lepidoptera). *Sericologia* **22,** 1, 25–31.

Quiot, J. M., and Paradis, S. (1984). Establishment of a cell line (SPCG1 13) from dorsal vessels of the cockroach: *Gromphadorhina laevigata,* Saussure and Zentner (Dictyoptera: Blattidae). *In Vitro* (in preparation).

Quiot, J. M., Vey, A., Vago, C., and Pais, M. (1980). Action antivirale d'une mycotoxine. Etude d'une toxine de l'hyphomycéte *Metarhizium anisopliae* (Metsch.) Sorok. en culture cellulaire. *C.R. Acad. Sci. (Paris), Ser. D.* **291,** 763–765.

Roberts, D. W. (1966a). Toxins from the entomogenous fungus *Metarhizium anisopliae.* I. Production in submerged and surface cultures, and in inorganic and organic nitrogen media. *J. Invert. Pathol.* **8,** 212–221.

Roberts, D. W. (1966b). Toxins from the entomogenous fungus *Metarhizium anisopliae.* II. Symptoms and detection in moribund hosts. *J. Invert. Pathol.* **8,** 222–227.

Suzuki, A., Kuyama, S., Kodaira, Y., and Tamura, S. (1966). Structural elucidation of destruxin A. *Agr. Biol. Chem. (Tokyo)* **30,** 517–518.

Suzuki, A., Kawakami, K., and Tamura, S. (1971). Detection of destruxins in silkworm larvae infected with *Metarhizium anisopliae. Agric. Biol. Chem. (Tokyo)* **35,** 1641–43.

Tamura, S., Kuyama, S., Kodaira, Y., and Higashi Kawa, S. (1964). The structure of destruxins B, a toxic metabolite of *Oospora destructor. Agr. Biol. Chem. (Tokyo)* **28,** 237–238.

Vey, A., and Quiot, J. M. (1975). Effet *in vitro* de substances toxiques produites par le champignon *Metarhizium anisopliae* (Metsch.) Sorok, sur la réaction hémocytaire du Coléoptére *Oryctes rhinoceros* L. *C.R. Acad. Sci. (Paris), Ser. D.* **280,** 931–934.

Vey, A., and Quiot, J. M. (1976). Action toxique du champignon *Mucor hiemalis* sur les cellules d'insectes en culture *in vitro. Entomophaga* **21,** (3), 275–279.

Vey, A. Bouletreau, M., Quiot, J. M., and Vago, C. (1975). Etude *in vitro* en microcinématographie des réactions cellulaires d'invertébré vis-á-vis d'agents bactériens et cryptogamiques. *Entomophaga* **20,** (4), 337–351.

Vey, A., Quiot, J. M., and Vago, C. (1984). Cytotoxic effect of the destruxins of the entomogenous deuteromycete, *Metarhizium anisopliae. J. Invert. Pathol.* (in preparation).

ADVANCES IN CELL CULTURE, VOL. 4

MICROCARRIERS IN CELL CULTURE: STRUCTURE AND APPLICATIONS

Shaul Reuveny[1]

Department of Biotechnology
Israel Institute for Biological Research
Ness-Ziona, Israel

I. INTRODUCTION

Animal and human cells can be propagated *in vitro* in two basically different modes: as anchorage-independent cells growing freely in suspension throughout the bulk of the culture, or as anchorage-dependent

[1]Present address: New Brunswick Scientific Co., Inc., 44 Talmadge Rd., Edison, New Jersey 08818.

ISBN 0-12-007904-6

cells (ADC) requiring attachment to solid substrate for their propagation (monolayer type of cell growth).

The cells from continuous established cell lines which have the ability to grow in suspension are the most obvious cell type to be used for large-scale production of cells and cell products. These cells can be relatively easily cultivated on a large scale in homogeneous culture, products can be produced at a lower cost and in a greater bulk, and as they grow indefinitely *in vitro* their availability is unlimited. However, these cell lines possess, generally, an abnormal chromosomal complement and tumorgenic properties which are mainly expressed by a high degree of undifferentiation, an infinite life span in culture, and the potential for causing tumors in immunodeficient mice. As a result of these characteristics, such cells have limited use for the production of biological products: U.S. Department of Health, Education and Welfare Regulations (1967) forbid their use in production of biologicals for human use because of the possibility of introducing potential oncogenic cell material into human beings. Further, viruses cultivated for vaccine production frequently cannot be effectively adapted to grow in suspension cultures. They tend to undergo rapid changes in specific viral markers, leading to reduction in immunogenicity as compared to viruses propagated on ADC (Bahnemann, 1980).

In contrast to cells that grow in suspension, ADCs are used extensively in production of different biological products (such as interferons and viral vaccines for human and veterinary applications): primary cells and cells from diploid strains are considered to be "normal" cells and, thus, allowed to be used for the production of biological products for human use, and cells which are anchorage dependent for their growth usually can support an efficient viral propagation as compared to cells growing in suspension.

Traditionally, ADCs were propagated on the bottom of small glass or plastic vessels. The restricted surface-to-volume ratio offered by these classical and traditional techniques has created a bottleneck in the production of cells and cells' products on a large scale. In an attempt to provide systems that offer large accessible surface for cell growth in small culture volume, a number of techniques have been proposed: the roller bottle system (Ubertini *et al.*, 1960), the multiplate propagator (Schleicher and Weiss, 1968), the spiral film bottles (House *et al.*, 1972), the artificial capillary propagators (Knazek *et al.*, 1972), the packed glass bead system (Whiteside *et al.*, 1979), and the membrane tubing reels (Jensen, 1977). These systems and their relative advantages and disadvantages were reviewed recently by Spier (1980).

In general, these systems suffer from the following shortcomings:

limited potential for scale-up, difficulties in taking cell samples, limited potential for measuring and controlling the system, and difficulty in maintaining homogeneous environmental conditions throughout the culture. In an effort to overcome these limitations, van Wezel (1967) developed the concept of the microcarrier (MC) culturing system. In this system cells are propagated on the surface of small solid particles suspended in the growth medium by slow agitation. These cells attach to the MC and grow gradually up to confluency of the MC's surface.

In fact, in this culture system both monolayer and suspension culture have been brought together, thus combining the necessary surface on which cells will grow with the advantages of the homogeneous suspension culture known to us from the traditional animal and bacterial cell-submerged cultures.

II. Advantages of the Microcarrier Culturing Systems

As a result of the unique character of the MC culture described above, this technique has the following advantages over other ADC large-scale cultivation methods:

1. High surface-to-volume ratio can be achieved which can be varied by changing the MC concentration. This leads to high cell yields per unit volume and potential for obtaining highly concentrated cell products. Cell yields of up to 9×10^6 per milliliter were achieved using MC perfused cultures (Butler *et al.*, 1983).
2. Cell propagation can be carried out in one high-productivity vessel instead of using many small low-productivity units, thus achieving a better utilization and a considerable saving of culture medium. A 2-liter MC cell culture using less than 5 liters of medium gave cell and virus yields equivalent to 250 Roux bottles using 25 liters of medium (Griffiths *et al.*, 1980). Cell propagation in a single reactor leads to a reduction in laboratory space and in the number of handling steps required per cell, thus reducing considerably labor cost and risk of contamination.
3. The well-mixed MC suspension culture in which cells are homogeneously distributed makes it possible to monitor and control different environmental conditions (e.g., pH, pO_2, and the concentration of medium components), thus leading to more reproducible cell propagation and product recovery.
4. It is possible to take a representative sample for microscopic ob-

servations, chemical testing, or enumeration, an option which is not available with most other techniques.

5. Since the MCs may be settled easily out of suspension, harvesting of cells and cell products can be done relatively easily.

6. The mode of ADC propagation on the MC makes it possible to use this system for new cellular manipulation such as cell transfer without use of proteolytic enzymes (Ryan *et al.*, 1980), cocultivation of cells (Davis and Kerr, 1982), perfusion of cell culture in columns (Bone and Swenne, 1982), large-scale mitotic cell recovery (Crespi and Thilly, 1982) and studies of cell differentiation in a three-dimensional matrix (Shahar *et al.*, 1983).

7. MC cultures can be relatively easily scaled up using conventional equipment (fermenters) used for the cultivation of bacteria and animal cells in suspension cultures.

III. The Development of the Microcarriers

van Wezel (1967, 1973, 1977) and van Wezel *et al.* (1978) examined the use of a commercial ion-exchange resin, diethylaminoethyl (DEAE)-Sephadex A-50 (Pharmacia, Uppsala, Sweden) as MC for cultivation of ADCs. This MC is composed of cross-linked dextran beads, charged with tertiary amine groups (DEAE), having an exchange capacity of 3.5 milliequivalent (mEq)/g dry materials. van Wezel (1967) demonstrated that primary cells and cells from a human diploid cell strain can be cultivated on these MCs and that poliomyelitis virus could be propagated. However, a toxic effect on cell growth was observed at bead concentration exceeding about 1 g of DEAE-Sephadex A-50 per liter culture, as indicated by increased inoculum losses (~50%), long lag periods, and diminished capacity for cell growth (van Wezel, 1977). Complete death of the cell inoculum was observed at bead concentrations exceeding 2 g/liter.

Various approaches were employed in order to decrease the "toxic effect." The main approach was to coat the MCs with serum proteins (Spier and Whiteside, 1969), nitrocellulose (van Wezel, 1973), or carboxymethylcellulose (Levine *et al.*, 1977a), but only partial improvement of the MC was achieved. Levine (1977) and Levine *et al.* (1977b, 1979a,b) found that this toxicity could be largely eliminated by reducing the exchange capacity of the commercial ion-exchange resin (DEAE-Sephadex A-50) to 1.5 mEq/g dry materials (~ 40% of the exchange capacity of the commercial preparation). Cells from a human diploid cell strain were propagated on the new low-charged DEAE-dextran MCs in concentrations of 5 g/liter, achieving cell concentrations of 4×10^6 cells/ml. These new developments and the increasing demand for large-scale propagation of ADCs have led to the commercial production of the low-charged DEAE-dextran MCs.

The new low-charged MCs were found to be suitable for cultivation of a wide variety of cell types including primary cells, cells from diploid cell strains, and established or transformed cell lines (Hirtenstein *et al.*, 1980). However, several problems connected with cell propagation and product production on these MCs (Reuveny, 1983a) have led commercial companies and researchers around the world to develop new MCs with different characteristics which make them more suitable for a specific cell propagation or product production.

The purpose of this review is to describe to the cell biologist the different commercially and noncommercially developed MCs, critical parameters which determine their suitability for supporting cell growth and production of biologicals, and the ways in which the MCs are used as production and research tools.

IV. Commercial Microcarriers

Several kinds of MCs suitable for cultivation of animal and human cells are now commercially available from a variety of sources. These MCs can be divided into 5 main groups.

1. Tertiary amino derivatized MCs (Biocarrier, Cytodex 1, Superbeads). These are beaded MCs having a porous hydrophilic matrix (dextran or polyacrylamide) with positively charged tertiary amino groups distributed throughout the whole matrix of the MC.
2. Surface-charged MCs (Cytodex 2). These are quaternary amine-derivatized beaded MCs in which the charged groups are distributed in a layer on the outer surface of the MC.
3. Gelatin or collagen coated MCs (Cytodex 3, Ventragel, Gelibeads). These are beaded MCs which are either covered with a thin layer of denaturated collagen or composed entirely of cross-linked gelatin.
4. Polystyrene MCs (Biosilon, Cytospheres). These MCs have a hydrophobic nonporous matrix (tissue culture-treated polystyrene) with a low negative charge on their surface.
5. DEAE-cellulose MCs (DE-53). These are cylindrical shaped MCs having a cellulose matrix and charged with tertiary amine groups throughout the matrix.

The detailed properties of the commercial MCs are described in Table I. These MCs and their relative advantages and disadvantages were reviewed by Reuveny (1983b).

TABLE I

COMMERCIAL MICROCARRIERS

Registered trade name and manufacturer	Main property	Charged groups and exchange capacity (or equivalent)	Matrix composition	Shape and dimensions	Surface area (cm^2/g dry weight MC)	Specific gravity	Porosity	Transparency	References
Cytodex 1 Pharmacia, Sweden	Positively charged groups distributed throughout the MC matrix	Diethylaminoethyl (DEAE) 1.5 mEq/g dry materials	Dextran	Beads 131–210 μm diameter	6000	1.03	+	+	Pharmacia (1982); Hirtenstein *et al.* (1980)
Superbeads Flow Labs, USA	Positively charged groups distributed throughout the MC matrix	Diethylaminoethyl (DEAE) 2.0 mEq/g dry dextran	Dextran	Beads 135–205 μm diameter	5000–6000	ND[a]	+	+	Flow (1978); Lewis and Volkers (1979)
Biocarrier Bio-Rad, USA	Positively charged groups distributed throughout the MC matrix	Dimethylaminopropyl 1.4 mEq/g dry materials	Polyacrylamide	Beads 120–180 μm diameter	5000	1.04	+	+	Bio-Rad (1979); Monthony *et al.* (1980)
Cytodex 2 Pharmacia, Sweden	Positively charged groups distributed in a layer on the outer surface of the MC	Trimethyl-2-hydroxyaminopropyl, 0.6 mEq/g dry materials	Dextran	Beads 114–198 μm diameter	5500	1.04	+	+	Pharmacia (1982); Gebb *et al.* (1982)

Cytodex 3 Pharmacia, Sweden	Type I denaturated collagen covalently linked to the outer surface of the MC	60 μg collagen/cm^2 MC surface	Dextran	Beads 133–215 μm diameter	4600	1.04	+	+	Pharmacia (1982); Gebb *et al.* (1982)
Ventragel Ventrex, USA	Gelatin (denaturated collagen) MC	Gelatin	Cross-linked gelatin	Beads 150–250 μm diameter	ND	ND	+	+	Ventrex (1984)
Gelibeads KC Biological, USA	Gelatin (denaturated collagen) MC	Gelatin	Cross-linked gelatin	Beads 115–235 μm diameter	3300–4300	1.03–1.04	+	+	Paris *et al.* (1983)
Biosilon Nunc, Denmark	Tissue culture treated polystyrene MC	Negative charges (tissue culture treatment)	Polystyrene	Beads 160–300 μm diameter	225	1.05	−	±	Johansson and Nielsen (1980); Nielsen and Johansson (1980); Nunc (1981)
Cytospheres Lux, USA	Tissue culture treated polystyrene MC	Negative charges (tissue culture treatment)	Polystyrene	Beads 160–300 μm diameter	250	1.04	−	±	Lux (1980)
DE-53 Whatman, UK	Positively charged groups distributed throughout the cylindrical cellulose matrix	DEAE 2.0 mEq/g dry materials	Microgranular cellulose	Cylinders 40–50 μm diameter, 80–400 μm length	ND	ND	+	−	Reuveny *et al.* (1980, 1982a,b)

[a] ND, No data available.

V. Noncommercial Microcarriers

Commercial MCs were designed to meet certain requirements needed mainly by industry for large-scale production of biological products (Reuveny 1983a,b). However, several researchers have found that these MCs do not fit several criteria essential for their cell culture system (Jacobson and Ryan, 1982; Damme and Billiau, 1981; Olson *et al.*, 1981). Therefore, efforts have been made to develop new MCs which will have different new characteristics. A list of these noncommercial developed MCs is given in Table II. However, it should be emphasized that there is no available MC which is optimized for use for all purposes. Each MC has its advantages and disadvantages and should be tested in the culturing system in which it is to be used. In the next section we will describe different parameters in the MC's structure and composition and their effect on the efficiency of MC use, thus helping cell biologists in choosing MC for their culture system.

TABLE II

Noncommercial Microcarriers

Microcarrier	Reference
Protein-coated DEAE-Sephadex MCs	David *et al.* (1981)
Amino acids (glycine) derivatized polystyrene-divinylbenzene MCs	Kuo *et al.* (1981)
Gelatin MCs	Nilsson and Mosbach (1980, 1981)
Negatively charged (sulfonated), positively charged (polyethylene-imine), or collagen, BSA, laminin, or fibronectin coated polystyrene-divinylbenzene MCs	Jacobson and Ryan (1982); Burke *et al.* (1983); Fairman and Jacobson (1983)
Polylysine, gelatin, or heparin coated MCs	Obrenovitch *et al.* (1982)
Hollow glass spheric MCs	Varani *et al.* (1983)
Primary amino-derivatized polyacryl-amide MCs	Reuveny *et al.* (1983a,b)
Liquid MCs	Keeze and Giaever (1983)
Positively charged (tertiary amine) polystyrene MCs	Reuveny *et al.* (1984)
Positively charged (polylysine, DEAE, or diaminohexane) or gelatin-derivatized agarose-polyaldehyde microspheres (Ag-Acrobead)	Lazar *et al.* (1984)

VI. Important Parameters in the Microcarrier Composition

The requirements for an optimal MC were described earlier by several groups (van Wezel, 1977; Hirtenstein *et al.*, 1980; Pharmacia, 1982; Nunc, 1981; Reuveny, 1983a,b).

In this section we will describe these optimal parameters and how they are reflected in the composition of the developed commercial and noncommercial MCs.

A. *Functional Groups on the Microcarrier*

ADCs can grow on the surface of MCs derivatized with several functional groups: positively charged (primary, tertiary, or quaternary amines), negatively charged (tissue culture-treated polystyrene or glass), or denaturated collagen (see Sections IV and V).

1. *Positively Charged Microcarriers*

Our group has systematically tested positively charged particles with different side chains as well as matrix composition for their ability to support mouse fibroblast cell growth on their surface. Cell attachment spreading and growth was observed on all the tested positively charged particles (Reuveny, 1983a). This phenomenon can be explained by the fact that all animal and human cells possess a net negative charge on their surface at physiological pH (Borysenko and Wood, 1979). Thus, these cells are attracted by electrostatic force to the positively charged particles. This strong electrostatic interaction created between the cells and the positively charged MCs is critical in MC culture, since the cells must withstand the shearing force created in the stirred culture without being dislodged from the MC.

a. Positively Charged Microcarriers—Type and Amount of Charged Groups. The type and amount of charged groups on the MC have a profound effect on cell attachment, spreading, and growth. Levine *et al.*, (1979b) have demonstrated that there is a discrete range of anion exchange capacity which is necessary for achieving optimal cell growth on DEAE-dextran MCs. A similar phenomenon was demonstrated for quaternary amine-derivatized dextran MCs (Gebb *et al.*, 1982), primary and tertiary amine-derivatized polyacrylamide MCs (Reuveny *et al.*, 1983a,b), and DEAE-cellulose MCs (Reuveny *et al.*, 1982a). At exchange capacity lower than the optimal range, usually, a low degree of cell attachment was observed, and some of the cells which did attach to the MCs grew to some degree and then withdrew from the MC surface and accumulated in large aggregates. At exchange capacities higher than optimal, the cells attached to the MC surface and spread; however, a "toxic effect" on cell growth was observed. These phenomena can be explained by the fact that a certain level of cell attraction and cell spreading on the growth surface is needed in order to achieve DNA synthesis and cell growth (Folkman and Moscona, 1978; Penman *et al.*, 1982). On the other hand, the toxic effect on cell growth at high exchange capacity can be attributed to an arrest of cell membrane movement as the result of the tight cell surface–MC interaction, a phenomenon shown by Ebbessen and Guttler (1979) and Wolpert and Gingell (1968).

The optimal exchange capacity for cell propagation is not a constant value, and it was shown to vary in accordance with the type of the charged group and the degree of hydrophobicity. Reuveny *et al.* (1983a,b) have shown that primary amino-derivatized MCs can support cell attachment spreading and growth on their surface at a significantly lower exchange capacity than tertiary amino-derivatized ones (0.2–0.5 as compared with 1.8–2.0 mEq/g dry polyacrylamide). Moreover, the rate at which cells attach and spread on the primary amino-derivatized MCs is faster than on tertiary amino-derivatized ones.

A threshold effect of the amount of the charge required for cell attachment, spreading, and growth was shown for primary amino-derivatized MCs, whereas a gradual increase rather than a threshold effect was observed with tertiary amino-derivatized MCs (Reuveny *et al.*, 1983a,b). It is pertinent to note that the threshold effect was previously observed for cell adhesion and spreading on surfaces carrying specific ligands such as carbohydrates (Weigel *et al.*, 1979), fibronectin (Huges *et al.*, 1979), and lectins (Aplin and Huges, 1981).

Quaternary amine-derivatized MCs were found to support maximal cell yields at a lower degree of substitution than tertiary amine-derivatized MCs, as shown by Gebb *et al.* (1982) for dextran MCs and Reuveny (1983a) for cellulose MCs.

b. Positively Charged Microcarriers—The Hydrophobicity of the Charged Group. Reuveny *et al.* (1983b) have demonstrated the effect of the hydrophobicity of the primary amino positively charged group on cell growth. The optimal degree of hydrophobicity for cell growth was found to be when the hydrocarbon side chain carrying the primary amino-charged group contained 4–6 carbons. However, the rate of cell attachment and spreading was increased gradually with increasing hydrophobicity. It was shown that there are hydrophobic interactions between the cell and the growth surface (Grinnell and Feld, 1982).

Apparently, an increase of hydrophobicity of the growth surface (MC) will increase cell-surface interaction. However, for optimal cell propagation a certain level of cell–MC interaction is needed (see Section VI,A,1,a), and, thus, there is a need for a certain degree of hydrophobicity for cell growth.

c. Positively Charged Microcarriers—Localization of the Charged Groups in the MC. Ideally MCs should be noncharged in order to reduce binding of proteins from the culture medium and to facilitate the removal of such proteins from the culture by a simple washing. Gebb *et al.* (1982) have found that no cell growth was obtained on MCs coated with a monolayer of quaternary amino positively charged groups, and a certain depth of charged layer is needed for supporting cell growth. Similar results were found with primary amino-derivatized polyacrylamide MCs (S. Reuveny, and A. Freeman, unpublished results). The reason for this may be that there is a need for a critical amount of positively charged groups on the outer parts of the MC in order to achieve the electrostatic attraction between the cells and the MC.

The only MC in which the positive charges are located at the outer parts of the MC are Cytodex 2 (see Table I). The exchange capacity of these MCs was lowered up to 0.6 mEq/g dry material (as compared with 1.5 mEq/g dry material with MCs which are charged throughout the matrix of the MC).

2. Negatively Charged Microcarriers

Traditionally, ADCs are propagated on negatively charged surfaces such as glass or "tissue culture-treated" polystyrene. It was demonstrated by Maroudas (1976) that optimal cell spreading was obtained at a density of $2–10 \times 10^{14}$ negative charges/cm^2. Negatively charged nonporous polystyrene MCs in which the negative charges are located only on the MC's surface are produced commercially (see Table I). However, cell attachment on these MCs is significantly lower as compared with positively charged MCs. The reason for this seems to be an electrical repulsion between the negatively charged cells and the MCs. This electrostatic repulsion may be overcome through ionic interaction, protein bridges, or local concentration of positive charges on the edges of the cell philopodia (Grinnell, 1978; Weiss and Harlos, 1972).

Fairman and Jacobson (1983) have found that porous MCs in which the negatively charged sulfonated groups are located in a 30-nm layer at the outer surface of the MC can support cell growth. However, if these negative charges are distributed throughout the MC matrix, no cell attachment or growth can be achieved (Reuveny, 1983a), presumably due to excessive electrical repulsion.

3. Nonionic Microcarriers

Biological molecules which play a role in cell attachment (collagen, fibronectin, laminin) were tried as a material for coating or preparing MCs (Gebb *et al.*, 1982; Nilsson and Mosbach, 1980; Fairman and Jacobson, 1983; Obrenovitch *et al.*, 1982). However, only collagen was found to be effective in inducing rapid cell attachment spreading and growth under stirring conditions.

Commercial MCs are available in two forms: dextran beads coated with a surface layer of cross-linked type I denaturated collagen and beads composed of cross-linked gelatin (see Table I).

The collagen MCs were found to be effective in supporting growth of cells with epithelial morphology (Gebb *et al.*, 1982) and cells with low plating efficiency (Gebb *et al.*, 1982, 1984). Cells can be harvested from these MCs by employing a specific protease, collagenase. Using this method, the harvested cells were reported to have maximal retention of the membrane. Gebb *et al.* (1984) have reported that these harvested cells have a higher plating efficiency as compared with cells harvested from ionic charged MCs. These collagen MCs absorb less serum proteins from the culture medium as compared with positively charged MCs (Gebb *et al.*, 1982).

B. The Microcarrier Matrix

The chemical composition, porosity, rigidity, and shape of the MC matrix have a profound effect on cell propagation. Dextran, polyacrylamide, cellulose, polystyrene, agarose, and polystyrene divinylbenzene have been used as matrices for MCs.

1. The Hydrophobicity of the Matrix

Reuveny *et al.* (1983b) have shown that by introducing hydrophobic elements into the polymeric backbone of primary amino-derivatized polyacrylamide MCs, a drastic decrease in cell growth on the MC surface occurs. Hydrophobic porous DEAE-polystyrene-divinylbenzene MCs were found to support lower cell attachment rate as compared with hydrophilic DEAE-dextran MC (Horng and McLimans, 1975; Reuveny, 1983a). These two observations indicate that higher rates of cell attachment and cell yields can be achieved on MCs having a hydrophilic matrix.

2. Matrix Porosity

Porous MCs can swell in aqueous media and absorb various substances from the growth medium. This is an unwanted phenomenon since it makes it difficult to remove medium component from the cell culture. Moreover, porous MCs have a tendency to shrink during cell culture preparation for microscopy and thus cause major deformation of the cultured cells (Sargent *et al.,* 1981). We have found that the degree of porosity of the primary amino-derivatized MCs has no effect on cell propagation and yields (S. Reuveny and A. Freeman, unpublished results).

The only commercial MCs which are nonporous are the polystyrene MCs (see Table I).

3. Matrix Rigidity

Maroudas (1973) has shown that fibroblast cells apply mechanical tension on their growth surface and have the ability to bend thin glass rods on which they are growing. Thus, a certain degree of rigidity of the growth surface is needed in order to make it possible for cells to spread on the surface (Maroudas, 1973, 1975; Harris, 1972).

Various MCs were reported to have different degrees of rigidity. Keeze and Giaever (1983) have described liquid nonrigid MCs while Varani *et al.* (1983) have described rigid glass MCs. There are no available data as to how this drastic change in the rigidity of the MCs affects cell growth on their surface. DEAE-cellulose MCs were found to have a higher rate of cell attachment kinetics as compared with DEAE-dextran MCs, and this was explained by the higher degree of rigidity of the cellulose MCs (Reuveny *et al.,* 1982a,b). However, there is a possibility that cells may be damaged due to collisions between rigid MCs under stirring conditions.

C. Transparency and Specific Density of the Microcarrier

Good optical quality of the MCs facilitates microscopic examination of the cells without or after staining. Usually porous MCs (dextran, polyacrylamide) are completely transparent so that observation of cells attached to the beads is achieved with great clarity. With other MCs (polystyrene) cells can be seen in the light microscope by applying a special illumination source (Johansson and Nielsen, 1980).

Specific density of the MCs should be slightly above the culture medium so that the MCs can be maintained in suspension at low stirring speeds. At a low specific density, the MCs will float to the surface, and, at a higher specific density, the stirring speeds needed to keep them in suspension will cause detachment of cells from the MCs by shearing forces.

D. Dimensions and Shape of the Microcarrier

The various available MCs have dimensions between 100 and 300 μm in diameter. Particles with this size can carry a few hundred cells on their surface and still be easily held in suspension. To guarantee the homogeneity of the culture, the size distribution of the MCs should be as narrow as possible. Since cells do not easily jump from one MC to the other during cultivation, each MC should be inoculated with 5–10 cells so that all the MCs will reach confluency at the same time.

Most of the available MCs have a beaded shape. The main reason for this is ease of production since these MCs are produced usually by an emulsion technique (Reuveny *et al.*, 1983a). The only nonbeaded MCs are the DEAE-cellulose ones. These MCs have an elongated cylindrical shape. This special MC shape has several advantages as compared with beaded MCs.

1. Cellulose MCs have a higher surface-to-volume ratio which leads to a lower degree of adsorption of ingredients from the culture medium onto the MC.
2. There is a tendency for the elongated MCs to attach to each other by forming cell bridges (Reuveny *et al.*, 1982a,b). As a result of this phenomenon, the cells can grow in aggregates in multilayered fashion (see Figs. 3 and 4). High cell yields with chick embryo fibroblasts were achieved in this way (Reuveny *et al.*, 1982a) and several viruses have been propagated on these cells (Reuveny *et al.*, 1982b; Fiorentini and Mizrahi, 1984). Neuronal and muscular cells grew as aggregates on the DEAE-cellulose MCs (Fig. 4) to a high degree of maturation (i.e., myelination and cross-striation) (Shahar *et al.*, 1983, 1984a,b; Shainberg *et al.*, 1983).
3. Cells which have a tendency to grow in small aggregates attaching to the flat growth surface in petri dishes (e.g., *Aedes aegypti* cell lines or human adenocarcinoma cells; Fig. 3) cannot grow on the beaded MCs since they peel off very quickly. These cells have the ability to grow in aggregates on the cylindrical cellulose MCs (Reuveny, 1983a,b).

VII. Applications of Microcarrier Cultures

A. Cell Propagation

A wide range of cells have been cultivated on various MCs (Table III, Figs. 1–4). Anchorage-dependent and -independent cells, cells from

TABLE III

CELL TYPES PROPAGATED IN MICROCARRIER CULTURE

Cell type	Reference
1. Primary or secondary cells	
Bovine kidney cells	Gebb *et al.* (1982)
Bovine endothelium cells	Busch *et al.* (1982); Busch and Owen (1982); Davis (1981, 1982); Folkman (1982); Ryan *et al.* (1980)
Bovine buccal cells	Beaudry *et al.* (1979); Frappa *et al.* (1979)
Bovine anterior pituitary cells	Horng and McLimans (1975)
Bovine ocular cells	Drayja *et al.* (1982)
Chick embryo fibroblast	Beaudry *et al.* (1979); Clark and Hirtenstein (1981b); Frappa *et al.* (1979); Giard *et al.* (1977); Griffiths *et al.* (1980); Levine (1977, 1979a,b); Mered *et al.* (1980); Nielsen and Johansson (1980); Nunc (1981); Pharmacia (1982); Reuveny *et al.* (1982a,b); Scattergood *et al.* (1983); Thilly and Levine (1980); Zavalnyi *et al.* (1980)
Chick embryo skeletal muscle	Pawlowski *et al.* (1979); Shahar *et al.* (1984b); Shainberg *et al.* (1983)
Chick heart fibroblasts	Norrgren *et al.* (1984)
Chick embryonic bone cells	Howard *et al.* (1983)
Dog kidney	Pharmacia (1981, 1982); van Wezel *et al.* (1978)
Human hepatocytes	Pharmacia (1981)
Human endothelial cells	Busch *et al.* (1982); Davis (1982)
Human amniotic cells	Bernstein (1980)
Human lymphocytes	Sundquist *et al.* (1980, 1981)
Japanese quail embryo cells	Bektemirov and Nagieva (1980a)
Monkey kidney	Pharmacia (1981, 1982); van Wezel (1972, 1973); van Wezel *et al.* (1978, 1980); Zavalnyi *et al.* (1980)
Mouse fibroblasts	Horst *et al.* (1980)
Mouse macrophages	Lewis and Volkers (1979); Ostlund *et al.* (1983); Ren (1982)
Porcine (newborn) kidney	Zavalnyi *et al.* (1980)
Porcine thyroid cells	Fayet and Hovsepian (1979)
Rat hepatocytes	Chessebeuf *et al.* (1983)
Rabbit endothelial cells	Ryan *et al.* (1982)
Rat embryo neuronal cells	Shahar *et al.* (1983, 1984a,b)
Rat embryo muscular cells	Shainberg *et al.* (1983)
Rat hepatocytes	Hout *et al.* (1979); Pharmacia (1981)

(continued)

TABLE III (*Continued*)

Cell type	Reference
Rat pancreatic islet cells	Bone and Swenne (1982); Spiess *et al.* (1982)
Rat pituitary cells	Smith and Vale (1980, 1981)
2. Established and transformed cell lines	
Canine	
Madin–Darby canine kidney cell line (MDCK)	Butler and Thilly (1982); Crespi and Thilly (1981); Imamura *et al.* (1982); Reuveny *et al.* (1980, 1982a,b, 1983a,b)
Feline	
Lung fibroblastic cell line	Bergman and Straman (1984)
Fish	
Various cell lines	Nicholson (1980)
Guinea pig	
Keratocytes	Griffiths *et al.* (1984)
Hamster	
Baby hamster kidney (BHK)	Duda (1982); Monthony *et al.* (1980); Pharmacia (1982); Reuveny *et al.* (1980, 1982a,b, 1983a,b); Spier and Whiteside (1969); Whiteside *et al.* (1979)
Chinese hamster ovary (CHO)	Crespi and Thilly (1981, 1982); Giard *et al.* (1977); Lai *et al.* (1980); Levine *et al.* (1979a); Mitchell and Wray (1979); Pharmacia (1981)
Human	
Lymphoblastoid cells (Namalva)	Talbot and Keen (1980)
Human cervix carcinoma (HeLa)	Billig *et al.* (1984); Carter and Ewell (1982); Levine *et al.* (1979a); Mitchell and Wray (1979); Talbot and Keen (1980)
Human pancreatic tumor cells	Kelly and Grant (1982)
Human carcinoma of the larynx (HEp-2)	Beaudry *et al.* (1979)
Monkey	
Kidney cell lines (vero, LLC-MK2, CV1, BSC1)	Billig *et al.* (1984); Carter and Ewell (1982); Levine *et al.* *al.* (1980); Gebb *et al.* (1982); Giard *et al.* (1977); Griffiths *et al.* (1982); Hirtenstein *et al.* (1980); Mered *et al.* (1980, 1981a,b); Monthony *et al.* (1980); Pharmacia (1981, 1982); Polsari *et al.* (1984); Seefried and Chun (1981)

TABLE III (*Continued*)

Cell type	Reference
Mouse	
Fibroblast (various lines)	Beaudry *et al.* (1979); Gebb *et al.* (1984); Giard *et al.* (1977); Levine (1977); Levine *et al.* (1977a); Nielsen and Johansson (1980); Reuveny *et al.* (1980, 1982a,b)
Hepatocyte cell line	Vickrey *et al.* (1982)
Macrophage cell line	Prestidge *et al.* (1981)
Porcine	
Kidney cell line (IBR 52)	Meignier (1979); Meignier *et al.* (1980)
3. Human diploid cell strains	
Embryonic lung cells (WI-38, MRC-5, IMR-90, HEL-299)	Clark *et al.* (1980, 1982); Giard *et al.* (1979); Griffiths and Thornton (1982); Hirtenstein *et al.* (1980); Levine *et al.* (1977a,b,c, 1979a,b); Lewis and Volkers (1979); Manousos *et al.* (1980); Morandi *et al.* (1982); Nunc (1981); Pharmacia (1981, 1982); Seefried and Chun (1981); Thilly and Levine (1980); Varani *et al.* (1983)
Foreskin (FS-4)	Edy *et al.* (1982); Fleischaker and Sinskey (1981); Giard and Fleischaker (1980); Giard *et al.* (1977, 1979, 1981, 1982); Levine *et al.* (1979a); Nunc (1981); Pharmacia (1982)

invertebrate, fish, bird, and mammalian origin, transformed and normal cells, fibroblastic or epithelial cells, and lately genetically modified cell lines have been cultivated on MCs.

Several cultivation systems were used for ADC propagation on MCs. In laboratory scale, ADCs were propagated on MCs in stationary culture in petri dishes (Reuveny *et al.*, 1983a,b) or in regular or specially designed stirred spinner flasks (Hirtenstein *et al.*, 1982; deBruyen and Morgan, 1981). Cells were propagated on MCs inside roller bottles (Ryan *et al.*, 1980), hollow fibers (Strand *et al.*, 1982), and on an industrial scale in fermenters up to 1000 liters (Montagnon *et al.*, 1984).

Culture media which are used for cell propagation in traditional monolayer culture were found suitable for MC cultures. For cells with low plating efficiency, a richer medium was reported to be used in the

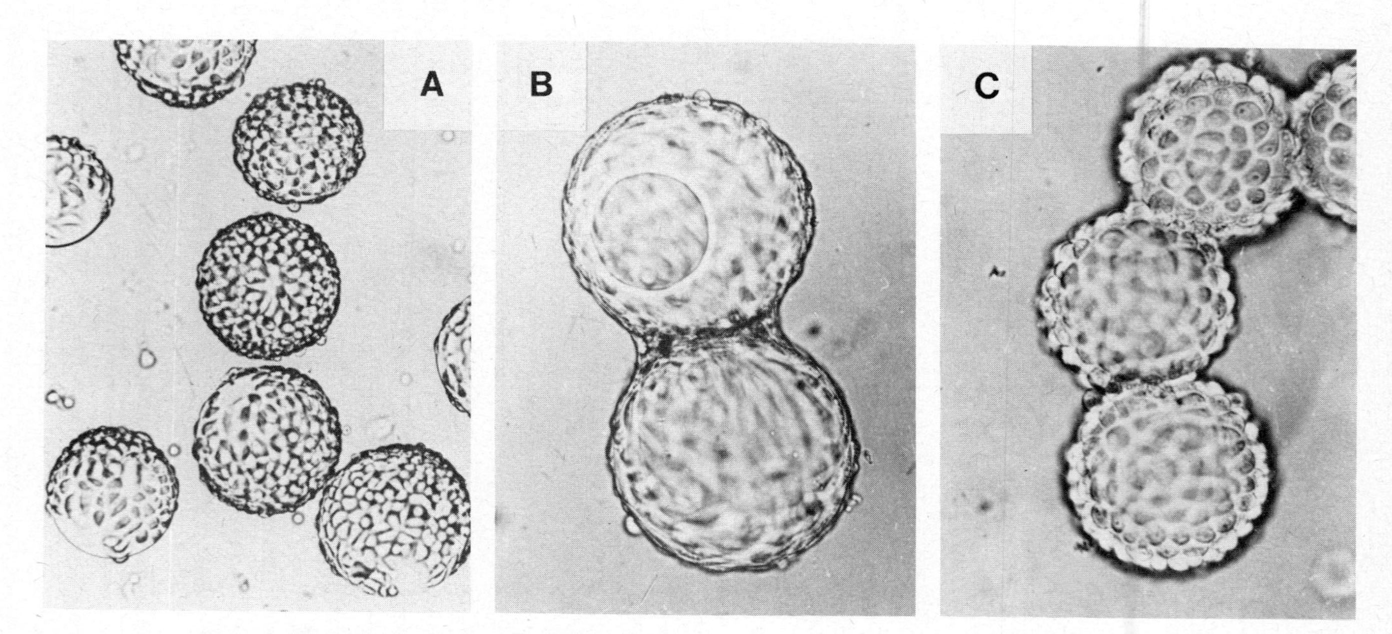

FIG. 1. Cell growth on beaded MCs. (A) Mouse fibroblasts (L-929). (B) Primary chick embryo fibroblasts. (C) MDCK (Madin–Darby canine kidney cell line) cells (epithelial morphology).

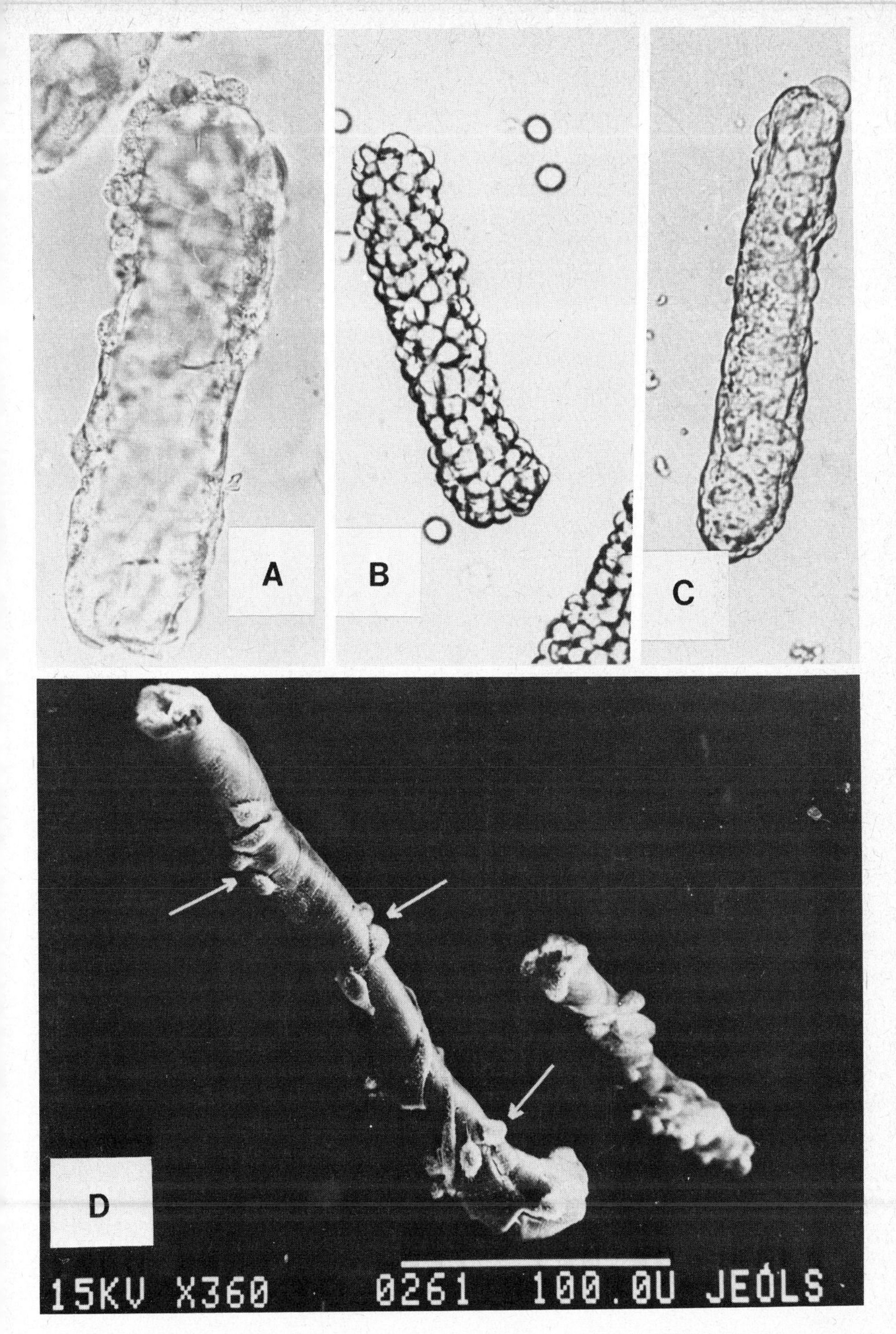

FIG. 2. Cell growth on cylindrical DEAE-cellulose MCs. (A) BHK (baby hamster kidney cell line) cells. (B) Mouse fibroblasts (L-929). (C) MDCK cells. (D) BHK cells (SEM, 2 days in culture), arrows indicating rounded cells in pairs, presumably after mitosis.

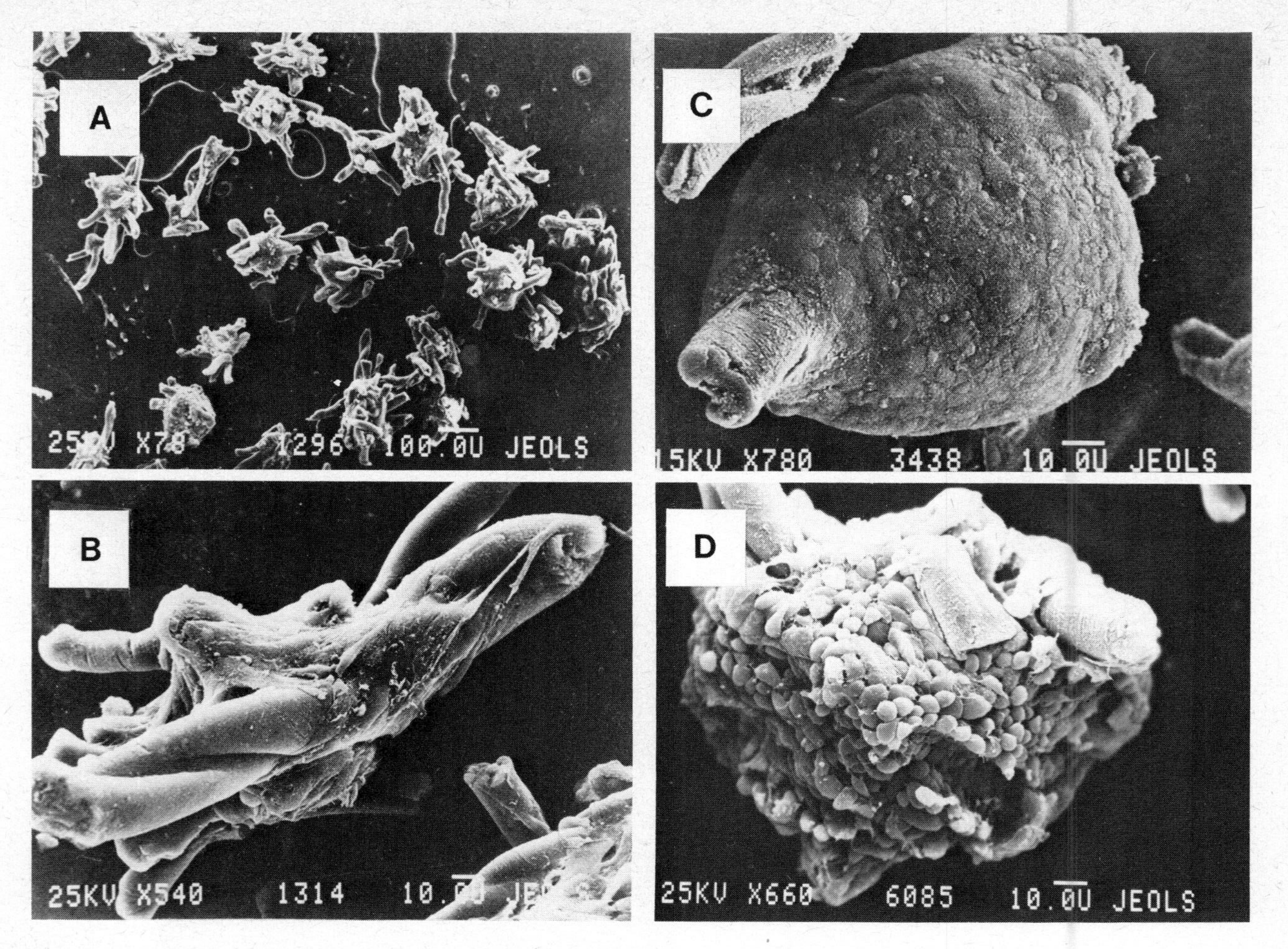

FIG. 3. Aggregate cell cultures on DEAE-cellulose MCs. (A,B) Primary chick embryo fibroblasts. Cell-MC aggregate culture and a single cell-MC aggregate. (C) A cell-MC aggregate of a human adenocarcinoma cell line. (D) A cell-MC aggregate of *Aedes*

initial stages of the MC culture (Clark *et al.*, 1980; Griffiths *et al.*, 1982). Clark *et al.* (1982) and Crespi *et al.* (1981) described growth of cells on MCs in specially designed serum-free media.

Along with the ability to achieve high cell densities in MC cultures special problems arise. These are high levels of lactic acid production with a drop in pH, rapid utilization of the growth medium, and difficulties in supplying oxygen for cell growth (especially in large tanks). In order to achieve high cell densities in culture with high MC concentrations (i.e., high surface growth area per unit volume), growth medium should be replenished during cell propagation, or perfusion systems can be used (Griffiths and Thornton, 1982).

B. Virus Propagation

Cells cultured on MCs can be used for the propagation of viruses. A wide range of viruses were propagated in MC cultures (Table IV). In

TABLE IV

VIRUSES WHICH HAVE BEEN GROWN IN MICROCARRIER CULTURE

Virus	Reference
Fish viruses	Nicholson (1980)
Foot-and-mouth disease virus	Meignier (1979); Meignier *et al.* (1980); Spier and Whiteside (1969)
Herpesviruses	Bektemirov and Nagieva (1980b); Griffiths *et al.* (1982); Griffiths and Thornton (1982); Polastri *et al.* (1984)
Influenza	Reuveny *et al.* (1982b)
Marek virus	Fiorentini and Mizrahi (1984)
Newcastle disease virus	Fiorentini and Mizrahi (1984)
Oncornaviruses	Manousos *et al.* (1980)
Papovavirus	Föhring *et al.* (1980)
Parvovirus	Bergman and Straman (1984)
Poliovirus	Giard *et al.* (1977); Mered *et al.* (1981); Montagnon *et al.* (1981, 1984); Seefried and Chun (1981); van Wezel *et al.* (1980)
Rabies virus	Nagieva *et al.* (1980); van Wezel and van Steenis (1978)
Rubella virus	van Hemert *et al.* (1969)
Simian virus 40	Pharmacia (1982)
Sindbis virus	Giard *et al.* (1977); Reuveny *et al.* (1982b); Tyo and Wang (1981)
Vesicular stomatitis virus	Crespi and Thilly (1981); Frappa *et al.* (1979); Giard *et al.* (1977)

spite of reduced viral yields per cells seen in some instances (Giard *et al.*, 1977), the great cellular productivity appears to have a great advantage for large-scale virus production. Poliomyelitis virus (van Wezel *et al.*, 1980) and foot-and-mouth disease virus (Meignier *et al.*, 1980) were propagated on DEAE-dextran MC cell cultures for vaccine production on an industrial scale. The conditions for optimal replication of viruses in culture differ from that for optimal cell propagation (van Hemert *et al.*, 1969; Tyo and Wang, 1981). Some virus infections destroy cells, but others do not cause cell breakdown, and thus, viruses (e.g., measles, rubella) are continually produced and released into the growth medium at a more or less constant rate for many days. These viruses can be harvested by fully continuous or semicontinuous methods (van Hemert *et al.*, 1969).

C. Production of Nonviral Biological Products

Several biological products have been reported to be produced in MC culture (Table V). These products can be produced in a batch mode (e.g., β-human interferon). However, if the product is produced spontaneously, sometimes a continuous or semicontinuous mode of product harvesting should be used (e.g., carcinoembryonic antigen).

The biological products reported to be produced on a large scale in MC culture are human (β)-fibroblast interferon (Edy *et al.*, 1982) and plasminogen activator (Kluft *et al.*, 1983). "Genetically engineered"

TABLE V

BIOLOGICALS PRODUCED IN MICROCARRIER CULTURE

Product	Reference
Carcinoembryonic antigen	Michel and Colette (1979); Reuveny *et al.* (1984)
Interferon	
Human (β)	Clark and Hirtenstein (1981a); Crespi *et al.* (1981); Damme and Billiau (1981); Edy *et al.* (1982); Giard *et al.* (1979, 1980, 1982); Pharmacia (1982)
Mouse	Bloch-Olszewska and Shurska (1979); Reuveny *et al.* (1980)
Interleukin-1	Prestidge *et al.* (1981); deVries *et al.* (1983)
Nerve growth promoter	Norrgren *et al.* (1984)
Plasminogen activators	Griffiths *et al.* (1984); Kluft *et al.* (1983); Scott *et al.* (1982)

animal cell lines which have been developed lately are now currently in use for production of viral antigen (e.g., hepatitis), tissue plasminogen activator, plasma proteins, and hormones.

D. Cell Transfer without Use of Proteolytic Enzymes

Subcultivation or cell transfer of ADC is done usually by using proteolytic enzymes (trypsin, collagenase, pronase, hyaluronidase) with or without the addition of chelating agents. These treatments cause a chemical trauma to the cells and affect the integrity of the plasma membrane (Anghileri and Dermietzel, 1976).

Often in MC culture, harvesting and subsequent separation of cells from the MC are not necessary. Cell transfer can be done directly from MC to MC, from MCs to a flat surface (petri dishes), or from petri dishes to the MCs, thus avoiding chemical trauma of the cells.

Davis (1981, 1982) demonstrated the continuous culture of endothelial cells using polystyrene MCs. Ryan *et al.* (1980) and Horst *et al.* (1980) observed that endothelial and mouse fibroblast cells can migrate from DEAE-dextran MCs and the surface of petri dishes. Crespi and Thilly (1981) succeeded in transferring cells from Chinese hamster ovary (CHO) and monkey kidney (LLC-MK2) from MC to MC under stirring conditions in media containing low calcium concentrations, thus enabling scaling up of the culture without the need for harvesting cells from the MC surface. Manousos *et al.* (1980) used this technique to initiate a new wave of cell proliferation and production of oncornoviruses.

E. Cryopreservation of Anchorage-Dependent Cells

Cells may be cryopreserved and stored while they are still attached to the surface of the MCs. The procedures for cryopreservation were described by Duda (1982) and the Pharmacia cell culture biology group (1982). This procedure eliminates the need for cell harvesting from the MC (see Section VII,D). Moreover, improved recovery of cells was reported.

F. Cell Membrane Isolation and Studies

Large-scale isolation of cell plasma membrane can be done by attaching whole cells to MCs and then shearing away the internal component by hypotonic lysis, agitation in a vortex mixer, or by sonication. In this way enzyme markers from the plasma membrane of mouse fibroblast cells were purified 10- to 20-fold relative to whole cell homogenates (Gotlib, 1982). Wright *et al.* (1982) reported a threefold increase in specific activity of plasma membrane glycoconjugates from mouse embryonic cells as compared with values obtained with whole cells. Lai *et al.* (1980) used the electron spin resonance (ESR) method to study membrane fluidity of Chinese hamster ovary (anchorage dependent) cells while still spreading on the surface of the MC. With this method they were able to monitor changes of membrane fluidity during cell growth. Hoeg *et al.* (1982) used MC culture in order to measure and characterize lipoprotein–receptor interaction in membrane of cells adhering to MCs.

G. *Obtaining Large Amounts of Mitotic Cells*

MC culture can be used as an efficient method for obtaining large quantities of cells in mitosis. Terasima and Tolmach (1963) have shown that ADCs round up during mitosis, and thus they can be detached from monolayers by shear forces which are insufficient to detach nonmitotic cells. Due to the large growth surface area per unit volume offered by the MC culture systems, high yields of mitotic cells can be obtained. By using a mitotic inhibitor (colcemid) and the right shearing force a yield of 4–6% of the total population is obtained with a mitotic index up to 95% (Ng *et al.*, 1980; Mitchell and Wray, 1979; Crespi *et al.*, 1981). Crespi and Thilly (1982) have used the mitotic cell population to study the cell cycle dependence of certain mutagenic chemicals.

H. *Study of Cell–Cell and Cell–Substrate Interactions*

MCs provide a dynamic attachment assay system in which the cell attachment ligand can be changed and the shearing force can be varied by increasing the speed at which confluent MCs are stirred. Vosbeck and Roth (1976) have used MCs to study the effect of different treatments on intercellular adhesion. They studied the adhesion of radioactively labeled cells to confluent MCs.

Gebb *et al* (1982) have shown that epithelial cells grow better on collagen-coated MCs as compared with DEAE-dextran MCs. Reuveny *et al* (1983a) have shown that epithelial cells grow better on diaminoethane-derivatized polyacrylamide MCs as compared with DEAE ones, while fibroblastic cells grow better on DEAE MCs. Thus we suggest that the different affinities of ADCs to MCs derivatized with different types of functional groups can be used as a means for enriching certain cell populations.

Jacobson and Ryan (1982), Burke *et al.* (1983), and Fairman and Jacobson (1983) have used polystyrene-divinylbenzene MCs derivatized with negatively and positively charged groups and cell attachment proteins in order to study cell attachment to different ligands under different degrees of shearing forces. Fairman and Jacobson (1983) and Varani *et al.* (1983) have investigated the morphology of cell attachment on different MCs such as those coated with "biological" and "nonbiological" substances or those of DEAE-dextran and glass. They have shown a difference in mode of cell attachment on the different MCs.

Kotler *et al.* (1984) have shown the ability of primary normal cells to grow on DEAE-cellulose MCs in multilayers (Fig. 3a,b, Fig. 4) in a pattern which differs from their growth in monolayers on petri dishes or DEAE-dextran MC (Fig. 1). They suggested that the ability of cells to grow in multilayers is determined not only by their state of transformation but also by the properties of the support on which they are cultured.

I. Use of Microcarriers for Electron Microscopy

Samples from MC cultures can be processed for electron microscopy without requiring large numbers of cells. This is a simple procedure as compared with the use of cell propagating on a planar surface. Sargent *et al.* (1981) have shown that polystyrene MCs are more suitable for electron microscopy than other porous MCs (e.g., DEAE-dextran), since they remain rigid during dehydration and dissolve in 1,2-epoxypropane. Thus, a good preservation of morphological details is obtained.

J. Cell Differentiation and Maturation

Traditionally, studies of cell function and differentiation *in vitro* are done in bidimensional monolayer cultures in polylysine- or collagen-coated petri dishes. The ability to propagate differentiating cells at high density in a three-dimensional mode provides a unique opportunity for studying cell differentiation and functions.

Different types of differentiating cells were propagated on beaded MCs. The system of choice is collagen-coated MCs, since these MCs allow growth of cell with low plating efficiency and only negligible amounts of growth medium protein are adsorbed on the beads.

Pawlowski *et al.* (1979) have propagated chick embryo muscle cells on DEAE-dextran MCs, achieving myogenesis. Ostlund *et al.* (1983) have propagated mouse peritoneal macrophages on different beaded MCs. Macrophages cultured on these MCs retain their ability to phagocytize latex beads. Chessebeuf *et al.* (1983) have propagated rat liver epithelial cells in serum-supplemented and serum-free media. These cells cultured on MCs maintain their hepatic functions. Differentiating bone cells (Howard *et al.* 1983), thyroid cells (Fayet and Hovsepian, 1979), pituitary cells (Proudman and Opel, 1981; Smith and Vale, 1980, 1981; McIntosh *et al.*, 1981), pancreatic β cells (Spiess *et al.*, 1982; Kelly and Grant, 1982), endothelial cells (Davis, 1981, 1982; Busch *et al.*, 1982; Busch and Owen, 1982), and neuronal cells (Shahar *et al.*, 1983) were also propagated on different beaded MCs.

Shahar *et al.* (1983, 1984a) and Shainberg *et al.* (1983) have found that DEAE-cellulose cylindrical MCs are better than beaded MCs in studying cell differentiation for several reasons.

1. Since cells grow on the cellulose MCs in multilayered form (in conglomerates, entrapping the cylindrical MCs) similar to their growth conditions *in vivo* (see Section VII,H), a higher degree of cell differentiation is achieved. This was demonstrated by myelination and high specific activity of acetylcholinesterase in neuronal cultures (Shahar *et al.*, 1983) and cross-striation, spontaneous contractions, parallel fiber orientation, and high activity of specific proteins in muscular cultures (Shainberg *et al.*, 1983).
2. Feeding of the cells inside the conglomerates occurred (presumably) by penetration of ingredients through the porous MCs into the center of the cell–MC conglomerates.
3. Neuronal and muscular cells usually peel off from their growth surface after 2–3 weeks of propagation in petri dishes. However, since cells attach to DEAE-cellulose MCs very firmly (Reuveny *et al.*, 1982a,b; see Section VI,B,3), neuronal and muscular cells can be cultured on these MCs for several months (Shahar *et al.*, 1983, 1984a, Shainberg *et al.*, 1983). Thus these cultures can be used as an effective tool in studying cell aging.
4. Moreover, because of the firm attachment of the cells to the cellulose MCs, these cultures can be washed thoroughly with physiological buffers (up to 10 washings, in order to conduct meaningful biochemical determinations). Similar washing of cultures on beaded MCs results in peeling off of cells (Shahar *et al.*, 1984a).

FIG. 4. Culture of primary rat neurons (A) and muscle fibers (B) on DEAE-cellulose MCs.

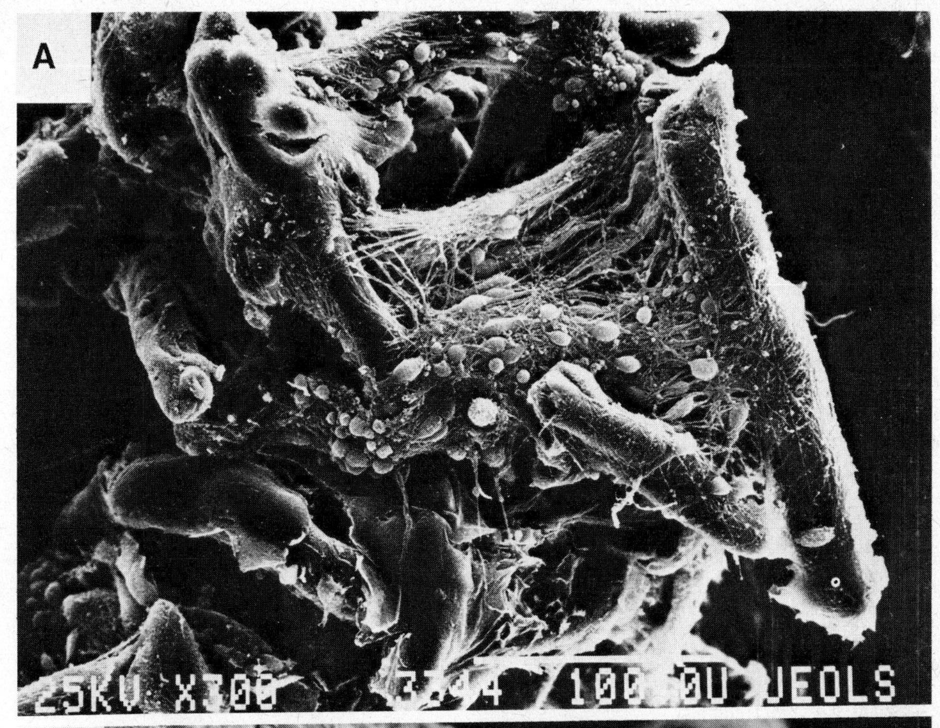
A

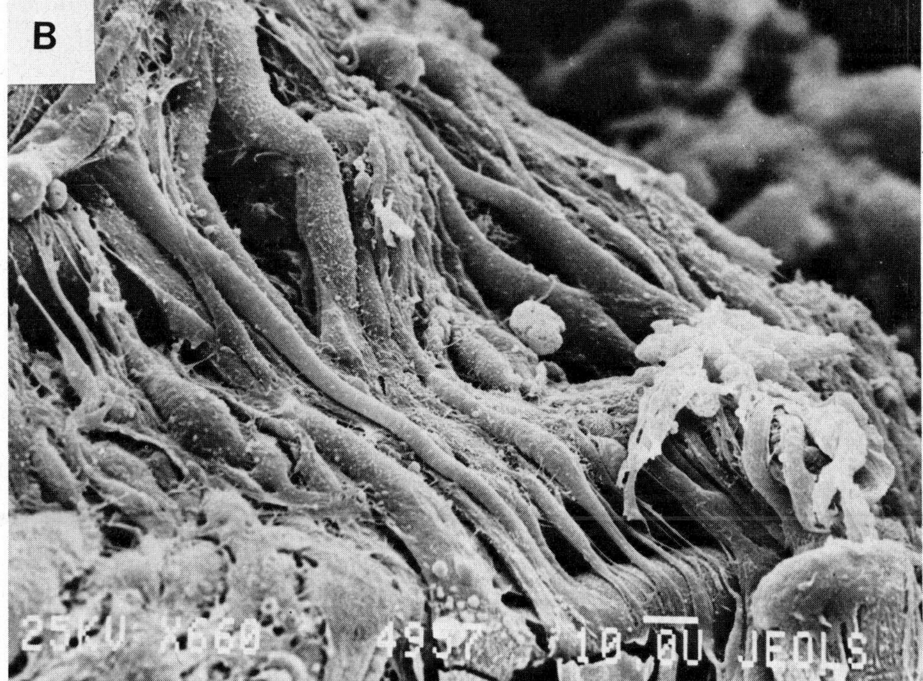
B

K. Perfusion of Microcarrier Cultures

Perfused MC cell cultures were used for 2 main purposes: (1) the large-scale production of ADCs (Butler *et al.*, 1983; Clark and Hirtenstein, 1981b; Feder and Tolbert, 1983; Griffiths and Thornton, 1982) and (2) the study of cell metabolism (Busch *et al.*, 1982; Smith and Vale, 1980, 1981).

A perfusion system of MC cell culture has several advantages. It allows a constant supply of nutrient and a constant removal of inhibitors which may accumulate in the culture medium. Thus, maintenance of growth in a constant environment (similar to *in vivo* conditions) is achieved.

In large-scale production in perfused systems, high cell yields in a relatively small culture vessel and concentrated cell products are obtained, thus reducing handling and labor costs as well as risks of contamination.

Griffiths and Thornton (1982) used the closed perfused system for maintaining constant conditions and especially for supplying oxygen in a large-scale human diploid cell (MRC-5) propagation system. A fluid lift system for large-scale propagation of cells was described by Clark and Hirtenstein (1981b). Butler *et al.* (1983) investigated the limiting factor in a perfusion system with a column separator. They have described multilayered cell growth of a canine kidney cell line (MDCK) in a perfused system, achieving cell yields of 9×10^6 cells/ml. Strand *et al.* (1982) used a hollow fiber unit to entrap MC in a perfused system for β-interferon production. Morandi and Valeri (1982) and Feder and Tolbert (1983) continuously dialyzed culture medium in hollow fiber or spin filter units (respectively) in order to achieve higher cell concentrations.

The contact between a large number of cells and a small volume of medium in a constant environment makes the perfusion system suitable for pharmacological studies in which the effect of certain chemicals or stimuli injected into the perfused medium can be observed. In practice MCs covered with cells were transferred into a column in which they were perfused. Bone and Swenne (1982) and Spiess *et al.* (1982) studied the release of insulin and other hormones from pancreatic cell-coated MCs in response to stimuli of glucose and other substrates. Smith and Vale (1980, 1981) studied hormone secretion from perfused rat pituitary cells. Davis (1981, 1982), Busch *et al.* (1982), and Busch and Owen (1982) perfused endothelial cells in MC cultures as a model for microcirculation of blood, and Hout *et al.* (1979) have perfused normal and transformed rat hepatocyte cells in order to study differences in biotransformation reactions.

L. Cocultivation of Cells

Interactions between cells of similar or different origins may be studied when one or both of the populations are cultivated on MCs. Davis and Kerr (1982) have cocultivated vascular endothelial cells and smooth muscle cells in order to study cell–cell interactions without cell–cell contact. One type of cell was propagated on MCs in a small chamber separated from the second type of cell by nylon mesh, while the second type was propagated in monolayers or on MC on the bottom of a petri dish.

Shahar *et al.* (1984b) have studied neuromuscular interactions by adding dissociated rat spinal cord cells to DEAE-cellulose MCs covered with differentiated muscular fibers.

M. Other Application

Different research groups have found MCs useful for various other applications in MC cell cultures. Sundquist and Wagner (1980, 1981) have shown that DEAE-dextran MCs can potentiate stimulation of lymphocytes by Con A. Ren (1982) used DEAE-dextran MCs for depletion of macrophages from spleen cell population. Bernstein (1980) used MCs for propagating human amniotic cells to be used in diagnosis of X-linked diseases. Yefenof *et al.* (1983) injected MC beads confluent with allogenic tumor cells in order to increase humoral and cellular immunity in mice. Bornman and Zachrisson (1982) have immobilized plant protoplast cells to DEAE-dextran MCs. We believe that many other applications to MCs in cell culture will be found in the near future.

VIII. Concluding Remarks

Recently significant progress in the development of MC culturing systems was made. New MCs with different properties were developed. Research regarding the special growth condition of cells in MC cultures was done. New cultivation vessels especially designed for propagation of cells on MCs were developed. MCs were used in industrial-scale operations (fermenters up to 1000 liters) for propagating cells in order to produce biological products (e.g., viral vaccines, interferons, plasminogen activators). In research laboratories, MCs were found to be an important tool in cell culture research, with various kinds of applications.

As a result of these developments we believe that in the future MCs will be widely used by cell biologists as a daily research tool similar to the way in which petri dishes, T-flasks, Roux bottles, and roller bottles are used today.

Acknowledgment

The author wishes to express his gratitude to Dr. L. Miller for her help and advice.

References

Anghileri, L. J., and Dermietzel, P. (1976). *Oncology* **33,** 11–23.
Aplin, J. D., and Huges, R. C. (1981). *J. Cell Sci.* **50,** 89–93.
Bahnemann, H. G. (1980). *Proc. ACS Meeting* **180,** 5.
Beaudry, Y., Quillan, J. P., Frappa, J., Deloince, R., and Fantanges, R. (1979). *Biotechnol. Bioeng.* **21,** 2351–2356.
Bektemirov, T. A., and Nagieva, F. G. (1980a). *Vopr. Virusol.* **4,** 400–403.
Bektemirov, T. A., and Nagieva, F. G. (1980b). *Vopr. Virusol.* **5,** 615–618.
Bergman, R., and Ströman, L. (1984). *Develop. Biol. Stand.* **55,** 77–78.
Bernstein, R. E. (1980). *Clin. Genet.* **18,** 147–150.
Billig, D., Clark, J. M., Ewell, A. J., Carter, C. M., and Gebb, C. (1984). *Develop. Biol. Stand.* **55,** 67–75.
Bio-Rad Laboratories (1979). *Bioradiation* **31,** 1. Trade publication.
Bloch-Olszewska, Z., and Shurska, Z. (1979). *Interferon Memoranda,* Nov. 1979.
Bone, A. J., and Swenne, I. (1982). *In Vitro* **18,** 141–148.
Bornman, C. H., and Zachrisson, A. (1982). *Plant Cell Rep.* **1,** 151–153.
Borysenko, J. S., and Wood, W. (1979). *Exp. Cell Res.* **118,** 215–227.
deBruyen, N. A., and Morgan, B. J. (1981). *Am. Lab., June,* 49–54.
Burke, D., Brown, M. J., and Jacobson, B. S. (1983). *Tissue Cell* **15,** 181–191.
Busch, C., and Owen, W. G. (1982). *J. Clin. Invest.* **69,** 726–729.
Busch, C., Cancilia, P. A., DeBault, L. E., Goldsmith, J. C., and Owen, W. G. (1982). *Lab. Invest.* **47,** 498–504.
Butler, M., and Thilly, W. G. (1982). *In Vitro* **18,** 213–219.
Butler, M., Imamura, T., Thomas, J., and Thilly, W. G. (1983). *J. Cell Sci.* **61,** 351–363.
Carter, C. M., and Ewell, A. J. (1982). *In Vitro* **18,** 312.
Chessebeuf, M., Mignot, G., and Padieu, P. (1983). *In Vitro* **19,** 290.
Clark, J. M., and Hirtenstein, M. D. (1981a). *J. Interferon Res.* **1,** 391–400.
Clark, J. M., and Hirtenstein, M. D. (1981b). *Ann. N. Y. Acad. Sci.* **369,** 33–46.
Clark, J., Hirtenstein, M., and Gebb, C. (1980). *Develop. Biol. Stand.* **46,** 117–124.
Clark, J., Gebb, C., and Hirtenstein, M. D. (1982). *Develop. Biol. Stand.* **50,** 81–92.
Crespi, C. L., and Thilly, W. G. (1981). *Biotechnol. Bioeng.* **23,** 983–994.
Crespi, C. L., and Thilly, W. G. (1982). *Mutation Res.* **106,** 123–125.
Crespi, L. C., Imamura, T., Leong, P. M., Fleischaker, R. J., Thilly, W. G., and Giard, D. J. (1981). *Biotechnol. Bioeng.* **23,** 2673–2689.
Damme, J. V., and Billiau, A. (1981). *Meth. Enzymol.* **78,** 101–111.
David, A., Segard, E., Braun, G., Berneman, A., and Horodiceanu, F. (1981). *Ann. N. Y. Acad. Sci.* 369, 61–64.

Davis, F. D. (1981). *Exp. Cell Res.* **134,** 367–373.
Davis, F. D. (1982). *Develop. Biol. Stand.* **50,** 125–136.
Davis, P. F., and Kerr, C. (1982). *Exp. Cell Res.* **141,** 455–459.
Drayja, T. P., D'Amico, D. J., Tyo, M. A., Croft, J. L., and Albert, D. M. (1982). *Invest. Ophthal. Vis. Sci.* **23,** 332–338.
Duda, E. (1982). *Cryo Lett.* **3,** 67–70.
Ebbessen, P., and Guttler, F. (1979). *J. Cell Sci.* **37,** 181–189.
Edy, V. G., Augenstein, D. C., Edwards, C. R., Gruttenden, V. F., and Lubiniecki, A. S. (1982). *Tex. Rep. Biol. Med.* **41,** 169–174.
Fairman, K., and Jacobson, B. S. (1983). *Tissue Cell* **15,** 167–180.
Fayet, G., and Hovsepian, S. (1979). *Biochimie* **61,** 923–930.
Feder, J., and Tolbert, W. R. (1983). *Sci. Am.* **248,** 24–31.
Fiorentini, D., and Mizrahi, A. (1984). *Develop. Biol. Stand.* (in press).
Fleischaker, R. J., and Sinskey, A. J. (1981). *Eur. J. Appl. Microbiol. Biotechnol.* **12,** 193–197.
Flow Laboratories (1978). Superbeads, Microcarriers. Trade publication.
Föhring, B., Tjia, S. T., Zenke, W. M., Sauer, G., and Doerfler, W. (1980). *Proc. Soc. Exp. Biol. Med.* **164,** 222–228.
Folkman, J. (1982). *J. Cell. Biol.* **95,** A458.
Folkman, J., and Moscona, A. (1978). *Nature (London)* **273,** 345–350.
Frappa, J., Beaudry, Y., Quillan, J. P., and Fantanges, R. (1979). *Develop. Biol. Stand.* **42,** 153–158.
Gebb, C., Clark, J. M., Hirtenstein, M. D., Lindgren G., Lindgren, U., Lindskog, U., Lundgren, B., and Vertblad, P. (1982). *Develop. Biol. Stand.* **50,** 93–102.
Gebb.C., Lundgren, B., Clark, J., and Lindskog, U. (1984). *Develop. Biol. Stand.* **55,** 57–65.
Giard, D. J., and Fleischaker, R. J. (1980). *Anti. Ag. Chem.* **18,** 130–136.
Giard, D. J., Thilly, W. G., Wang, D. I. C., and Levine, D. W. (1977). *Appl. Environ. Microbiol.* **34,** 668–672.
Giard, D. J., Loeb, D. H., Thilly, W. G., Wang, D. I. C., and Levine, D. W. (1979). *Biotechnol. Bioeng.* **21,** 433–442.
Giard, D. J., Fleischaker, R. J., Sinskey, A. J., and Wang, D. I. C. (1981). *Develop. Indust. Microb.* **22,** 299–309.
Giard, D. J., Fleischaker, R. J., and Sinskey, A. J. (1982). *J. Interferon Res.* **2,** 471–480.
Gotlib, L. J. (1982). *Biochem. Biophys. Acta* **685,** 21–26.
Griffiths, B., and Thornton, B. (1982). *J. Chem. Tech. Biotechnol.* **32,** 324–329.
Griffiths, B., Thornton, B., and McEntee, I. (1980) *Eur. J. Cell Biol.* **22,** 606.
Griffiths, B., Thornton, B., and McEntee, I. (1982). *Develop. Biol. Stand.* **50,** 103–110.
Griffiths, B., Atkinson, T., Electricwala, A., Latter, A., McEntee, I., Riley, P. A., and Sutton, P. M. (1984). *Develop. Biol. Stand.* **55,** 31–36.
Grinnell, F. (1978). *Int. Rev. Cytol.* **53,** 65–144.
Grinnell, F., and Feld, M. K. (1982). *J. Biol. Chem.* **257,** 4888–4892.
Harris, A. K. (1972). *Ciba Foundation Symp.* **14,** 3–12.
van Hemert, P., Kilburn, D. G., and van Wezel, A. L. (1969). *Biotechnol. Bioeng.* **11,** 875–885.
Hirtenstein, M., Clark, J., Lindgren, G., and Vretblad, P. (1980). *Develop. Biol. Stand.* **46,** 109–116.
Hirtenstein, M. D., Clark, J. M., and Gebb, C. (1982). *Develop. Biol. Stand.* **50,** 73–80.
Hoeg, J. M., Osborne, J. C., and Brewer, H. B. (1982). *J. Biol. Chem.* **257,** 2125–2128.
Horng, C., and McLimans, W. (1975). *Biotechnol. Bioeng.* **17,** 713–732.

Horst, J., Kern, M., and Ulmer, E. (1980). *Eur. J. Cell Biol.* **22,** 599.
House, W., Shearer, M., and Maroudas, G. (1972). *Exp. Cell Res.* **71,** 293–296.
Hout, J., Canillard, M., Gudserley, H., Gagnon, D., and Heisler, S. (1979). *In Vitro* **15,** 226.
Howard, G. A., Turner, R. T., Fuzas, J. E., Nichols, F., and Baylink, D. J. (1983). *JAMA* **249,** 258–259.
Huges, R. C., Pena, S. D. J., Clark, J., and Dourmashkin, R. R. (1979). *Exp. Cell Res.* **121,** 307–314.
Imamura, T., Crespi, C. L., Thilly, W. G., and Brunengraber, H. (1982). *Anal. Biochem.* **125,** 353–358.
Jacobson, B. S., and Ryan, U. S. (1982). *Tissue Cell* **14,** 69–84.
Jensen, M. D. (1977). *In* "Cell Culture and Its Application" (R. T. Acton and J. D. Lynn, eds.), pp. 589–602. Academic Press, New York.
Johansson, A., and Nielsen, V. (1980). *Develop. Biol. Stand.* **46,** 125–129.
Keeze, C. R., and Giaever, I. (1983). *Science* **219,** 1448–1449.
Kelly, S. A., and Grant, A. G. (1982). *Cell Biol. Int. Rep.* **6,** 733–739.
Kluft, C., van Wezel, A. L., van der Velden, C. A. M., Emeis, J. J., Verheijen, J. H., and Wijngoards, A. (1983). *In* "Advances in Biotechnological Processes" (A. Mizrahi and A. L. van Wezel, eds.), Vol 2, pp. 97–100. Alan R. Liss, New York.
Knazek, R. A., Kohler, P., and Dedrick, R. (1972). *Science* **178,** 65–67.
Kotler, M., Reuveny, S., Mizrahi, A., and Shahar, A. (1984). *Develop. Biol. Stand.* (in press).
Kuo, M. J., Lewis, Jr., R. A., Martin, A., Miller, R. E., Schoenfeld, R. A., Schuck, J. M., and Wildi, B. S. (1981). *In Vitro* **17,** 901–912.
Lai, C. S., Hopwood, L. E., and Swartz, H. H. (1980). *Exp. Cell Res.* **130,** 437–442.
Lazar, A., Silberstein, L., Margel, S., and Mizrahi, A. (1984). *Develop. Biol. Stand.* (in press).
Levine, D. W. (1977). Ph. D. Thesis, Massachusetts Institute of Technology.
Levine, D. W., Wang, D. I. C., and Thilly, W. G. (1977a). *In* "Cell Culture and Its Application" (R. T. Acton and J. D. Lynn, eds.), pp. 191–216. Academic Press, New York.
Levine, D. W., Wong, J. S., Wang, D. I. C., and Thilly, W. G. (1977b). *Somatic Cell Genet.* **3,** 149–155.
Levine, D. W., Wang, D. I. C., and Thilly, W. G. (1977c). U.S. Patent 4,036,693.
Levine, D. W., Thilly, W. G., and Wang, D. I. C. (1979a). *Develop. Biol. Stand.* **42,** 159–164.
Levine, D. W., Wang, D. I. C., and Thilly, W. G. (1979b). *Biotechnol. Bioeng.* **21,** 821–845.
Lewis, D. H., and Volkers, S. A. S. (1979). *Develop. Biol. Stand.* **42,** 147–152.
Lux (1980). Cytospheres. Trade publication.
Manousos, M., Ahmed, M., Torchio, C., Wolfi, G., Shibley, G., Stephens, R., and Mayyasi, S. (1980). *In Vitro* **15,** 507–515.
Maroudas, N. G. (1973). *Nature (London)* **244,** 353–354.
Maroudas, N. G. (1975). *Nature (London)* **254,** 695–701.
Maroudas, N. G. (1976). *J. Cell Physiol.* **90,** 511–520.
McIntosh, R. P., McIntosh, J. E. A., and Berridge, M. V. (1981). *Endocr. Soc. Aust.* **24,** 66.
Meignier, B. (1979). *Develop. Biol. Stand.* **42,** 141–145.
Meignier, B., Mougeot, H., and Favre, H. (1980). *Develop. Biol. Stand.* **46,** 249–256.
Mered, B., Albrecht, P., and Hopps, H. E. (1980). *In Vitro* **16,** 859–865.
Mered, B., Albrecht, P., Hopps, H. E., Petricciani, J. C., and Salk, J. (1981a). *J. Biol. Stand.* **9,** 137–146.

Mered, B., Albrecht, P., Hopps, H. E., Petricciani, J. C., and Salk, J. (1981b). *Develop. Biol. Stand.* **47,** 41–54.
Michel P., and Colette, D. (1979). *J. Cell Biol.* **83,** 302A.
Mitchell, K. J., and Wray, W. (1979). *Exp. Cell Res.* **123,** 452–455.
Montagnon, B. J., Fanget, B., and Nicholas, A. J. (1981). *Develop. Biol. Stand.* **47,** 55–64.
Montagnon, B. J., Vincent-Falquet, J. C., and Fanget, B. (1984). *Develop. Biol. Stand.* **55,** 37–42.
Monthony, J. P., Schwartz, N. D., Hollis, D. F., and Polsari, G. D. (1980). U.S. Patent 4,237,218.
Morandi, M., and Valeri, A. (1982). *Biotechnol. Lett.* **4,** 465–486.
Morandi, M., Bandinelli, L., and Valeri, A. (1982). *Experientia* **38,** 668–670.
Nagieva, F. G., Bektemirov, M. S., Matevosyan, K. S., Bektemirov, T. A., and Pille, E. R. (1980). *Vopr. Virusol.* **4,** 429–431.
Ng, J. J. Y., Crespi, C. L., and Thilly, W. G. (1980). *Anal. Biochem.* **109,** 231–238.
Nicholson, B. L. (1980). *Appl. Environ. Microbiol.* **39,** 394–397.
Nielsen, V., and Johansson, A. (1980). *Develop. Biol. Stand.* **46,** 131–136.
Nilsson, K., and Mosbach, K. (1980). *FEBS Lett.* **118,** 145–150.
Nilsson, K., and Mosbach, K. (1981). *Proc. 2nd European Congress of Biotechnology, England.*
Norrgren, G., Ebendal, T., Gelb, C., and Wikstrom, H. (1984). *Develop. Biol. Stand.* **55,** 43–51.
Nunc (1981). Biosilon Bulletin No. 1: Cultivation Principles and Working Procedure. Trade publication.
Obrenovitch, A., Sene, C., Maintier, C., Boschetti, E., and Monsigny, M. (1982). *Biol. Cell.* **45,** 28–35.
Ostlund, C., Clark, J., and Kruse, M. (1983). *In Vitro* **19,** 279.
Paris, M. S., Eaton, D. L., Sempolinski, D. E., and Sharma, B. P. (1983). *In Vitro* **19,** 262.
Pawlowski, R., Krajcik, R., Loyed, R., and Przybylski, R. (1979). *J. Cell Biol.* **83,** 115A.
Olson, R. A., Lubiniecki, A. S., and Edy, V. G. (1981). *Interferon Memoranda,* June 1981, 5.
Penman, S., Fulton, A., Copo, D., Ben Zéev, A., Wittelsberger, S., and Tse, C. F. (1982). *Cold Spring Harbor Symp. Quant. Biol.* **44,** 1013–1028.
Pharmacia Fine Chemicals (1981). Microcarrier Cell Culture: Technical Notes. Trade publication.
Pharmacia Fine Chemicals (1982). Microcarrier Cell Culture: Principle and Method. Trade Publication.
Polastri, G. D., Friesen, H. J., and Mauler, R. (1984). *Develop. Biol. Stand.* (in press).
Prestidge, R. L., Sandlin, G. M., Koopman, W. J., and Bennett, J. C. (1981). *J. Immunol. Meth.* **46,** 197–204.
Proudman, J. A., and Opel, H. (1981). *Poultry Sci.* **60,** 1714.
Ren, E. C. (1982). *J. Immunol. Meth.* **49,** 105–111.
Reuveny, S. (1983a). Ph.D. Thesis, The Hebrew University, Jerusalem, Israel.
Reuveny, S. (1983b). *In* "Advances in Biotechnological Processes" (A. Mizrahi and A. L. van Wezel, eds.), Vol. 2, pp. 1–32. Alan R. Liss, New York.
Reuveny, S., Bino, T., Rosenberg, H., and Mizrahi, A. (1980). *Develop. Biol. Stand.* **46,** 137–145.
Reuveny, S., Silberstein, L., Shahar, A., Freeman, E., and Mizrahi, A. (1982a). *In Vitro* **18,** 92–98.
Reuveny, S., Silberstein, L., Shahar, A., Freeman, E., and Mizrahi, A. (1982b). *Develop. Biol. Stand.* **50,** 115–124.

Reuveny, S., Mizrahi, A., Kotler, M., and Freeman, A. (1983a). *Biotechnol. Bioeng.* **25,** 469–480.

Reuveny, S., Mizrahi, A., Kotler, M., and Freeman, A. (1983b). *Biotechnol. Bioeng.* **25,** 2969–2980.

Reuveny, S., Corett, R., Mizrahi, A., Freeman, A., and Kotler, M. (1984). *Proc. Biotech. 84, England,* 1984.

Ryan, U. S., Mortara, M., and Whitaker, C. (1980). *Tissue Cell* **12,** 619–636.

Ryan, U. S., White, L. A., Lopez, M., and Ryan, J. W. (1982). *Tissue Cell* **14,** 597–605.

Sargent, G. F., Sims, T. A., and McNeish, A. S. (1981). *J. Microsc.* **122,** 209–212.

Scattergood, E. M., Schlabach, A. J., Macleer, W. J., and Hilleman, M. R. (1983). *Ann. N. Y. Acad. Sci.* **413,** 332–339.

Schleicher, J. B., and Weiss, R. E. (1968). *Biotechnol. Bioeng.* **10,** 617–624.

Scott, R. W., Eaton, D. L., and Baker, J. B. (1982). *J. Cell Biol.* **95,** 121A.

van Seefried, A. V., and Chun, J. H. (1981). *Develop. Biol. Stand.* **47,** 25–34.

Shahar, A., Reuveny, S., Amir, A., Kotler, M., and Mizrahi, A. (1983). *J. Neurosci. Res.* **9,** 339–348.

Shahar, A., Mizrahi, A., Reuveny, S., Zinman, T., and Shainberg, A. (1984a). *Develop. Biol. Stand.* (in press).

Shahar, A., Amir, A., Reuveny, S., Silberstein, L., and Mizrahi, A. (1984b). *Develop. Biol. Stand.* **55,** 25–30.

Shainberg, A., Isac, A., Reuveny, S., Mizrahi, A., and Shahar, A. (1983). *Cell Biol. Int. Rep.* **7,** 727–734.

Smith, M. A., and Vale, W. W. (1980). *Endocrinology* **107,** 1425–1431.

Smith, M. A., and Vale, W. W. (1981). *Endocrinology* **108,** 752–759.

Spier, R. E. (1980). *Adv. Biochem. Eng.* **14,** 120–162.

Spier, R. E., and Whiteside, J. P. (1969). *Biotechnol. Bioeng.* **18,** 659–667.

Spiess, Y., Smith, M. A., and Vale, W. (1982). *Diabetes* **31,** 189–193.

Strand, J. M., Quarles, J. M., and McConnell, S. (1982). *In Vitro* **18,** 311.

Sundquist, K. G., and Wagner, L. (1980). *Immunology* **41,** 883–890.

Sundquist, K. G., and Wagner, L. (1981). *Immunology* **43,** 573–580.

Talbot, P., and Keen, M. J. (1980). *Develop. Biol. Stand.* **46,** 147–149.

Terasima, J., and Tolmach, J. (1963). *Exp. Cell Res.* **30,** 344–362.

Thilly, W. G., and Levine, D. W. (1980). *Meth. Enzymol.* **50,** 184–194.

Tyo, M. A., and Wang, D. I. C. (1981). *In* "Advances in Biotechnology" (E. Moo-Young, C. W. Robinson, and C. Venzina, eds.), Vol. I. pp. 141–147.

Ubertini, B. L., Nardelli, L., Santero, G., and Panina, G. (1960). *J. Biochem. Microbiol. Technol. Eng.* **2,** 237–342.

U.S. Department of Health, Education and Welfare (1967). Public Health Service: Regulation for the Manufacture of Biological Products (42, part 73), DHEW Publ. No. (NIH)-71-161. Formerly PHS Publ. No. 437 (revised 1971).

Varani, J., Dame, M., Beals, T. F., and Wass, J. A. (1983). *Biotechnol. Bioeng.* **25,** 1359–1372.

Ventrex Labs (1984). The Gelatin Microcarrier. Trade publication.

Vickrey, H. M., Thilly, W. G., Barngrover, D., Joseph, R. R., and McCann, D. S. (1982). *J. Cell Biol.* **95,** A458.

Vosbeck, K., and Roth, S. (1976). *J. Cell Sci.* **22,** 657–670.

deVries, N. E., Vyth-Dreese, F. A., Figdor, C. G., Spitz, H., Leemans, J. M., and Bont, W. S. (1983). *J. Immunol.* **131,** 201–206.

Weigel, P. H., Schonoar, R. L., Kuhleneshmidt, M. S., Schmell, E., Lee, R. T., Lee, Y. S., and Roseman, S. (1979). *J. Biol. Chem.* **254,** 10830–10836.

Weiss, L., and Harlos, J. P. (1972). *Prog. Surface Sci.* **1,** 304–355.
van Wezel, A. L. (1967) *Nature (London)* **216,** 64–65.
van Wezel, A. L. (1972) *Prog. Immunobiol. Stand.* **5,** 187–192.
van Wezel, A. L. (1973). *In* "Tissue Culture: Methods and Applications" (P. F. Kruse and M. K. Patterson, eds.), pp. 372–377. Academic Press, New York.
van Wezel, A. L. (1977). *Develop. Biol. Stand.* **37,** 143–147.
van Wezel, A. L. (1982). *J. Chem. Tech. Biotechnol.* **32,** 318–323.
van Wezel, A. L., and van der Velden-de-Groot, C. A. M. (1978). *Process Biochem.* **13,** 6–8.
van Wezel, A. L., and van Steenis, G. (1978). *Develop. Biol. Stand.* **40,** 69–75.
van Wezel, A. L., van Steenis, G., Hannik, C. A., and Cohen, A. (1978). *Develop. Biol. Stand.* **41,** 159–168.
van Wezel, A. L., van der Velden-de-Groot, C. A. M., and van Herwaarden, J. A. M. (1980). *Develop. Biol. Stand.* **46,** 151–158.
Whiteside, J. P., Whiting, B. R., and Spier, R. E. (1979). *Develop, Biol. Stand.* **42,** 113–120.
Wolpert, L., and Gingell, D. (1968). *Symp. Soc. Exp. Biol.* **23,** 169–175.
Wright, J. T., Elmer, W. A., and Dunlop, A. T. (1982). *Anal. Lab.* **125,** 100–104.
Yefenof, E., Schwartz, S., and Katz-Gross, A. (1983). *Immunol. Lett.* **6,** 39–43.
Zavalnyi, M. A., Grochev, V. P., Denisova, T. N., Popva, V. L., and Mironova, L. L. (1980). *Vopr. Virusol.* **5,** 583–590.
XF

ADVANCES IN CELL CULTURE, VOL. 4

THE ESTABLISHMENT OF CELL LINES FROM HUMAN SOLID TUMORS

Albert Leibovitz

Arizona Cancer Center
University of Arizona College of Medicine
Tucson, Arizona

I. Introduction

When George Gey established his HeLa cell line, he changed the course of my life as I am sure he did that of countless others. He "opened the door" to modern medical diagnostic virology. In the U.S. Army I was given a crash course in the use of tissue culture in diagnostic medical virology for three months under Trygve Berge at the Sixth Army Medical Laboratory in Sausalito, California and was then sent to St. Louis to establish a virology laboratory from scratch. After an empty warehouse was converted into a full-fledged laboratory that was stocked with HeLa and HEp II cell lines and Eagle's medium, I was ready for business. It was not long before my technicians complained that they felt like robots in refeeding hundreds of tubes three time a week as these rapidly growing cell lines lowered the pH of the medium to toxic levels. I agreed with them, and started my research in media composition which finally led to the development of medium L-15 (1) which negated the refeeding of tubes inoculated with suspected virus-containing specimens.

I retired from the Army in 1970 to join the staff of the Scott and White Clinic in Temple, Texas as head of their Microbiology Section.

ISBN 0-12-007904-6

As in most hospitals in those days, virology was not included in this Section for economic reasons. Being naive as to the complexities of establishing cell lines from human tumors, I assured the Board of Directors that I could establish a virology laboratory at nominal cost by initiating my own cell lines rather than by buying them commercially. Thus, my initial venture into establishing cell lines from human tumors was not to develop a quick cure for cancer but to get cheap cell lines. The Scott and White Clinic now possesses an excellent medical diagnostic virology laboratory.

I had ideal conditions at the Scott and White Clinic from 1971 to 1978 for trying to establish cell lines from all sources. Dr. James C. Stinson, Director of the Pathology Division, was keenly interested in this program and helped to establish the tissue culture laboratory. His Pathology Section was encircled by 10 operating rooms. All resected tissues suspected of containing carcinoma were immediately transferred to a pathologist for rapid diagnosis by frozen section. Within several minutes of resection, a tentative diagnosis was made and the tumor tissue was immediately sent to the tissue culture laboratory.

During this period, I succeeded in establishing about 100 cell lines from a wide variety of human solid tumors. But as this was the result of about 2000 primary explants, the overall success rate was 5%. Thus, 95% of my efforts were largely fruitless. Although short-term cultures were readily attained, most died within a few weeks, some survived for several months, but relatively few developed into permanent cell lines.

During this 8-year span, I developed some insight into what a tumor cell must endure to survive *in vitro* and my success rate improved significantly. My simplistic reasoning was that tumor cells had to survive in a hostile environment. The tumor mass is often necrotic. On the average, only 10 to 20% of the cells obtained by mechanical means (2,3) were viable. The dead and dying cells were releasing powerful enzymes that would destroy the viable cells unless prompt steps were taken to neutralize or remove them (3).

II. Establishment of Cell Lines

A. Transport Medium

The first step was to develop a transport medium that would sustain the viable cells. Although it is preferable to process specimens within 1 hour after excision, circumstances often necessitated longer delays or even overnight holding of specimens received late in the day. The transport medium I devised is shown in Table I.

TABLE I

TISSUE TRANSPORT MEDIUM

Additive	Per liter
L-15 culture medium	1 liter
PVP-360	10.00 g
Penicillin G	200.00 units/ml
Streptomycin sulfate	200.00 μg/ml
Fungazone (amphotericin B)	1.25 μg/ml

L-15 medium is an ideal transport medium. As it is bicarbonate free, there is minimal fluctuation in pH during transit even when the specimen is exposed to the atmosphere. Polyvinylpyrrolidone (PVP) has been noted by investigators (4,5) to help retain cell viability when tissues are exposed to harsh treatment. Presumably, its role is similar to the large serum protein molecules which provide physical protection to cells (6). Antibiotics help to eliminate overt contamination.

B. Debridement and Processing

The second step was proper debridement. The specimen was placed in a sterile plastic petri dish and completely covered with a "detoxification medium" containing antibiotics. All necrotic areas, adherent stromal tissue, fat, and blood clots were removed. The debrided tumor tissue was washed in a second dish containing detoxification medium and then placed in a third dish with the same medium for mincing (3). During that period, I preferred the spillout (3,7) and the spinner spillout (3) techniques as they yielded tumor cells with minimal stromal cell contamination. Cell viability was determined by the Trypan Blue exclusion test; one could also determine at this time whether the yield resembled tumor cells. This was especially true of adenocarcinoma specimens as these tumor cells were usually seen in grape-like clusters as well as single cells. If insufficient tumor cells were obtained by these methods, collagenase digestion of the remaining minces enhanced the tumor cell yield and also increased stromal cell contamination. This problem is somewhat alleviated by the newer growth media which require less serum. We found that collagenase disaggregation was most helpful in obtaining outgrowth of hypernephromas, but was not superior to the spillout techniques in establishing cell lines from other carcinomas. However, when large numbers of tumor cells are required, as for chemotherapeutic screens, collagenase digestion may be necessary. This aspect will be discussed in greater detail below.

C. Media

For short-term cultures, almost any good growth medium will suffice if the specimen is properly transported and processed. This becomes obvious on reviewing the literature. However, most reports of established cell lines refer to singular successes and no mention is made of the number of specimens processed before efforts are fruitful. This may lead one to believe that such successes are fortuitous.

1. Medium L-15

Medium L-15 with certain additives is useful in establishing cell lines from tumor specimens. The medium is bicarbonate and glucose free. It is buffered by the free-base amino acids, especially L-arginine, and contains galactose, sodium pyruvate, and dl-alanine as its source of carbohydrate. Amino acids are present in maximum amounts that are nontoxic to tissue cells (1). This permits greater flexibility in feeding the cells; once a week is usually sufficient for rapidly growing cells. However, the buffering system must be modified if one desires to use this medium in a CO_2 incubator.

2. Detoxification and Growth Media

When we developed detoxification growth media at Scott and White, our success rate in establishing cell lines from colorectal adenocarcinoma specimens rose from 3 to 16% (3).

Detoxification reagents are incorporated to neutralize the powerful enzymes being released by dead and dying cells as well as peroxides produced by exposure of the cell cultures to visible light (8). PVP-360 and methocel (methylcellulose) protect tissue cells similar to the large molecular weight serum proteins (6). Catalase, glutathione, and selenium decompose peroxides. Selenium is essential for glutathione peroxidase activity, a selenoenzyme that catalyzes the decomposition of peroxides (9).

Numerous growth factors have been uncovered during the past few years which enhance specific tissue cell growth. Many of these have come from Dr. Sato's laboratory, the most significant reagent being transferrin (10,11). This iron-transport protein, in conjunction with insulin and hydrocortisone, permits optimal growth of most tissue cells with markedly reduced amounts of fetal bovine serum. Media M-3 and M-4 (Table II) have 5% fetal bovine serum for optimal growth, al-

TABLE II

CULTURE MEDIA M-3 AND M-4

Medium M-3	
A. Basic medium	1 liter pkg.
L-15 medium powdered	
B. Detoxification reagents	
Polyvinylpyrrolidone-360	1.00 g[a]
Methylcellulose, 15 cps	200.00 mg
Catalase, 11,000 U/mg	5.00 mg
Glutathione, reduced	15.00 mg
Selenious acid	0.79 mg
C. Growth factors	
Insulin	10.00 mg
Transferrin	10.00 mg
Glutamine	292.00 mg
Hydrocortisone	3.61 mg
Orotic acid	15.00 mg
Myoinositol	15.00 mg
Ornithine	15.00 mg
2-Mercaptoethanol	0.80 mg
Fetal bovine serum	50.00 ml
Antibiotics, 100×	10.00 ml
Buffers: Hepes and Hepes, Na in equal parts	4.80 g
Medium M-4	
Above ingredients plus	
17β-Estradiol	2.72 mg
Testosterone	2.88 mg

[a] Per liter.

though excellent growth can be attained with as little as 2% serum. Mercaptoethanol may be necessary for some cell lines to utilize the essential amino acid cystine (12). The other ingredients are added as possible precursors for essential nutrients for tissue cells in general. Fetal bovine serum functions both as a detoxifying agent and as a source of essential nutrients, enzymes, and growth factors.

Medium M-3 is free of steroid hormones except for the minimal amounts present in the low percentage of fetal bovine serum added to the medium. This medium is designed for cancer cells obtained from the nonendocrine organs as steroid hormones may retard the growth of some of these cells (13). M-4 contains 17β-estradiol and testosterone for cells derived from the endocrine organs. However, cancer cells from all sites are explanted in both media to determine whether these hormones are growth stimulating or inhibiting or have no obvious effects.

III. Chemotherapeutic Screens

After trying to retire for 3 years, I was most happy to accept the kind invitation of Sydney Salmon and Jeffrey Trent to join their staff at the University of Arizona Cancer Center. Dr. Salmon and his colleagues (14,15) have developed the stem cell clonogenic soft agar technique into a precise tool for studying tumor biology and anticancer drug sensitivity. The procedure is still in its infancy, but already has attracted worldwide interest, and intensive investigations are in progress in many laboratories to enhance its usefulness (16–18). Usually less than 0.1% of tumor cells isolated from primary tissues or body fluids are clonogenic; millions of viable cells are necessary for testing a battery of potentially useful chemotherapeutic agents, and not all tumors have stem cells that will form colonies in soft agar with our present media.

A. Plastic vs Soft Agar

For purposes of developing cell lines it is simpler to work on media improvement on plastic than in soft agar. I therefore compared outgrowth of tumor cells on plastic to aliquots of the same specimen in soft agar for colony formation. Of 128 specimens tested from all sources, there was no significant difference in short-term growth. With minor exceptions, growth or no growth was highly correlated in both systems ($p = 0.005$). I could thus use the plastic system to develop media that would enhance tumor cell growth, and am using the development of cell lines as the ultimate test of a medium's value. Since developing my "M" series (hopefully to establish myeloma cell lines as well as those from solid tumors), cell lines from a wide variety of tumors were established during the past 2 years, as noted in Table III.

B. Hypoosmotic Medium to Disaggregate Cell Clumps

As millions of viable cells are necessary for testing batteries of potentially useful chemotherapeutic agents, many laboratories are using enzymatic methods to obtain large cell yields per gram of tissue (16–18). We are using a collagenase–DNase method which yields large number of viable cells that are clonogenic in soft agar (19), but often a large share of the yield cannot be used because of cell clumps that cannot be dispersed into single cells. As single cells are essential to the chemotherapeutic screen test, methods have been sought to disrupt these clumps while still retaining viable cells. A recent article by Julia Carter and colleagues (5) indicated that single cells could be isolated

TABLE III

CELL LINES ESTABLISHED IN DETOXIFICATION GROWTH MEDIA

Primary site of carcinoma	Number of cell lines
Endometrium	1
Fallopian tube	1
Lung	2
Myeloma	1
Mammary	2
Melanoma	3
Ovary	5
Seminal	1
Uterine	3

from hamster small intestinal segments with the use of a hypoosmotic medium to disrupt the cells at their desmosome connections and PVP to help retain cell viability. This promising avenue has been pursued and a disaggregate medium (Table IV) and techniques have been devised for the disruption of cell clumps (19).

The medium developed has an osmolarity of about 200 mOsm and includes bovine serum albumin and methocel as well as PVP-40 to help retain cell viability. Cell clumps when rotated in this medium in equal parts with collagenase-DNase for 1.5 hours are sufficiently disrupted to greatly enhance the single-cell yield and significantly increase colony growth in soft agar (see Figs. 1–4).

Cell viability decreases when cells are exposed to this treatment for longer than 6 hours, although some survive for as long as 24 hours.

TABLE IV

HYPOOSMOLAR MEDIUM

Additives	ml/liter
Fraction V	
Bovine serum albumin, 0.8%[a]	688
(in Dulbecco's BSS, w/o Ca^{2+} or Mg^{2+})	
PVP-40, 5.0%	200
Methylcellulose (15 cps), 2.0%	100
EDTA $(4Na_2)\cdot 2H_2O$, 10.0%	10
Osmolarity: 200 mOsm, pH 7.0	

[a] Sigma, #A8022, pH 5.2.

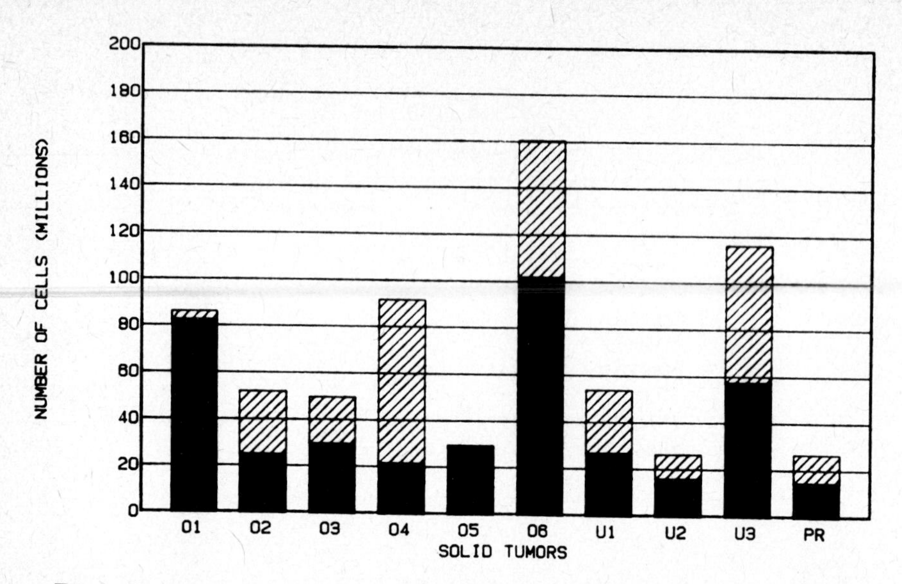

FIG. 1. Process and viable cells. Black area: single-cell harvest in millions following 0.15% collagenase–0.015% DNase digestion of tumor tissue. Hatched area: additional single-cell yield in millions following disruption of cell clumps by the hypoosmolar treatment. Specimens 1–6 are ovarian carcinoma tissues; specimens U1–3 are adenocarcinomas of unkown primary site; and the PR specimen is from a prostate carcinoma.

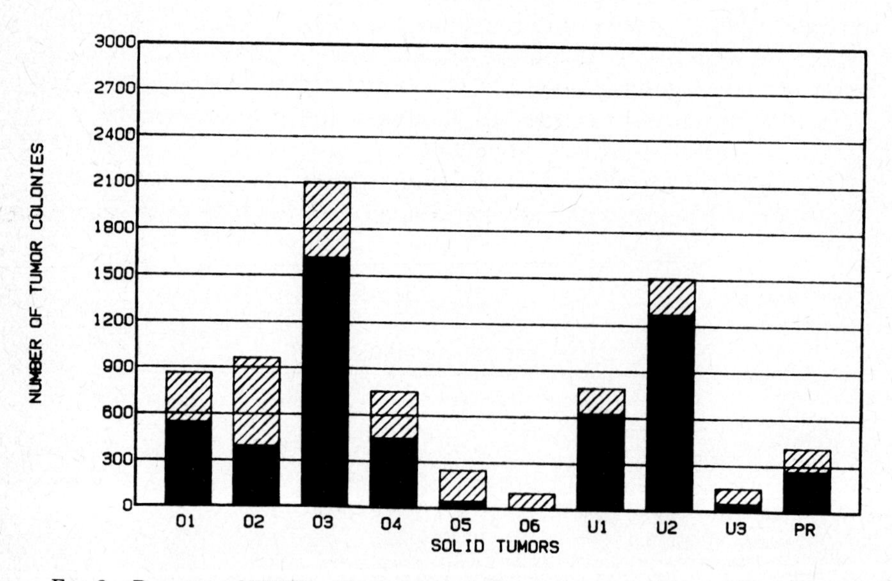

FIG. 2. Process and TCFUs. Black area: colonies produced in soft agar by the single cells obtained in the collagenase–DNase digestion of specimens in Fig. 1. Hatched area: colonies produced in soft agar from the single cells obtained by disruption of cell clumps by the hypoosmolar treatment.

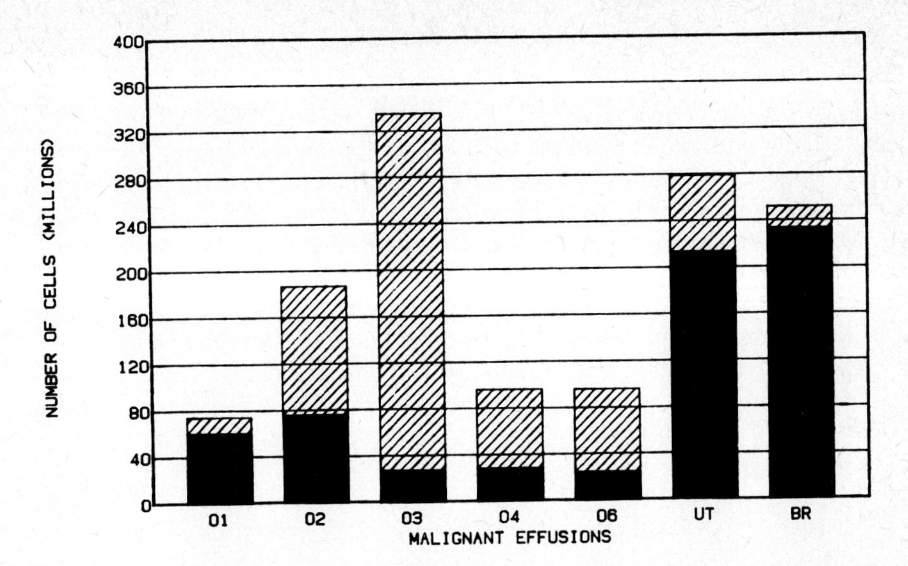

FIG. 3. Process and viable cells. Black area: single cells harvested from malignant effusions in millions. Hatched area: single cells derived from cell clumps in malignant effusions following hypoosmolar treatment. Specimens 1–6 are from ovarian carcinomas; UT is from a uterine carcinoma; and BR is from a mammary carcinoma.

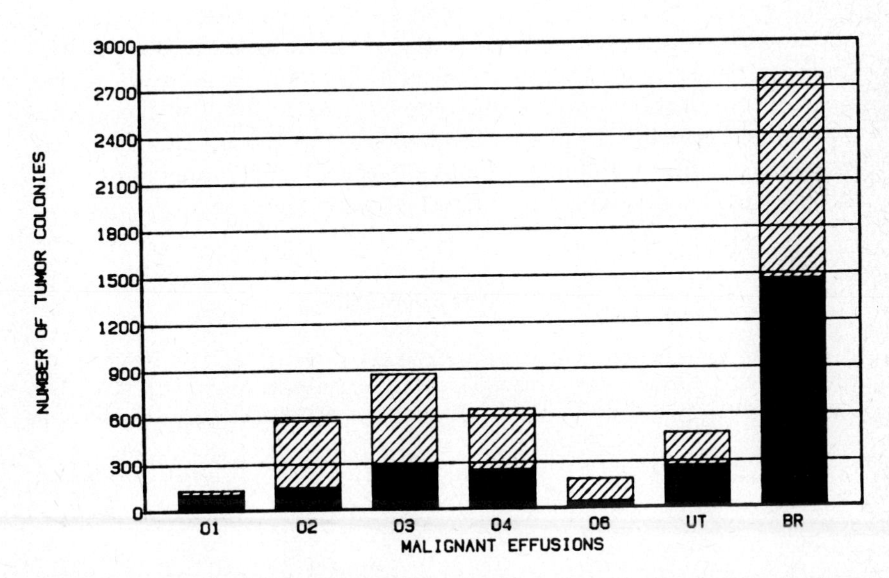

FIG. 4. Process and TCFUs. Black area: colonies formed in soft agar from the single cells isolated from malignant effusions with no treatment. Hatched area: colonies formed in soft agar from single cells derived from cell clumps dispersed by the hypoosmolar treatment. Specimens are the same as in Fig. 3.

C. *Hypoosmolar Medium for Cell Passage*

The hypoosmolar medium with 0.001% porcine 1× crystalline trypsin (already contains 0.1% versene) is significantly less traumatic than the crude 0.25% trypsin–0.1% versene in common use for cell passage. This may be important in attempting to pass newly established tumor cell growth as the crude trypsin often kills the cells (3). Monolayers exposed to the hypoosmolar trypsin–versene medium are usually dislodged from the plastic within 10 minutes. Cell clumps can be disaggregated into viable single cells with incubation at 37°C for 45 minutes. Remaining clumps can usually be dispersed by pipetting back and forth about 6 times.

IV. Summary and Future Perspectives

In summary, recent advances in our knowledge of the growth, nutrition, and metabolic requirements of cancer cells lend hope that cell lines can be established from all carcinomas in reasonable numbers. Tissue culture is unmasking the heterogeneity of cancer cells derived from any given site (3). While some cell lines continue to undergo clonal evolution after they have been established, others appear to be stable with respect to drug sensitivity over numerous passages. Such lines can be useful for studies of cancer biology and for analysis of drug resistance. Additionally, the ability to have batteries of such cell lines available for the testing of specific chemotherapeutic agents, biological response modifiers, and monoclonal antibodies (20,21) may play a crucial role in the development of effective cancer therapy.

Acknowledgments

This work was supported by Grants CA-21839, CA-17094, CA-23074, and contract no. N01-CM17497 from the National Institutes of Health, Bethesda, Maryland 20205. It was presented, in part, at the 34th Annual Meeting of the Tissue Culture Association in Orlando, Florida, June 12–16, 1983.

References

1. Leibovitz, A. (1963). The growth and maintenance of tissue cell cultures in free gas exchange with the atmosphere. *Am. J. Hyg.* **78,** 173–180.
2. Leibovitz, A. (1975). Development of media for isolation and cultivation of human

cancer cells. *In* "Human Tumor Cells *in Vitro*" (J. Fogh, ed.), pp. 23–50. Plenum, New York.

3. Leibovitz, A., Stinson, J. C., McCombs III, W. B., McCoy, C. E., Mazur, K. C., and Mabry, N. D. (1976). Classification of human colorectal adenocarcinoma cell lines. *Cancer Res.* **36,** 4562–4569.
4. Lechner, J. F., Shankar Narayan, K., Ohnuki, Y., Babcock, M. S., Jones, L. W., and Kaighn, M. E. (1978). Replicative epithelial cell cultures from normal prostate gland. *J. Natl. Cancer Inst.* **60,** 797–799.
5. Carter, J. H., Carter, H., Nussbaum, J., and Eichholz, A. (1982). Isolation of hamster intestinal epithelial cells using hypoosmotic media and PVP. *J. Cell. Physiol.* **111,** 55–67.
6. Sanford, K. K., and Evans, V. J. (1982). A quest for the mechanism of "spontaneous" malignant transformation in culture and associated advances in culture technology. *J. Natl. Cancer Inst.* **68,** 895–913.
7. Lasfargue, E. Y., and Ozzello, O. (1958). Cultivation of human breast carcinomas. *J. Natl. Cancer Inst.* **21,** 1131–1147.
8. Parshad, R., Sanford, K. K., Taylor, W. G., Tarone, R. E., Jones. G. M., and Baeck, A. E. (1979). Effect of intensity and wavelength of fluorescent light on chromosome damage in cultured mouse cells. *Photochem. Photobiol.* **29,** 971–975.
9. Parshad, R., Taylor, W. G., Sanford, K. K., Camalier, R. F., Gantt, R., and Tarone, R. E. (1980). Fluorescent light-induced chromosome damage in human IMR-90 fibroblasts: Role of hydrogen peroxide and related free radicals. *Mutat. Res.* **73,** 115–124.
10. Barnes, D., and Sato, G. (1980). Methods for growth of cultured cells in serum-free medium. *Anal. Biochem.* **102,** 255–270.
11. Barnes, D., van der Bosch, J., Masui, H., Miyazaki, K., and Sato, G. (1981). The culture of human tumor cells in serum-free medium. *Meth. Enzymol.* **79,** 368–391.
12. Ishii, T., Bannai, S., and Sugita, Y., (1981). Mechanism of growth stimulation of L1210 cells by 2-mercaptoethanol *in vitro*. Role of the mixed disulfide of 2-mercaptoethanol and cystine. *J. Biol. Chem.* **256,** 12387–12892.
13. Murakami, H., and Masui, H. (1980). Hormonal control of human colon carcinoma growth in serum-free medium. *Proc. Natl. Acad. Sci. U.S.A.* **77,** 3464–3468.
14. Hamburger, A. W., and Salmon, S. E. (1977). Primary bioassay of human tumor stem cells. *Science* **197,** 461–463.
15. Salmon, S. E. (1980). Background and overview. *In* "Cloning of Human Tumor Stem Cells" (S. E. Salmon, ed.), pp. 3–13. Alan R. Liss, New York.
16. Slocum, H. K., Pavelic, Z. P., Kanter, P. M., Nowak, N. J., and Rustum, Y. M. (1981). The soft agar clonogenicity and characterization of cells obtained from human solid tumors by mechanical and enzymatic means. *Cancer Chemother. Pharmacol.* **6,** 219–225.
17. Kern, D. H., Campbell, M. A., Cochran, A. J., Burk, M. W., and Morton, D. L. (1982). Cloning of human solid tumors in soft agar. *Int. J. Cancer* **30,** 725–729.
18. Kedar, E., Ikejiri, B. L., Bonnard, G. D., and Herberman, R. B. (1982). A rapid technique for isolation of viable tumor cells from solid tumors: Use of the tumor cells for induction and measurement of cell-mediated cytotoxic responses. *Eur. J. Cancer Clin. Oncol.* **18,** 911–1000.
19. Leibovitz, A., Lui, R., Hayes, C., and Salmon, S. E. (1983). A hypoosmotic medium to disaggregate tumor cell clumps into viable and clonogenic single cells for the human tumor stem cell clonogenic assay. *Int. J. Cell Cloning* **1,** 478–485.
20. Herlyn M., Steplewski, Z., Herlyn, D., and Koprowski, H. (1979). Colorectal car-

cinoma-specific antigen: Detection by means of monoclonal antibodies. *Proc. Natl. Acad. Sci. U.S.A.* **76,** 1438–1442.

21. Herlyn, D. M., Steplewski, Z., Herlyn, M., and Koprowski, H. (1980). Inhibition of growth of colorectal carcinoma in nude mice by monoclonal antibody. *Cancer Res.* **40,** 717–721.

INDEX

D

E

F

G

V

X